Communications
in Computer and Information Science 22

Joaquim Filipe   Boris Shishkov
Markus Helfert   Leszek A. Maciaszek (Eds.)

Software and Data Technologies

Second International Conference, ICSOFT/ENASE 2007
Barcelona, Spain, July 22-25, 2007
Revised Selected Papers

 Springer

Volume Editors

Joaquim Filipe
Polytechnic Institute of Setúbal – INSTICC
Av. D. Manuel I, 27A - 2. Esq., 2910-595 Setúbal, Portugal
E-mail: jfilipe@insticc.org

Boris Shishkov
Interdisciplinary Institute for Collaboration and
Research on Enterprise Systems and Technology – IICREST
P.O. Box 104, 1618 Sofia, Bulgaria
E-mail: b.b.shishkov@ewi.utwente.nl

Markus Helfert
Dublin City University
School of Computing
Dublin 9, Ireland
E-mail: markus.helfert@computing.dcu.ie

Leszek A. Maciaszek
Macquarie University
Department of Computing
Sydney, NSW, 2109, Australia
E-mail: leszek@ics.mq.edu.au

Library of Congress Control Number: Applied for

CR Subject Classification (1998): D.2, E.1, D.3, I.2.4, C.1.4

ISSN   1865-0929
ISBN-10   3-540-88654-0 Springer Berlin Heidelberg New York
ISBN-13   978-3-540-88654-9 Springer Berlin Heidelberg New York

Springer is a part of Springer Science+Business Media

springer.com

© Springer-Verlag Berlin Heidelberg 2008
Printed in Germany

Typesetting: Camera-ready by author, data conversion by Scientific Publishing Services, Chennai, India
Printed on acid-free paper   SPIN: 12549450   06/3180   5 4 3 2 1 0

Preface

This book contains the best papers of the Second International Conference on Software and Data Technologies (ICSOFT 2007), held in Barcelona, Spain. It was organized by the Institute for Systems and Technologies of Information, Communication and Control (INSTICC), co-sponsored by the Workflow Management Coalition (WfMC), in cooperation with the Interdisciplinary Institute for Collaboration and Research on Enterprise Systems and Technology (IICREST).

The purpose of ICSOFT 2007 was to bring together researchers and practitioners interested in information technology and software development. The conference tracks were "Software Engineering," "Information Systems and Data Management," "Programming Languages," "Distributed and Parallel Systems" and "Knowledge Engineering."

Being crucial for the development of information systems, software and data technologies encompass a large number of research topics and applications: from implementation-related issues to more abstract theoretical aspects of software engineering; from databases and data warehouses to management information systems and knowledge-base systems; next to that, distributed systems, pervasive computing, data quality and other related topics are included in the scope of this conference.

ICSOFT 2007 received 292 paper submissions from more than 56 countries in all continents. To evaluate each submission, a double-blind evaluation method was used: each paper was reviewed by at least two internationally known experts from the ICSOFT Program Committee. Only 41 papers were selected to be published and presented as full papers, i.e., completed work (8 pages in proceedings / 30-min oral presentations), 74 additional papers describing work-in-progress were accepted as short papers for 20-min oral presentation, leading to a total of 115 oral paper presentations. Another 76 papers were selected for poster presentation. The full-paper acceptance ratio was thus 14%, and the total oral paper acceptance ratio was 39%.

ICSOFT was organized in conjunction with ENASE—the Evaluation of Novel Approaches to Software Engineering working conference. ENASE provides a yearly forum for researchers and practitioners to review and evaluate new software development methodologies, practices, architectures, technologies and tools. The background body of knowledge for ENASE is formed by novel approaches to software engineering with emphasis on software product and process improvement.

This book includes a special section with the three best papers from ENASE.

We hope that you will find these papers interesting and we hope they represent a helpful reference in the future for all those who need to address any of the research areas mentioned above.

August 2008

Joaquim Filipe
Boris Shishkov
Markus Helfert
Leszek Maciaszek

Conference Committee

Conference Chair

Joaquim Filipe   Polytechnic Institute of Setúbal / INSTICC, Portugal

Program Co-chairs

Cesar Gonzalez-Perez   IEGPS, CSIC, Spain (ENASE)
Markus Helfert   Dublin City University, Ireland (ICSOFT)
Leszek A. Maciaszek   Macquarie University - Sydney, Australia (ENASE)
Boris Shishkov   University of Twente, The Netherlands (ICSOFT)

Organizing Committee

Paulo Brito   INSTICC, Portugal
Marina Carvalho   INSTICC, Portugal
Helder Coelhas   INSTICC, Portugal
Vera Coelho   INSTICC, Portugal
Andreia Costa   INSTICC, Portugal
Bruno Encarnação   INSTICC, Portugal
Luís Marques   INSTICC, Portugal
Vitor Pedrosa   INSTICC, Portugal
Vera Rosário   INSTICC, Portugal
Mónica Saramago   INSTICC, Portugal

ICSOFT Program Committee

Francisco Abad, Spain
Sergey Jemal Abawajy, Australia
Silvia Abrahão, Spain
Muhammad Abulaish, India
Hamideh Afsarmanesh, The Netherlands
Jacky Akoka, France
Rafa Al Qutaish, Canada
Markus Aleksy, Germany
Tsanka Petrova Angelova, Bulgaria
Keijiro Araki, Japan
Alex Aravind, Canada

Colin Atkinson, Germany
Juan Carlos Augusto, UK
Elisa Baniassad, China
Luciano Baresi, Italy
Joseph Barjis, USA
Bernhard Beckert, Germany
Noureddine Belkhatir, France
Fevzi Belli, Germany
Alexandre Bergel, Germany
Sue Black, UK
Maarten Boasson, The Netherlands

Wladimir Bodrow, Germany
Marcello Bonsangue, The Netherlands
Pere Botella, Spain
Lisa Brownsword, USA
Gerardo Canfora, Italy
Cinzia Cappiello, Italy
Antonio Cerone, China
W.K. Chan, Hong Kong
Shiping Chen, Australia
T.Y. Chen, Australia
Kung Chen, Taiwan, R.O.C.
Samuel Chong, UK
Peter Clarke, USA
Rolland Colette, France
Rem Collier, Ireland
Kendra Cooper, USA
Alfredo Cuzzocrea, Italy
Bogdan Czejdo, USA
Mehdi Dastani, The Netherlands
Sergio de Cesare, UK
Clever de Farias, Brazil
Rogerio de Lemos, UK
Andrea De Lucia, Italy
Serge Demeyer, Belgium
Steven Demurjian, USA
Elisabetta Di Nitto, Italy
Massimiliano Di Penta, Italy
Nikolay Diakov, The Netherlands
Oscar Dieste, Spain
Jan L.G. Dietz, The Netherlands
Jin Song Dong, Singapore
Jing Dong, USA
Brian Donnellan, Ireland
Juan C. Dueñas, Spain
Jürgen Ebert, Germany
Paul Ezhilchelvan, UK
Behrouz Far, Canada
Massimo Felici, UK
Rudolf Ferenc, Hungary
Juan Fernandez-Ramil, UK
Bernd Fischer, UK
Gerald Gannod, USA
Jose M. Garrido, USA
Dragan Gasevic, Canada
Nikolaos Georgantas, France
Paola Giannini, Italy

John Paul Gibson, France
Holger Giese, Germany
Karl Goeschka, Austria
Swapna Gokhale, USA
Jose Ramon Gonzalez de Mendivil,
   Spain
Jesus M. Gonzalez-Barahona, Spain
Daniela Grigori, France
Klaus Grimm, Germany
Yann-Gaël Guéhéneuc, Canada
Tibor Gyimothy, Hungary
Michael Hanus, Germany
Naohiro Hayashibara, Japan
Reiko Heckel, UK
Christian Heinlein, Germany
Markus Helfert, Ireland
Rattikorn Hewett, USA
Jang-Eui Hong, Republic of Korea
Shinichi Honiden, Japan
Ilian Ilkov, The Netherlands
Ivan Ivanov, USA
Stephen Jarvis, UK
Damir Kalpic, Croatia
Krishna Kavi, USA
Taghi Khoshgoftaar, USA
Roger (Buzz) King, USA
Paul Klint, The Netherlands
Alexander Knapp, Germany
Mieczyslaw Kokar, USA
Rainer Koschke, Germany
Jens Krinke, Germany
Padmanabhan Krishnan, Australia
Martin Kropp, Switzerland
Tei-Wei Kuo, Taiwan, R.O.C.
Yvan Labiche, Canada
Michele Lanza, Switzerland
Eitel Lauria, USA
Insup Lee, USA
Jonathan Lee, Taiwan, R.O.C.
Yu Lei, USA
Hareton Leung, Hong Kong
Kuan-Ching Li, Taiwan, R.O.C.
Man Lin, Canada
Panos Linos, USA
Hua Liu, USA
Chengfei Liu, Australia

David Lorenz, USA
Christof Lutteroth, New Zealand
Jianhua Ma, Japan
Broy Manfred, Germany
Tiziana Margaria, Germany
Katsuhisa Maruyama, Japan
Johannes Mayer, Germany
Tommaso Mazza, Italy
Fergal McCaffery, Ireland
Hamid Mcheick, Canada
Massimo Mecella, Italy
Karl Meinke, Sweden
Simão Melo de Sousa, Portugal
Emilia Mendes, New Zealand
Manoel Mendonça, Brazil
Raffaela Mirandola, Italy
Hristo Mirkov, USA
Prasenjit Mitra, USA
Dimitris Mitrakos, Greece
Birger Møller-Pedersen, Norway
Mattia Monga, Italy
Sandro Morasca, Italy
Maurizio Morisio, Italy
Markus Müller-Olm, Germany
Paolo Nesi, Italy
Alan O'Callaghan, UK
Rory O'Connor, Ireland
Pasi Ojala, Finland
Claus Pahl, Ireland
Witold Pedrycz, Canada
Steve Peters, The Netherlands
Mario Piattini, Spain
Martin Pinzger, Switzerland
Lori Pollock, USA
Andreas Polze, Germany
Peter Popov, UK
Wenny Rahayu, Australia
Jolita Ralyte, Switzerland
Anders P. Ravn, Denmark
Marek Reformat, Canada
Arend Rensink, The Netherlands
Werner Retschitzegger, Austria
Gustavo Rossi, Argentina
Guenther Ruhe, Canada

Stefano Russo, Italy
Mortaza S. Bargh, The Netherlands
Shazia Sadiq, Australia
Francesca Saglietti, Germany
Bernhard Schätz, Germany
Douglas Schmidt, USA
Andy Schürr, Germany
Isabel Seruca, Portugal
Samir Shah, USA
Boris Shishkov, The Netherlands
Harvey Siy, USA
Jacob Slonim, Canada
George Spanoudakis, UK
Peter Stanchev, USA
Nenad Stankovic, Japan
Larry Stapleton, Ireland
Richard Starmans, The Netherlands
Leon Sterling, Australia
Junichi Suzuki, USA
Ramayah T., Malaysia
Yarar Tonta, Turkey
Mark van den Brand, The Netherlands
Marten van Sinderen, The Netherlands
Enrico Vicario, Italy
Aurora Vizcaino, Spain
Christoph von Praun, USA
Christiane Gresse von Wangenheim,
    Brazil
Bing Wang, UK
Edgar Weippl, Austria
Danny Weyns, Belgium
Ing Widya, The Netherlands
Dietmar Wikarski, Germany
Hongwei Xi, USA
Haiping Xu, USA
Hongji Yang, UK
Tuba Yavuz-Kahveci, USA
Rym Zalila Mili, USA
Kang Zhang, USA
Du Zhang, USA
Xiaokun Zhang, Canada
Jianjun Zhao, China
Hong Zhu, UK
Andrea Zisman, UK

ENASE Program Committee

Pekka Abrahamsson, Finland
Witold Abramowicz, Poland
Mehmet Aksit, The Netherlands
Colin Atkinson, Germany
Giuseppe Berio, Italy
Lawrence Chung, USA
Jens Dietrich, New Zealand
Schahram Dustdar, Austria
Jonathan Edwards, USA
Maria João Ferreira, Portugal
Bogdan Franczyk, Germany
Ulrich Frank, Germany
Steven Fraser, USA
Paolo Giorgini, Italy
Cesar Gonzalez-Perez, Spain
Brian Henderson-Sellers, Australia
Zbigniew Huzar, Poland
Stefan Jablonski, Germany
Ryszard Kowalczyk, Australia

Philippe Kruchten, Canada
Michele Lanza, Switzerland
Xabier Larurcea, Spain
Pericles Loucopoulos, UK
Leszek Maciaszek, Australia
Lech Madeyski, Poland
Janis Osis, Latvia
Jeffrey Parsons, Canada
Mario Piattini, Spain
Bala Ramesh, USA
Gil Regev, Switzerland
Francisco Ruiz, Spain
Motoshi Saeki, Japan
Keng Siau, USA
Dave Thomas, Canada
Rainer Unland, Germany
Jean Vanderdonckt, Belgium
Jaroslav Zendulka, Czech Republic

ICSOFT Auxiliary Reviewers

Jonatan Alava, USA
David Arney, USA
Louise Avila, USA
Djuradj Babich, USA
Tibor Bakota, Hungary
Nurlida Basir, UK
Massimo Canonico, Italy
Glauco Carneiro, Brazil
Su-Ying Chang, Taiwan, R.O.C.
Shih-Chun Chou, Taiwan, R.O.C.
Daniela Cruzes, Brazil
Marco D'Ambros, Switzerland
Florian Deissenböck, Germany
Daniele Theseider Duprè, Italy
Lavinia Egidi, Italy
Ekaterina Ermilove, The Netherlands
Hua-Wei Fang, Taiwan, R.O.C.
Massimo Ficco, Italy
Christina von Flach, Brazil
Rita Francese, Italy
Lajos Fulop, Hungary
Lajos Jenő Fülöp, Hungary

Udo Gleich, Germany
Leonardo Grassi, Italy
Andreas Griesmayer, Macau SAR China
Ralph Guderlei, Germany
Michael Haupt, Germany
Stefan Henkler, Germany
Martin Hirsch, Germany
Florian Hoelzl, Germany
Bernhard Hohlfeld, Germany
Endre Horváth, Hungary
Ping-Yi Hsu, Taiwan, R.O.C.
Judit Jasz, Hungary
Joop de Jong, The Netherlands
Elrmar Juergens, Germany
Madhan Karky, Australia
Steven van Kervel, The Netherlands
Tariq M. King, USA
Peter Lammich, Germany
Massimiliano de Leoni, Italy
Martin Leucker, Germany
Yun-Hao Li, Taiwan, R.O.C.
Adrian Lienhard, Switzerland

Ruopeng Lu, Australia
Heng Lu, Hong Kong
Viviane Malheiros, Brazil
Sergio Di Martino, Italy
Michael Meisinger, Germany
Samar Mouchawrab, Canada
Simon S. Msanjila, The Netherlands
Sudarsanan Nesmony, Australia
Joseph Okika, Denmark
Rocco Oliveto, Italy
Jennie Palmer, UK
Ignazio Passero, Italy
Gustavo Perez, USA
Christian Pfaller, Germany
Roberto Pietrantuono, Italy
Dan Ratiu, Germany
Giancarlo Ruffo, Italy
Ruggero Russo, Italy
Laís Salvador, Brazil

Valeriano Sandrucci, Italy
Giuseppe Scanniello, Italy
Siraj Shaikh, Macau SAR China
Marwa Shousha, Canada
Istvan Siket, Hungary
Carine Souveyet, France
Michael Sowka, Canada
Bas Steunebrink, The Netherlands
Tatiana Tavares, Brazil
Matthias Tichy, Germany
Carlo Torniai, Italy
Kun-Yi Tsai, Taiwan, R.O.C.
Laszlo Vidacs, Hungary
Stefan Wagner, Germany
Doris Wild, Germany
Tao Yue, Canada
Zhenyu Zhang, Hong Kong
Xiaohui Zhao, Australia

ENASE Auxiliary Reviewers

Monika Starzecka, Poland
Dominique Stein, Germany

Jong Woo, USA

Invited Speakers

Jan Dietz   Delft University of Technology, The Netherlands
David Lorge Parnas   University of Limerick, Ireland
Kalle Lyytinnen   Case Western Reserve University, Canada
Stephen Mellor   Australia
Bart Nieuwenhuis   K4B Innovation / University of Twente, The Netherlands
Tony Shan   Bank of America, USA
Brian Fitzgerald   Lero - the Irish Software Engineering Research Centre,
                             Ireland

Table of Contents

Invited Papers

Software and Data Technologies

Part I: Programming Languages

Part II: Software Engineering

Part III: Distributed and Parallel Systems

Part IV: Information Systems and Data Management

Part V: Knowledge Engineering

Evaluation of Novel Approaches to Software Engineering

Invited Papers

Benefits of Enterprise Ontology for the Development of ICT-Based Value Networks

Antonia Albani and Jan L.G. Dietz

Delft University of Technology
Chair of Information Systems Design
PO Box 5031, 2600 GA Delft, The Netherlands
{a.albani,j.l.g.dietz}@tudelft.nl

Abstract. The competitiveness of value networks is highly dependent on the cooperation between business partners and the interoperability of their information systems. Innovations in information and communication technology (ICT), primarily the emergence of the Internet, offer possibilities to increase the interoperability of information systems and therefore enable inter-enterprise cooperation. For the design of inter-enterprise information systems, the concept of business component appears to be very promising. However, the identification of business components is strongly dependent on the appropriateness and the quality of the underlying business domain model. The ontological model of an enterprise – or an enterprise network – as presented in this article, is a high-quality and very adequate business domain model. It provides all essential information that is necessary for the design of the supporting information systems, and at a level of abstraction that makes it also understandable for business people. The application of enterprise ontology for the identification of business components is clarified. To exemplify our approach, a practical case is taken from the domain of strategic supply network development. By doing this, a widespread problem of the practical application of inter-enterprise information systems is being addressed.

1 Introduction

Driven by drastic changing market conditions companies are facing an increasingly complex competitive landscape. Decisive factors such as globalization of sales and sourcing markets, shortened product life cycles, innovative pressure on products, services and processes and customers' request for individual products are forcing companies to undergo a drastic transformation of business processes as well as organizational and managerial structures [1]. The shift from a function-oriented to a process-oriented organization with a strong customer focus is essential in order to better adapt to fast changing market requirements and to become more flexible while meeting individual customer demands. Within an enterprise the core business processes need to be identified, improved and (partly) automated, while at the same time other processes are outsourced to business partners [2, 3]. As a consequence, business processes concerning e.g. product development, market research, sales, production,

J. Filipe et al. (Eds.): ICSOFT/ENASE 2007, CCIS 22, pp. 3–22, 2008.
© Springer-Verlag Berlin Heidelberg 2008

delivery and services are affected and have to be adjusted and integrated not only within a single company but also over multiple tiers of suppliers. Corporate networks, so called *value networks* [4-6], are formed to better fulfill specific customer requests providing customized products on time in the right quality and at a competitive price. In order to enhance the competitive advantages of value networks, the selection, development, management and integration of respective suppliers, located not only in tier-1 but also in the subsequent tiers, is of great relevance for enabling effective *cooperation* between enterprises. Modern information and communication technologies (ICT) – like the Internet, semantic standards, distributed applications, component based and service-oriented architectures – are necessary in order to enable the creation and management of dynamic corporate networks [7]. However, the deployment of information and communication technologies for enabling *interoperability* between business partners does not always meet expectations. At present, ICT-enabled value networks can be largely found in the form of rather small, flexible alliances of professionalized participants. The support of large value networks with multiple tiers of suppliers still causes considerable difficulties. The high degree of complexity resulting from dynamic changes in value networks is the main reason for the lack of practical implementations [8].

Looking at the potential of enabling and improving inter-enterprise cooperation by applying ICT, this article contributes to that area by illustrating the benefits of enterprise ontology for the development of ICT-based value networks. For developing intra- and inter-enterprise information systems it is necessary to use a suitable *methodology* for modeling the business domain satisfying defined quality criteria and for eliciting the relevant business requirements. Additionally, information systems need to be modeled on a level of abstraction that is understood also by the business people, who are defining the requirements and using the respective systems. The application of *business components* [9] for the design of high-level information systems appears to be adequate since they directly implement the business activities in an enterprise [9]. All other components are considered either to deliver services to these business components or to offer some general functionality. The identification of business components thus is the first step in the development of an information system according to current standards (component-based, object-oriented, etc.). Needless to say that this step is a very crucial one and that it should be performed at the highest possible level of quality. In the area of business components identification the research initiatives are still scarce see [10-12]. Vitharana [13] recognized that more formal approaches are needed to make the component based paradigm into an effective development tool. The starting point is the set of requirements that have been elicited from the business domain. Requirements engineering is still a weak link, although considerable improvements can be achieved if it is based on an appropriate and high-quality business domain model. Such a model offers a more objective starting point for extracting requirements and a more objective criterion for evaluating them than the traditional 'waiter strategy' [14]. In a very true sense however, this new approach to engineering requirements shifts the problem only to an earlier stage instead of solving it. The crucial factors now are the appropriateness and the quality of the business domain model. In [15] some quality criteria are proposed regarding a business domain model, which we adopted for our current research:

- It should make a clear and well-founded distinction between the *essential* business and infological actions. For example, delivering services to customers is essential, but computing the price to be paid is infological (it doesn't produce a new fact).
- It should have the right granularity or level of detail. "Right" in this respect means: finding the actions that are *atomic* from the business point of view. They may be composite only in their implementations. For example, the request to a supplier is atomic from the business point of view, but to perform a request by postal mail, a number of other (non-essential) actions have to be taken like sending the order form, transporting it and delivering it to the supplier.
- It should be *complete*, i.e. it should contain everything that is necessary and it should not contain anything that is irrelevant. As will be shown in the following, this requirement is probably the hardest to satisfy since it is common practice in most organizations to perform several kinds of coordination acts tacitly, according to the rule "no news is good news".

We call a business domain model that satisfies these requirements an *enterprise ontology*. Several domain modeling techniques exist propagating the *business specific* aspect of enterprise modeling, but most of them do not satisfy the quality criteria just mentioned. Looking at business processes modeling techniques, being a relevant part of domain modeling, next to the traditional flow charts, there exist e.g. Petri Net [16, 17], Event Driven Process Chains (EPC) [18] and Activity Diagrams [19]. However, in these techniques the notion of business processes is not well defined and there exists no distinction between business and infological actions. Consequently the difference between business processes and other kinds of process remains unclear. This leads to the conclusion that they do not specifically address business processes but can be used for any discrete event process. Other approaches, as e.g. from the Language/Action Perspective (LAP), claim to offer a solution for the mismatch between social perspectives and technical perspectives by explicitly focusing on business specific communication patterns, where social beings achieve changes in the (object) world by means of communication acts [20]. The *enterprise ontology* [21, 22] methodology is an approach that incorporates LAP and that additionally distinguishes between essential (business), infological and datalogical acts and facts, ultimately resulting in the distinction between three aspect organizations: the B-organization (B from business), the I-organization (I from infological), and the D-organization (D from datalogical). The apparent advantages of enterprise ontology – in particular the huge reduction of complexity [23] – led us to choose this methodology for modeling the business domain.

This article builds on previous work regarding enterprise ontology [21, 22] and business components [24-26]. To exemplify the usability of the proposed enterprise ontology and the identification of business components the domain of *strategic supply network development* [27-29] is introduced as a practical application of inter-enterprise information systems. The domain extends the frame of reference in strategic sourcing from a supplier centric to a supply network perspective and addresses a widespread problem in the practical application of inter-enterprise information systems. It is structured as follows: in section 2 the domain of strategic supply network development (SSND) is introduced. SSND is used throughout the article as an example domain for inter-enterprise cooperation. In section 3 we present the method that is used to arrive at a right ontological model of an enterprise. To illustrate it, the ontology of the SSND case is developed. In section 4, the method for identifying business components is applied to the SSND case. Discussions of the results as well as the conclusions that can be drawn are provided in section 5.

2 The Domain of Strategic Supply Network Development

The relevance of the purchasing function in enterprises has increased steadily over the past two decades. Till the 70's, purchasing was widely considered an operational task with no apparent influence on long term planning and strategy development [30]. This narrow perspective was broadened by research that documented the positive influence that a targeted supplier cooperation and qualification could bring to a company's strategic options [31]. In the 80's, trends spurred the recognition of the eminent importance of the development and management of supplier relationships for gaining competitive advantages. Such trends were e.g., the growing globalization, the focus on core competencies in the value chain with connected in-sourcing and out-sourcing decisions, as well as new concepts in manufacturing. As a result, purchasing gradually gained strategic relevance on top of its operational and tactical relevance [32].

Based on these developments, purchasing has become a core function in the 90's. Current empiric research shows a significant correlation between the establishment of a strategic purchasing function and the financial success of an enterprise, independent from the industry surveyed [33, p. 513]. One of the most important factors in this connection is the buyer-supplier-relationship. At many of the surveyed companies, a close cooperation between buyer and supplier led to process improvements and resulting cost reductions that were shared between buyer and suppliers [33, p. 516].

In practice, supplier development is widely limited to suppliers in tier-1, i.e. the direct suppliers. With respect to the superior importance of supplier development, as mentioned above, we postulated the extension of the traditional frame of reference in strategic sourcing from a supplier-centric to a supply-network scope [27-29]. It means the further development of the strategic supplier development to a strategic supply *network* development. This shifted the perspective in the field of strategic sourcing to analyze multi-tier supplier networks instead of single suppliers.

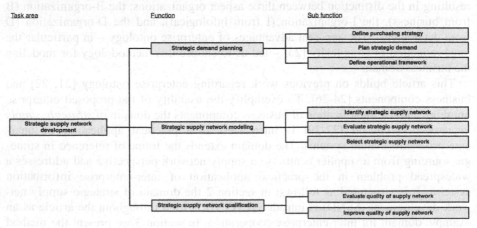

Fig. 1. Functional decomposition diagram for strategic supply network development

The tasks within the strategic supply network development can be grouped into 3 main areas as illustrated in Fig. 1: *strategic demand planning*, *strategic supply network modeling* and *strategic supply network qualification* [34].

Within the function *strategic demand planning*, a corporate framework for all purchasing-related activities is defined. This framework consists of a consistent and corporate-wide valid purchasing strategy (*define purchasing strategy*), a strategic demand planning and demand bundling function (*plan strategic demand*), and the definition of methods and tools to control performance and efficiency of purchasing and to establish a conflict management concept (*define operational framework*).

The function *strategic supply network modeling* provides a methodology for the identification (*identify strategic supply network*), evaluation (*evaluate strategic supply network*) and selection (*select strategic supply network*) of potential suppliers, not only located in tier-1 but also in the subsequent tiers. Using evaluation criteria such as lowest cost, shortest delivery time or best quality, and corresponding evaluation methods, the identified supply networks are evaluated. If there is a positive result on the evaluation, the supply network is selected and contractually linked to the company.

Within the function *strategic supply network qualification*, the quality of a performing supplier network is evaluated using evaluation criteria and evaluation methods (*evaluate quality of supply network*). Dependent on the result of the evaluation, sanctions may be used to improve the quality of the supply network (*improve quality of supply network*).

In this article we focus on the modeling of strategic supply networks and specifically on the identification sub-task. The reason for it is that compared to the traditional strategic purchasing the supplier selection process undergoes the most evident changes in the shift to a supply network perspective. The expansion of the traditional frame of reference in strategic sourcing requires more information than merely data on existing and potential suppliers in tier-1. Instead, the supply networks connected with those suppliers have to be identified and evaluated, e.g., by comparing alternative supply networks. Since the identification of strategic supply networks builds the basis for the evaluation and selection, we introduce shortly the preliminary work done in the area of identification and modeling of strategic supply networks [27, 29, 34].

To model and visualize the network in a structured way, a specific strategic demand for a product to be manufactured is communicated from the OEM to existing and/or potential suppliers. Fig. 2 illustrates an example for an identification process and its results. In the example the OEM is connected to a potential network of suppliers as shown in the left part of Fig. 2. It is assumed that the OEM needs to order two products externally, product 1 and product 2. During the identification process the OEM sends out demands for these products to its strategic suppliers in tier-1. In the example it is assumed that supplier 1-1 and supplier 1-2 get the demand for product 1 while supplier 1-3, supplier 1-4 and supplier 1-5 receive the demand for product 2. These suppliers check whether they can fulfill the demand internally and, if not, send out subsequent demands to their respective suppliers. Each node executes the same process as described until the demand has reached the last tier. The requested information is then returned to the OEM, aggregated and finally visualized as a supply network, in which each participant of the supply network constitutes a network node.

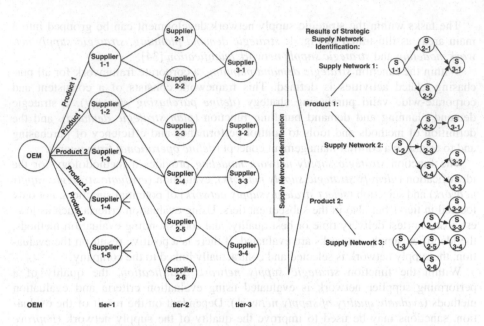

Fig. 2. Example for result of identification process

This process may result in the identification of several possible supply networks as shown in the right part of Fig. 2, where e.g. product 1 can be provided by two supply networks, supply network 1 (root node S1-1) and supply network 2 (root node S1-2), whereas product 2 can only be provided by supply network 3. It is now up to the OEM to decide which of the potential strategic supply networks (in the example above for product 1) will be selected to fulfill its original demand. Therefore an evaluation process is necessary which has been introduced by the authors in [35]; it will not be elaborated in detail in this article.

3 Enterprise Ontology and Its Application to the SSND Case

As motivated in the introduction, we apply enterprise ontology for modeling the business domain. The specific methodology that is used for constructing the ontology of an enterprise is DEMO (Design & Engineering Methodology for Organizations) [20, 21, 36, 37]. As is explained in [21, 36, 37] a distinction is made between production acts and facts and coordination acts and facts. The transaction axiom aggregates these acts/facts into the standard pattern of the (business) transaction. Consequently, two worlds are distinguished: the *production world* (P-world) and the *coordination world* (C-world). The complete ontological model of an organization in DEMO consists of four aspect models (Fig. 3).

The Construction Model (CM) specifies the composition, the environment and the structure of the organization. It contains the identified *transaction types* and the associated *actor roles* and describes the *links* to *production* or *coordination banks*, where all production and coordination facts are stored. The Process Model (PM) details each single transaction type of the CM according to the *universal transaction pattern* [21]. In addition, the *causal* and *conditional relationships* between transactions are exhibited. The Action Model (AM) specifies the *action rules* that serve as guidelines for the actors in dealing with their agenda. It contains one or more action rules for every agendum type. The State Model (SM) specifies the *object classes, fact types* and ontological *coexistence rules* in the production world.

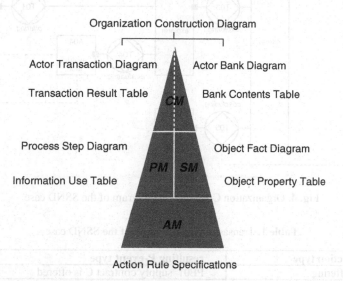

Fig. 3. The four aspect models

In Fig. 3 the CM triangle is split by a dashed line in a left and a right part. This is due to the logical sequence of producing the aspect models. First, the left part of the CM can be made straight away after having applied the elicitation procedure as discussed in [38]. It contains the active influences among actor roles, being initiator or executor of a transaction type. The CM is expressed in an Actor Transaction Diagram and a Transaction Result Table. Next, the Process Step Diagram, which represents a part of the PM, is produced and after that the AM, which contains the Action Rule Specifications. The action rules are expressed in a pseudo-algorithmic language, by which an optimal balance between readability and preciseness is achieved. Then the SM is produced, expressed in an Object Fact Diagram and an Object Property Table. Next, the right part of the CM is produced. It consists of an Actor Bank Diagram and a Bank Contents Table. Usually the Actor Bank Diagram is drawn as an extension of the Actor Transaction Diagram. Together they constitute the Organization Construction Diagram. After that we are able to complete the PM with the Information Use Table, showing which information is used in which process steps.

Based on this method, the ontology for the SSND case has been constructed. Space limitations prohibit us to provide a more extensive account of how the models in the figures below are developed. Also, we will not present and discuss the action rules. Fig. 4 exhibits the Organization Construction Diagram. The corresponding Transaction Result Table is shown in Table 1.

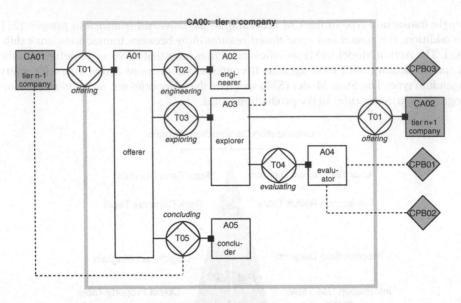

Fig. 4. Organization Construction Diagram of the SSND case

Table 1. Transaction Result Table of the SSND case

transaction type	resulting P-event type
T01 offering	PE01 supply contract C is offered
T02 engineering	PE02 the BoM of assembly A is determined
T03 exploring	PE03 supply contract C is a potential contract
T04 evaluating	PE04 supply contract C is evaluated
T05 concluding	PE05 supply contract C is concluded

The top or starting transaction type is the offering transaction T01. Instances of T01 are initiated by the environmental actor role CA01, which is a company in tier n-1 and executed by actor role A01 (offerer). This company asks the direct supplier (company in tier-n) for an offer regarding the supply of a particular product P. In order to make such an offer, A01 first initiates an engineering transaction T02, in order to get the bill of material of the requested product P. This is a list of (first-level) components of P, produced by A02 (engineerer). Next, A01 asks A03 (explorer) for every such component to get offers from companies that are able to supply the component. So, a number of exploring transactions T03 may be carried out within one T01, namely as many as there are components of P that are not produced by the tier-n company.

In order to execute each of these transactions, A03 has to ask companies for an offer regarding the supply of a component of P. Since this is identical to the starting transaction T01, we model this also as initiating a T01. Now however, the executor of the T01 is a company in tier (n+1). Consequently, the model that is shown in Fig. 4 must be understood as to be repeated recursively for every tier until the products to be supplied are elementary, i.e. non-decomposable. Note that, because of the being recursive, an offer (the result of a T01) comprises the complete bill of material of the concerned component of P.

Every offer from the companies in tier n+1 is evaluated in a T04 transaction by the actor role A04 (evaluator). So, there is a T04 for every 'output' T01, whereby each company can use their own evaluation and decision rules. The result of a T04 is a graded offer for some component of P. So, what A03 delivers back to A01 is a set of graded offers for every component of P. Next, A01 asks A05 (concluder), for every component of P, to select the best offer. The result is a set of concluded offers, one for every component of P. This set is delivered to A01. Lastly, A01 delivers a contract offer for supplying P, together with the set of concluded offers for delivering the components of P. Because of the recursive character of the whole model, this offer includes the complete bill of material of P, regardless its depth.

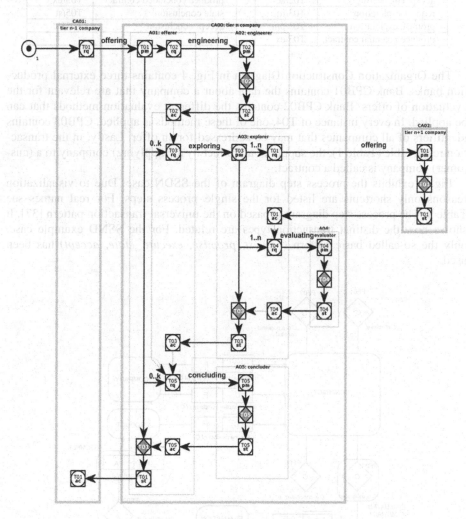

Fig. 5. Process Step Diagram of the SSND case

Table 2. Assignment of process steps to shortcuts

process steps names	shortcuts		process steps names	Shortcuts
request offering	T01/rq		state exploration	T03/st
promise offering	T01/pm		accept exploration	T03/ac
produce contract offering	T01/ex		request evaluation	T04/rq
state offering	T01/st		promise evaluation	T04/pm
accept offering	T01/ac		produce contract evaluation	T04/ex
request engineering	T02/rq		state evaluation	T04/st
promise engineering	T02/pm		accept evaluation	T04/ac
produce BoM explosion	T02/ex		request conclusion	T05/rq
state engineering	T02/st		promise conclusion	T05/pm
accept engineering	T02/ac		produce concluded contract	T05/ex
request exploration	T03/rq		state conclusion	T05/st
promise exploration	T03/pm		accept conclusion	T05/ac
produce potential contract	T03/ex			

The Organization Construction Diagram in Fig. 4 contains three external production banks. Bank CPB01 contains the data about a company that are relevant for the evaluation of offers. Bank CPB02 contains the different evaluation methods that can be applied. In every instance of T04, one of these methods is applied. CPB03 contains identifiers of all companies that may be addressed for an offer. Lastly, in the transaction result table (Table 1), the supply of a product by a (supplying) company to a (customer) company is called a contract.

Fig. 5 exhibits the process step diagram of the SSDN case. Due to visualization reasons only shortcuts are listed for the single process steps. For real names see Table 2. The process step diagram is based on the universal transaction pattern [37]. It shows how the distinct transaction types are related. For the SSND example case, only the so-called basic pattern (*request, promise, execute, state, accept*) has been used.

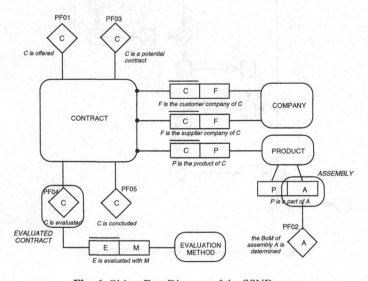

Fig. 6. Object Fact Diagram of the SSND case

Table 3. Object Property Table of the SSND

property type	object class	scale
< company information >	COMPANY	< aggregated data >
< contract terms >	CONTRACT	< aggregated data >
sub_products(*)	PRODUCT	set of PRODUCT
#sub_products(*)	PRODUCT	NUMBER
suppliers(*)	CONTRACT * PRODUCT	set of COMPANY
sub_contracts(*)	CONTRACT	set of CONTRACT
evaluation_mark	CONTRACT	NUMBER
evaluation_marks(*)	CONTRACT	set of NUMBER

From the state T01/pm (promised) a number of transactions T03 (possibly none) and a number of transactions T05 (possibly none) are initiated, namely for every first-level component of a product. This is expressed by the cardinality range 0..k. Likewise, from the state T03/pm, a number of transactions T01 and a number of transactions T04 are initiated, namely for every offer or contract regarding a first-level component of a product. The dashed arrows, from an accept state (e.g. T02/ac) to some other transaction state, represent waiting conditions. So, for example, the performance of a T03/rq has to wait for the being performed of the T02/ac.

Fig. 6 exhibits the Object Fact Diagram and Table 3 the Object Property Table. Together they constitute the State Model of the example case.

The Object Fact Diagram is a variant of the Object Role Modeling (ORM) model [39]. One of the major differences is that the OFD contains diamond-shaped unary fact types for representing the results of transactions, i.e., the production fact types. They correspond with the transaction results in Table 1. A roundangle around a fact type or a role defines a concept in an extensional way, i.e. by specifying the object class that is its extension. For example, the roundangle around the production fact type "C is evaluated" defines the concept of evaluated contract. Lastly, the roundangle around the role "A" of the fact type "P is a part of A" defines all assemblies, i.e. all products that do have parts. Properties are binary fact types that happen to be pure mathematical functions, of which the range is set of, usually ordered, values, called a scale. Instead of including them in an Object Fact Diagram they can be more conveniently represented in an Object Property Table. The property types marked by "(*)" in the Object Property Table are derived fact types. The derivation rules are as follows:

sub_products (P)   = {X | X ∈ PRODUCT **and** X is a part of P};
#sub_products(P)   = <u>card</u>(sub_products (P));
suppliers(F,P)   = {X | X ∈ COMPANY **and** (∃Y ∈ CONTRACT: P is the
                            product of Y **and** X is the supplying company of Y **and**
                            F is the customer company of Y)};
sub_contracts(C)   = {X | X ∈ CONTRACT **and** the product of X is Z
                            and the product of C is Y **and** Z is a part of Y};
evaluation_marks (C)   = {X | X is an evaluation mark of C};

The information items as defined in the SM, including the derived fact types, constitute all information that is needed to develop a supply network for a particular product.

4   Identification of Business Components in the SSND Case

The set of ontological models of an enterprise, as discussed above, constitutes the basis for the identification of business components. A business component provides a set of services out of a given business domain through well-defined interfaces and hides its implementation [40]. The specification of a business component is defined as a "complete, unequivocal and precise description of its external view" [41]. The enterprise ontology, satisfying the quality criteria (introduced in section 1) of providing a *complete* domain model, being described at the *right granularity or level of detail* and modeling only the *essential business actions* provides the necessary basis for the realization of business components. The enterprise ontology for the SSND case contains all essential information for that business domain.

The business components identification (BCI) method [24-26] applied in this section aims at grouping business tasks and their corresponding information objects into business components satisfying defined metrics. The metrics used – being loosely coupling and tight cohesion – are the basic metrics for the component-based development of inter-enterprise business applications focusing on the deployment of components that can be used in different enterprise systems [42-44]. According to the classification of business components identification methods by [44], the BCI method combines Clustering Analysis and CRUD (Create, Read, Update, Delete) based methods. With the BCI method we satisfy the recommendation of Wang et al. of combining current component identification methods in order to achieve satisfactory results.

Since the identification of business components is strongly dependant on the underlying business model, the BCI method uses the object classes and fact types from the State Model and the process steps from the Process Model, including their relationships. One can distinguish between three types of relationships necessary for the identification of business components. The relationship between single process steps, the relationship between information objects and the relationship between process steps and information objects. A relationship type distinguishes between subtypes expressing the significance of a relationship. E.g. the relationship between single process steps expresses – based on their cardinality constraints – how often a process step is executed within a transaction and therefore how close two process steps are related to each other in that business domain. The relationship between information objects defines how loosely or tightly the information objects are coupled, and the relationship between process steps and information objects defines whether a corresponding information object is used, created or altered while executing the respective process step. All types of relationship are of great relevance in order to define which information object and process steps belong to which component.

The relationships are modeled in the BCI method using a weighted graph. As the nodes represent information objects and process steps, the edges characterize the relationships between the nodes. Weights are used to define the different types and subtypes of relationships and build the basis for assigning nodes and information objects to components. Due to display reasons the graph is visualized in a three-dimensional representation having the process steps and information objects arranged in circles, and without showing the corresponding weights (see Fig. 7).

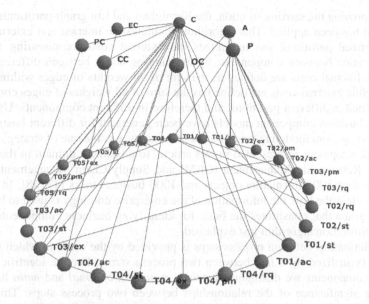

Fig. 7. Relationships between process steps and information objects

The graph shows all process steps and information objects as well as the relationships between them. For the purpose of clarity, only the shortcuts of all nodes are shown. For the real names, please see Table 2 and Table 4.

Table 4. Assignment of information object names to shortcuts

information object name	shortcut
Product	P
Assembly	A
Contract	C
Evaluated Contract	EC
Potential Contract	PC
Concluded Contract	CC
Offered contract	OC

In order to ensure optimal grouping satisfying the defined metrics of loosely coupling and tight cohesion, an optimization problem needs to be solved for which a genetic algorithm has been developed. The algorithm starts with a predefined solution (specific assignment of process steps and information objects to components) and generates better solutions through iterations. The starting solution is generated using a greedy graph-partitioning algorithm [45]. Using a model that provides a complete description of the relevant business domain implicates that a better starting solution is generated. The quality of the starting solution determines the quality of the resulting business components. The enterprise ontology provides an adequate basis for gaining all relevant information for building an adequate starting solution.

For improving the starting solution, the Kernighan and Lin graph-partitioning algorithm [46] has been applied. This algorithm calculates the internal and external costs of the defined partitions and minimizes the external costs, representing minimal communication between components, by exchanging nodes between different graph partitions. Internal costs are defined as the sum of all weights of edges within a component, while external costs are defined as the sum of all weights of edges connecting nodes located in different partitions, and therefore in different components. Using this algorithm business component models have been generated for different business domains, starting from modeling small business domains, e.g. the one of strategic supply network development as presented in this article, to a business domain in the area of Customer Relationship Management (CRM) and Supply Chain Management (SCM) with more then 500 information objects and 1000 business tasks [47, 48]. In the following we describe how the information of the enterprise ontology is used to build the weighted graph that constitutes the basis for identifying business components, based on the optimization algorithm just explained.

The relationship between process steps is provided by the order in which they are executed (visualized by a *link* between two process steps). For the identification of business components we distinguish between *standard*, *optional* and *main* links, defining the significance of the relationships between two process steps. This is expressed in Fig. 5 by the cardinality constraints assigned to the links. If a link has the cardinality range 0..1 we call it an *optional* link, since the following process step does not need to be executed. If the cardinality range is set to 0..n then we speak from a *main* link, indicating that the following process step is executed at least once. If no cardinality range is assigned to a link, we call it a *standard* link having the following process step executed exactly once. The different types of links are modeled by different weights in the graph. The process steps and their relationships are shown in Fig. 8, positioning all process steps as nodes in a circle and inserting edges for the relationship between two business services.

Let us have a closer look at the information objects relevant for the information system to be built. Fig. 6 and Table 3 introduce different types of potential information objects, namely the object classes, fact types and property types. Object classes

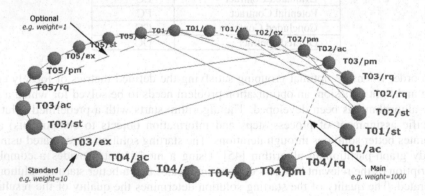

Fig. 8. Relationship between process steps

that are provided by external information systems, are traded in a special way in BCI method. They concern the data that are contained in the external production banks (see Fig. 4). Due to space limitations we omit this kind of information here. The remaining information objects relevant for the component identification of the SSND case and their relationships are shown in Fig. 9. Regarding the different types of relationships between single information objects we distinguish between *part-of, state-of* and *related-to* expressed by means of edge weights. In the example case a 'Contract' (C) that has been evaluated is a *state-of* 'Contract' (C). Additionally, we have 'Product' (P) being *part-of* 'Assembly' (A), whereas 'Contract' (C) is *related-to* a specific 'Product' (P).

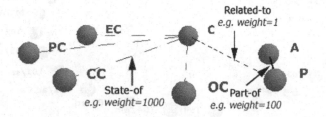

Fig. 9. Relationship between information objects

Next, the relationships between information objects and business services are examined. The information to define those relationships can be gained from the Create/Use Table, see Table 5, which is an extension of the Information Use table of the enterprise ontology.

Table 5. Create/Use Table of the SSND case

object class or fact type	process steps
PRODUCT	T01/rq T01/pm *T02/ex* T02/rq T02/pm T02/st T02/ac T03/rq T03/pm
product P is a part of product A	*T02/ac*
the BoM of assembly A is determined	*T02/ac*
ASSEMBLY	*T02/ex* T03/ex
CONTRACT	T01/ex T01/st T01/ac T03/ex T03/st T03/ac T04/ex T04/rq T04/pm T04/st T04/ac T05/ex T05/rq T05/pm T05/st T05/ac
supply contract C is offered	*T01/ac*
supply contract C is a potential contract	*T03/ac*
supply contract C is evaluated with method M	*T04/ac*
supply contract C is concluded	*T05/ac*

It contains object classes and fact types that are assigned to specific process steps by differentiating objects that are *used* and objects that are *created*. The process steps in which an instance of an object class or property type is created are printed in italic, as are the production fact types (transaction results). For the SSND example, the relationships between information objects and process steps and their weights are visualized in Fig. 10.

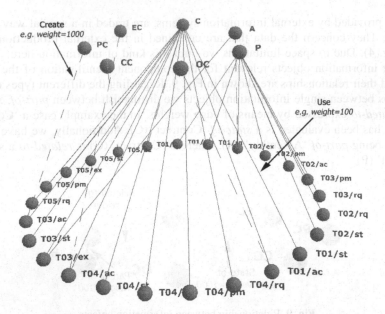

Fig. 10. Relationship between process steps and information objects

Combining the three dimensions of node dependencies – between information objects, between business services and between information objects and business services (shown in Fig. 8, Fig. 9 and Fig. 10) – results in the three dimensional model as shown in Fig. 7. This model is the basis for applying the genetic algorithms in order to cluster related nodes within single components. Applying the BCI method to that graph results in the following graph clustering (see Fig. 11).

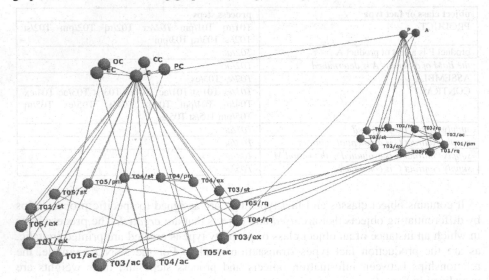

Fig. 11. Identified business components

Now that all the nodes in the pictures are ordered and grouped as resulting from the optimization algorithm, two business components can be identified immediately: one containing the business tasks and information objects related to *product management* and one containing the business tasks and information objects related to *contract management*. The two components are therefore named *Product Manager* and *Contract Manager* and are shown in Fig. 12 according to the UML 2.0 notation [19].

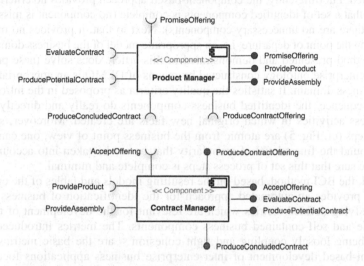

Fig. 12. Identified business components with their provided and required services

It is evident that the derivation of the components is strongly dependent on the weights chosen. Different results could be achieved in assigning different weights to the edges in the graph. In applying the BCI method to many real world cases, we aim at deriving reference values for the different weights.

Since the identified business components provide business services to or require business services from other business components either within an enterprise or between enterprises, the single business component services need to be defined. These services are related to the process steps executed in a business transaction. From the diagram in Fig. 5 – describing the process steps of the transaction types introduced in Fig. 4 – the business services are gained and modeled in Fig. 12 either as required services (composed of a semicircle attached to the related component by a line) or as provided services (composed of a circle attached to the component by a line).

Even if for this small example only two business components were found, the derivation of the services and information objects assigned to the components would not have been possible without a detailed analysis and modeling of the business domain. This information is essential and provides the basis for the implementation of the business components. Place limitations prevented us to derive components from a larger example domain.

5 Conclusions

In this article, we have addressed the impact of using an enterprise ontology for identifying business components, defined as the highest level software components, directly supporting business activities. Although component-based software development may be superior to more traditional approaches, it leaves the problem of requirements engineering unsolved. Put differently, the component-based approach provides no criteria for determining that a set of identified components is complete (no component is missing) and minimal (there are no unnecessary components). Next to that, it provides no means for determining the point of departure, i.e. an appropriate model of the business domain.

The method presented and demonstrated in this article does solve these problems. First, the enterprise ontology constructed by means of DEMO is an appropriate model of the business domain. It satisfies the quality criteria as proposed in the introduction. As a consequence, the identified business components do really and directly support the business activities in which original new facts are created. Moreover, since the process steps (cf. Fig. 5) are atomic from the business point of view, one can be sure to have found the finest level of granularity that has to be taken into account. Also, one can be sure that this set of process steps is complete and minimal.

Second, the BCI method, based on the resulting models and tables of the enterprise ontology, provides an automated approach for the identification of business components satisfying defined metrics which are relevant for the development of reusable, marketable and self-contained business components. The metrics introduced in this article – being loosely coupling and tight cohesion – are the basic metrics for the component-based development of inter-enterprise business applications focusing on the deployment of components that can be on different enterprise systems.

The usability of the proposed enterprise ontology and the identification of business components is illustrated by means of a practical application of the domain of strategic supply network development. The approach seams to be very promising, but for the identification of the business components additional simulations with different weights and different optimization algorithms are necessary in order to make it applicable in practice.

References

1. Burtler, P., et al.: A Revolution in Interaction. The McKinsey Quarterly 1/97, 4–23 (1997)
2. Davenport, T.H.: The Future of Enterprise System-Enabled Organizations. Information Systems Frontiers 2(2), 163–180 (2000)
3. Davenport, T.H., Short, J.E.: The New Industrial Engineering: Information Technology and Business Process Redesign. Sloan Management Review 31(4), 11 (1990)
4. Malone, T.W., Lautbacher, R.J.: The Dawn of the E-Lance Economy. Harvard Business Review, 145–152 (September-October 1998)
5. Pine, B.J., Victor, B., Boynton, A.C.: Making Mass Customization Work. Havard Business Review 36(5), 108–119 (1993)
6. Tapscott, D., Ticoll, D., Lowy, A.: Digital Capital: Harnessing the Power of Business Webs, Boston (2000)
7. Kopanaki, E., et al.: The Impact of Inter-organizational Information Systems on the Flexibility of Organizations. In: Proceedings of the Sixth Americas Conference on Information Systems (AMCIS), Long Beach, CA (2000)

8. Lambert, D.M., Cooper, M.C.: Issues in Supply Chain Management. Industrial Marketing Management 29(1), 65–83 (2000)
9. Barbier, F., Atkinson, C.: Business Components. In: Barbier, F. (ed.) Business Component-Based Software Engineering, pp. 1–26. Kluwer Academic Publishers Group, Dordrecht (2003)
10. Jang, Y.-J., Kim, E.-Y., Lee, K.-W.: Object-Oriented Component Identification Method Using the Affinity Analysis Technique. In: Konstantas, D., Léonard, M., Pigneur, Y., Patel, S. (eds.) OOIS 2003. LNCS, vol. 2817, pp. 317–321. Springer, Heidelberg (2003)
11. Levi, K., Arsanjani, A.: A Goal-driven Approach to Enterprise Component Identification and Specification. Communications of the ACM 45(10) (2002)
12. Réquilé-Romanczuk, A., et al.: Towards a Knowledge-Based Framework for COTS Component Identification. In: International Conference on Software Engineering (ICSE 2005), St. Louis, Missouri, USA. ACM Press, New York (2005)
13. Vitharana, P., Zahedi, F., Jain, H.: Design, Retrieval, and Assembly in Component-based Software Development. Communications of the ACM 46(11) (2003)
14. Dietz, J.L.G., Barjis, J.A.: Petri net expressions of DEMO process models as a rigid foundation for requirements engineering. In: 2nd International Conference on Enterprise Information Systems, Escola Superior de Tecnologia do Instituto Politécnico, Setúbal (2000); ISBN: 972-98050-1-6
15. Dietz, J.L.G.: Deriving Use Cases from Business Process Models. In: Song, I.-Y., Liddle, S.W., Ling, T.-W., Scheuermann, P. (eds.) ER 2003. LNCS, vol. 2813, pp. 131–143. Springer, Heidelberg (2003)
16. van der Aalst, W.M.P., van Hee, K.M.: Workflow Management: Models, Methods and Tools. MIT Press, MA (2001)
17. Jensen, K.: Coloured Petri Nets. Basic Concepts, Analysis Methods and Practical Use. Monographs in Theoretical Computer Science, Basic Concepts, vol. 1. Springer, Heidelberg (1997)
18. Scheer, A.-W.: ARIS - Business Process Modeling, 2nd edn. Springer, Berlin (1999)
19. UML, OMG Unified Modelling Language, Version 2.0. n.d, http://www.omg.org/technology/documents/modeling_spec_catalog.htm#UML
20. van Reijswoud, V.E., Mulder, J.B.F., Dietz, J.L.G.: Speech Act Based Business Process and Information Modeling with DEMO. Information Systems Journal (1999)
21. Dietz, J.L.G.: Enterprise Ontology - Theory and Methodology. Springer, Heidelberg (2006)
22. Dietz, J.L.G., Habing, N.: A meta Ontology for Organizations. In: Meersman, R., Tari, Z., Corsaro, A. (eds.) OTM-WS 2004. LNCS, vol. 3292, pp. 533–543. Springer, Heidelberg (2004)
23. Dietz, J.L.G.: The Deep Structure of Business Processes. Communications of the ACM 49(5) (2006)
24. Albani, A., Dietz, J.L.G., Zaha, J.M.: Identifying Business Components on the basis of an Enterprise Ontology. In: Interop-Esa 2005 - First International Conference on Interoperability of Enterprise Software and Applications, Geneva, Switzerland (2005)
25. Albani, A., Dietz, J.L.G.: The benefit of enterprise ontology in identifying business components. In: IFIP World Computing Conference. Santiago de Chile (2006)
26. Albani, A., et al.: Domain Based Identification and Modelling of Business Component Applications. In: Kalinichenko, L.A., Manthey, R., Thalheim, B., Wloka, U. (eds.) ADBIS 2003. LNCS, vol. 2798, pp. 30–45. Springer, Heidelberg (2003)
27. Albani, A., et al.: Dynamic Modelling of Strategic Supply Chains. In: Bauknecht, K., Tjoa, A.M., Quirchmayr, G. (eds.) EC-Web 2003. LNCS, vol. 2738, pp. 403–413. Springer, Heidelberg (2003)
28. Albani, A., Müssigmann, N., Zaha, J.M.: A Reference Model for Strategic Supply Network Development. In: Loos, P., Fettke, P. (eds.) Reference Modeling for Business Systems Analysis. Idea Group Inc. (2006)

29. Albani, A., Winnewisser, C., Turowski, K.: Dynamic Modelling of Demand Driven Value Networks. In: Meersman, R., Tari, Z. (eds.) OTM 2004. LNCS, vol. 3290, pp. 408–421. Springer, Heidelberg (2004)
30. McIvor, R., Humphreys, P., McAleer, E.: The Evolution of the Purchasing Function. Journal of Strategic Change 6(3), 165–179 (1997)
31. Ammer, D.: Materials Management, 2nd edn. Homewood (1968)
32. Kaufmann, L.: Purchasing and Supply Management - A Conceptual Framework. In: Hahn, D., Kaufmann, L. (eds.) Handbuch Industrielles Beschaffungsmanagement, pp. 3–33. Wiesbaden (2002)
33. Carr, A.S., Pearson, J.N.: Strategically managed buyer - supplier relationships and performance outcomes. Journal of Operations Management 17, 497–519 (1999)
34. Albani, A., et al.: Component Framework for Strategic Supply Network Development. In: Benczúr, A.A., Demetrovics, J., Gottlob, G. (eds.) ADBIS 2004. LNCS, vol. 3255, pp. 67–82. Springer, Heidelberg (2004)
35. Albani, A., Müssigmann, N.: Evaluation of Strategic Supply Networks. In: Meersman, R., Tari, Z., Herrero, P. (eds.) OTM-WS 2005. LNCS, vol. 3762, pp. 582–591. Springer, Heidelberg (2005)
36. Dietz, J.L.G.: The Atoms, Molecules and Fibers of Organizations. Data and Knowledge Engineering 47, 301–325 (2003)
37. Dietz, J.L.G.: Generic recurrent patterns in business processes. In: Mira, J., Álvarez, J.R. (eds.) IWANN 2003. LNCS, vol. 2687. Springer, Heidelberg (2003)
38. Dietz, J.L.G.: The notion of business process revisited. In: Meersman, R., Tari, Z. (eds.) OTM 2004. LNCS, vol. 3290, pp. 85–100. Springer, Heidelberg (2004)
39. Halpin, T.A.: Information Modeling and Relational Databases. Morgan Kaufmann, San Francisco (2001)
40. Fellner, K., Turowski, K.: Classification Framework for Business Components. In: Proceedings of the 33rd Annual Hawaii International Conference On System Sciences, Maui, Hawaii. IEEE, Los Alamitos (2000)
41. Ackermann, J., et al.: Standardized Specification of Business Components, Gesellschaft fÜr Informatik, Working Group 5.10.3, Technical Report, Augsburg (2002)
42. Jain, H., Chalimeda, N.: Business Component Identification - A Formal Approach. In: Proceedings of the Fifth International Enterprise Distributed Object Computing Conference (EDOC 2001). IEEE Computer Society, Los Alamitos (2001)
43. Kim, S.D., Chang, S.H.: A Systematic Method to Identify Software Components. In: 11th Asia-Pacific Software Engineering Conference (APSEC) (2004)
44. Wang, Z., Xu, X., Zhan, D.: A Survey of Business Component Identification Methods and Related Techniques. International Journal of Information Technology 2(4), 229–238 (2005)
45. Jungnickel, D.: The Greedy Algorithm. In: Jungnickel, D. (ed.) Graphs, Networks and Algorithms, pp. 123–146. Springer, Berlin (2005)
46. Kernighan, B.W., Lin, S.: An efficient heurisitc procedure for partitioning graphs. Bell Systems Technical Journal 49, 291–307 (1970)
47. Selk, B., Klöckner, K., Albani, A.: Enabling interoperability of networked enterprises through an integrative information system architecture for CRM and SCM. In: International Workshop on Enterprise and Networked Enterprises Interoperability (ENEI 2005), Nancy, France (2005)
48. Selk, B., et al.: Experience Report: Appropriateness of the BCI-Method for Identifying Business Components in large-scale Information Systems. In: Conference on Component-Oriented Enterprise Applications (COEA 2005) in conjunction with the Net.Objectdays, Erfurt, Germany (2005)

SOA Pragmatism

Tony C. Shan

IBM
10712 Hellebore Rd, Charlotte NC 28213, USA
tonycshan@yahoo.com

Abstract. This paper presents a pragmatic approach composed of Methodology, Automation, Patterns, and Strategy (MAPS), to effectively manage the architecture design practices and solution development lifecycle of information systems in a service-oriented paradigm. The key challenges in SOA are discussed, such as architecture complexity, evolving technologies, immature governance, fragmented specification efforts, and disparate visual notations. This comprehensive framework aims to provide a mature integration of appropriate knowledge and capabilities to filter the inessential from the essential. In the Methodology dimension, a hybrid method, SOA philosophy, and a methodical approach are the key components. The Automation dimension covers tools, service lifecycle, and COTS mapping. The prominent elements of the Patterns dimension are data caching patterns, reference model, and open source reference implementation. Finally, the Strategy dimension addresses the strategy metamodel, technology architecture planning, and strategy roadmapping. In addition, a 9-point list of SOA wisdom is articulated, which gives best-practice guidelines to adopt and implement SOA pragmatically in large organizations from a practitioner's perspeoctive.

Keywords: Service-oriented, computing, framework, model, architecture, process, integration, environment, technology, development, management, roadmap, infrastructure, standards, practice, pattern, method, taxonomy, strategy.

1 Introduction

Information Technology (IT) differs itself from other industries in that it is made of bits rather than atoms. Over the years, IT has achieved remarkable success, becoming pervasive in people's daily life. However, the IT uniqueness also brings about the poor project execution and defect rate that is unimaginable in other vertical sectors like manufacturing. A recent Chaos Report [1] by the Standish Group revealed some astonishing findings – almost half of IT projects in 2006 were behind schedule, over budget, or do not meet all the requirements, whereas close to one fifth of projects completely failed, i.e. being canceled or not deployed to the field. There have been a number of high-profile project failures in the past few years, wasting hundreds of millions of dollars. For example, the Virtual Case File System [2] by FBI was canceled with an estimated loss of $170 million in 2005. The Department of Homeland Security had to close the Emerge2 program [3] worth $229 million due to project failure in September 2006.

J. Filipe et al. (Eds.): ICSOFT/ENASE 2007, CCIS 22, pp. 23–28, 2008.

One of the root causes of these immature happenings is the architecture complexity that has grown exponentially in the evolution of several generations of the computing paradigm. A great many Enterprise Architecture (EA) programs in large firms are actually in a primitive shape, generally lack of best-practice disciplines. The IT architecture management and governance are inefficient in many cases, coupled with the organization, culture, and behavior challenges, which leads to inability to deal with the dynamic nature of today's IT responsiveness.

2 SOA Challenges

From a systems development lifecycle (SDLC) standpoint, it is not uncommon that the design efforts of software-intensive systems are disjointed, similar to the old tale of "The Blind Men and the Elephant." As the architecture complexity continuously grows dramatically, the limitation of the fragmented IT development practices tend to become more visible and devastating. Naturally, SOA is anticipated to be a miracle cure to solve all these problems, which could lead to a more catastrophic failure in the enterprise computing environments, if the expectation is not set right and the impact of SOA is mis-interpreted. In the real-world scenarios, the SOA activities without appropriate mindsets, well-defined objectives, rationalized reasoning, fact-based justification criteria, leadership support, and effective operating models often result in even worse outcome than without SOA.

A service-oriented solution can be viewed from different dimensions. In general, there are three key aspects as described below:

- Business Aspect
    - o   Business architecture
    - o   Value chain model
    - o   Business solution priorities
    - o   Modeling and taxonomy, e.g. BMM, BPMN, CBM
- Technology Aspect
    - o   Functionality expressed as services
    - o   Map to business processes
    - o   Based on industry standards (ACORD, iXRetail, eTOM, IFW, IFX, SCOR, HL7, NHI)
- Operations Aspect
    - o   Quality of services
    - o   Middleware and integration
    - o   Platforms and frameworks
    - o   Network and storage
    - o   Virtualization and utility computing

Facing the magnitude of the complexity in a service-oriented solution, some of the key SOA challenges are increasing dynamics, growing integration points, disparate visual notations, fragmented WS-* specification efforts, and barriers to deploy functional governance, which collectively further widen the gap of communications. For example, a typical large-size enterprise system today deals with the composition and integration of enterprise portal, enterprise services, user authentication, entitlement services, customer applications, business process management, service registry/ repository, service management, enterprise service bus, business activity monitoring, legacy systems, EAI, reporting, application frameworks, enterprise information integration, business intelligence, storage, infrastructure, network, quality of services, non-functional requirements, tools, etc. Many practitioners agree that a critical factor to the success of SOA adoption is the effectiveness of handling the complexity of the large-scale architecture as the size and number of details become virtually uncontrollable or unmanageable via traditional methods.

A widely-used approach to coping with complexity is divide-and-conquer, in an attempt to decompose the architecture to the level that is simple enough to address the design concerns and provide solutions in a straightforward way. The abstraction technique is also commonly applied to create the simplified models by hiding and filtering details that are irrelevant to the level of considerations. Complexity can be classified into *logical* – proof of proof of correctness becomes difficult, long, and impossible, *psychological* – comprehensibility to understand, and *structural* – number of elements. Various complexity metrics exist such as the graph-theoretic measure. Nevertheless, Fred Brooks pointed out that the essential complexity is irreducible [4]. As Grady Booch put it [5], the most fruitful path of investigation ties architectural complexity to entropy, i.e., the measure of the disorganization of a system and/or the size of the state space of the system. I can "feel" that one system or implementation is more or less complex than another, but I can't empirically defend that feeling in many cases. In reality, handling architectural complexity is essentially in an art format. We argue that wisdom should be leveraged for better analyzability in this regard.

3 Pragmatic Approach

As defined in the Merriam-Webster dictionary, wisdom is accumulated philosophic or scientific learning-knowledge and ability to discern inner qualities and relationships-insight. According to Doc Childre and Bruce Cryer, wisdom implies a mature integration of appropriate knowledge, a seasoned ability to filter the inessential from the essential [6].

In plain English, SOA wisdom is a whole array of better-than-ordinary ways of dealing with SOA pragmatically, making the best use of available knowledge and often incomplete information – a trait developed by experience, but not taught. We need to take advantage of SOA wisdom to move from chaos to coherence.

A pragmatic SOA approach is proposed, which consists of Methodology, Automation, Patterns, and Strategy (MAPS). The methodology dimension is with regard to the disciplined ways to execute SOA. The automation dimension deals with repeatable and objective processes. The patterns dimension documents the best practices and lesson learned, to avoid reinventing the wheel. Finally, the strategy dimension lays out the plans to run SOA in a controlled mode.

Figure 1 illustrates the key characteristics of each dimension in the MAPS model. The *Methodology* dimension addresses the SOA methods that deal with process, people, technology, and information. Standard notations are advocated in modeling and model transformation in round-trip engineering as a discipline in the end-to-end sustainable lifecycle management. Agile methods are leveraged for incremental iterations in project execution. Common terminology, taxonomy and ontology are defined to facilitate effective communications among the stakeholders of different background in a semantic fashion. Key indicators are specified to quantify the progress and growth in terms of the maturity of the SOA program. Service governance practices are enforced to properly manage the changes and issues related to culture, organization, and behavior, with roles and responsibilities clearly identified in a RACI matrix. In addition, streamlined methods in a lightweight format are applied to medium- and small-size projects, to promote pragmatism and practicality.

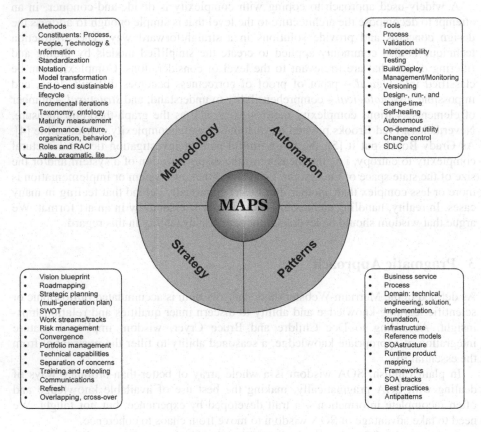

- Methods
- Constituents: Process, People, Technology & Information
- Standardization
- Notation
- Model transformation
- End-to-end sustainable lifecycle
- Incremental iterations
- Taxonomy, ontology
- Maturity measurement
- Governance (culture, organization, behavior)
- Roles and RACI
- Agile, pragmatic, lite

- Tools
- Process
- Validation
- Interoperability
- Testing
- Build/Deploy
- Management/Monitoring
- Versioning
- Design-, run-, and change-time
- Self-healing
- Autonomous
- On-demand utility
- Change control
- SDLC

- Vision blueprint
- Roadmapping
- Strategic planning (multi-generation plan)
- SWOT
- Work streams/tracks
- Risk management
- Convergence
- Portfolio management
- Technical capabilities
- Separation of concerns
- Training and retooling
- Communications
- Refresh
- Overlapping, cross-over

- Business service
- Process
- Domain: technical, engineering, solution, implementation, foundation, infrastructure
- Reference models
- SOAstructure
- Runtime product mapping
- Frameworks
- SOA stacks
- Best practices
- Antipatterns

Fig. 1. MAPS Model

The *Automation* dimension makes use of both commercial and open source tools for various tasks in the SOA project development process, including design-time, run-time and change-time, such as service versioning, testing, build/deploy, service management, business activity monitoring, compliance validation, and service repository. Conformance to the service and platform interoperability is crucial for a heterogeneous environment, using industry standards such as WS-I Basic Profile and Basic Security Profile. Advanced computing techniques are taken advantage of to provide service location transparency in an autonomous on-demand fashion, using workload management, virtualization, grid computing, self-healing, fault-tolerance, real-time optimization, and so on. Automatic tools help facilitate the change control more efficiently in the SDLC.

The *Patterns* dimension captures and documents the reusable best practices for business services and processes. Patterns can be grouped into different domains: technical, engineering, solution, implementation, foundation, and infrastructure. Reference models are built as an abstract SOA pattern, on which SOA stacks are based. SOA frameworks are constructed accordingly, where runtime products are mapped to individual capabilities to form a combination of SOA logical model and physical model, called SOAstructure. Traditional patterns such as GoF design patterns, Java EE patterns, integration patterns, and analysis patterns are aggregated to be incorporated in the SOA pattern library. Moreover, antipatterns are recorded to keep track of pitfalls and lessons learned in the project executions.

The *Strategy* dimension lays out a vision blueprint and alignment from a strategic viewpoint. A multi-generation plan is created to roadmap a variety of work tracks and streams to accomplish long-term and short-term SOA goals and realize technical capabilities. Different priorities, timelines, funding, constraints, overlapping, and drivers from various stakeholders and portfolios are thoroughly analyzed via techniques like SWOT to justify the tradeoffs and drive out a balanced approach with minimum risk of failures. A key to the success of the strategic SOA planning is the separation of concerns and duties. The deliberations and decisions must be well-communicated and vetted. Appropriate trainings need to be planned to retool the skillset of the project teams. The strategy contents must be refreshed periodically to reflect the changes of the conditions and the advance in the field.

4   Pragmatism Guidelines

The following is a list of best practice guidelines to implement pragmatism in real-world SOA projects:

1. **Strategy:** set a solid plan as a foundation and develop a SOA blueprint
2. **Ontology:** define a taxonomy with both business and technology semantics as well as service metadata
3. **Agreement:** enforce design by contract, and leverage repositories for automated interface and contract management
4. **Web 2.0/3.0:** migrate to social and collaborative computing with rich user experience via Wiki, RIA, mashups, and semantic web
5. **Integration:** focus on interoperability and portability in composite applications and systems integration
6. **Standardization:** build architecture based on industry standards and open platforms
7. **Discovery:** enable dynamic binding and better findability with registry and repository
8. **Optimization:** achieve evolutionary improvement through incremental advances and maturation
9. **Management:** leverage policy-driven governance in SOA lifecycle, including design-time, run-time, and change-time

As a reference, a case study of a pragmatic SOA implementation in the financial industry is presented in a separate publication [7].

5 Conclusions

Given the fact that the complexity, size and number of details in IT architecture has been growing at an unprecedented pace, we propose that the SOA wisdom should be leveraged to effectively manage the architecting practices in a disciplined manner. The pragmatic approach defined in this paper consists of four dimensions – Methodology, Automation, Patterns, and Strategy (MAPS). The key challenges in SOA are discussed, such as increasing integration, dynamics, disparate visual notations, fragmented WS-* specification efforts, and barriers to deploy useful governance, which collectively further widen the gap of communications. The MAPS model is a comprehensive framework aiming to enable a mature integration of appropriate knowledge and capabilities to filter the inessential from the essential. In the Methodology dimension, a hybrid method, SOA philosophy, and a methodical engineering framework are the key components. The Automation dimension covers tools, service lifecycle, and COTS mapping. The prominent elements of the Patterns dimension are data caching patterns, reference model, and open source reference implementation. Last but not least, the Strategy dimension addresses the strategy metamodel, technology architecture planning, and strategy roadmapping. In addition, a 9-point list of SOA wisdom is articulated, which gives best-practice guidelines to adopt and implement SOA pragmatically in large organizations from a practitioner's perspective.

References

1. The Standish Group (2006), http://www.standishgroup.com
2. Goldstein, H.: Who Killed the Virtual Case File? IEEE Spectrum (September 2005)
3. Federal Computer Week (2006), http://www.fcw.com/online/news/102253-1.html
4. Brooks, F.P.: No Silver Bullet - essence and accident in software Engineering. In: Proceedings of the IFIP Tenth World Computing Conference, pp. 1069–1076 (1986)
5. Booch, G.: Blog (Handbook of Software Architecture) (2007), http://booch.com/architecture/blog.jsp?archive=2007-02.html
6. Childre, D., Cryer, B.: From Chaos to Coherence, HeartMath, Boulder Creek (2000)
7. Shan, T.C., Hua, W.: A Service-Oriented Solution Framework for Internet Banking. International Journal of Web Services Research 3(1), 29–48 (2006)

Software and Data Technologies

Part I
Programming Languages

A Simple Language for Novel Visualizations of Information

Wendy Lucas[1] and Stuart M. Shieber[2]

[1] Computer Information Systems Department, Bentley College, Waltham, MA, USA
wlucas@bentley.edu
[2] Division of Engineering and Applied Sciences, Harvard University, Cambridge, MA, USA
shieber@deas.harvard.edu

Abstract. While information visualization tools support the representation of abstract data, their ability to enhance one's understanding of complex relationships can be hindered by a limited set of predefined charts. To enable novel visualization over multiple variables, we propose a declarative language for specifying informational graphics from first principles. The language maps properties of generic objects to graphical representations based on scaled interpretations of data values. An iterative approach to constraint solving that involves user advice enables the optimization of graphic layouts. The flexibility and expressiveness of a powerful but relatively easy to use grammar supports the expression of visualizations ranging from the simple to the complex.

1 Introduction

Information visualization tools support creativity by enabling discoveries about data that would otherwise be difficult to perceive [8]. Oftentimes, however, the standard set of visualizations offered by commercial charting packages and business intelligence tools is not sufficient for exploring and representing complex relationships between multiple variables. Even specialized visual analysis tools may not offer displays that are relevant to the user's particular needs. As noted in [9], the creation of effective visual representations is a labor-intensive process, and new methods are needed for simplifying this process and creating applications that are better targeted to the data and tasks. To that end, we introduce a language for specifying informational graphics from first principles. One can view the goal of the present research as doing for information visualization what spreadsheet software did for business applications. Prior to Bricklin's VisiCalc, business applications were built separately from scratch. By identifying the first principles on which many of these applications were built — namely, arithmetic calculations over geographically defined values — the spreadsheet program made it possible for end users to generate their own novel business applications. This flexibility was obtained at the cost of requiring a more sophisticated user, but the additional layer of complexity can be hidden from naive users through prepackaged spreadsheets.

Similarly, we propose to allow direct access to the first principles on which (a subset of) informational graphics are built through an appropriate specification language. The advantages are again flexibility and expressiveness, with the same cost in terms of user sophistication and mitigation of this cost through prepackaging.

J. Filipe et al. (Eds.): ICSOFT/ENASE 2007, CCIS 22, pp. 33–45, 2008.

For expository reasons, the functionality we provide with this language is presented by way of example, from simple scatter plots to versions of two quite famous visualizations: Minard's depiction of troop strength during Napoleon's march on Moscow and a map of the early ARPAnet from the ancient history of the Internet.[1] We hope that the reader can easily extrapolate from the provided examples to see the power of the language. We then describe how constraints are specified and how the constraint satisfaction process is coupled with user advice to reduce local minima. This is followed by descriptions of the primary constructs in the language and the output generation process for our implementation, which was used to generate the graphics shown in this paper.

2   The Structure of Informational Graphics

In order to define a language for specifying informational graphics from first principles, those principles must be identified. For the subset of informational graphics that we consider here, the underlying principles are relatively simple:

- Graphics are constructed based on the rendering of generic graphical objects taken from a small fixed set (points, lines, polygons, text, etc.).
- The graphical properties of these generic graphical objects are instantiated by being tied to values taken from the underlying data (perhaps by way of computation).
- The relationship between a data value and a graphical value is mediated by a function called a *scale*.
- Scales can be depicted via generic graphical objects referred to as *legends*. (A special case is an *axis*, which is the legend for a location scale.)
- The tying of values is done by simple constraints, typically of equality, but occasionally of other types.

For example, consider a generic graphical object, the *point*, which has graphical properties like horizontal and vertical position, color, size, and shape. The standard scatter plot is a graphic where a single generic object, a point, is instantiated in the following way. Its horizontal and vertical position are directly tied by an equality constraint to values from particular data fields of the underlying table. The other graphical properties may be given fixed (default) values or tied to other data fields. In addition, it is typical to render the scales that govern the mapping from data values to graphical locations using axes.

Suppose we have a table Table with fields f, g, and h as in Table 1. We can specify a scatter plot of the first two fields in just the way that was informally described above:

```
{make p:point with
   p.center = Canvas(record.f, record.g)
| record in SQL("select f, g from Table1")};
make a:axis with
   a.aorigin = (50,50),
   a.ll = (10,10),
   a.ur = (160,160),
   a.tick = (40,40);
```

---

[1] See Figures 4 and 7 for our versions of these visualizations.

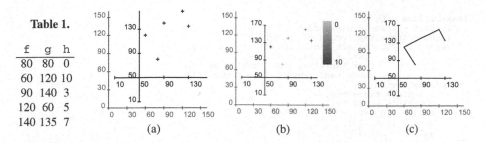

f	g	h
80	80	0
60	120	10
90	140	3
120	60	5
140	135	7

Table 1.

(a)   (b)   (c)

Fig. 1. Data table and simple scatter plots defined from first principles

The make keyword instantiates a generic graphical object (point) and sets its attributes. The set comprehension construct ({· | ·}) constructs a set with elements specified in its first part generated with values specified in its second part. Finally, we avail ourselves of a built-in scale, Canvas, which maps numeric values onto the final rendering canvas. One can think of this as the assignment of numbers to actual pixel values on the canvas. A depiction of the resulting graphic is given in Figure 1(a). (For reference, we show in light gray an extra axis depicting the canvas itself.)

Other graphical properties of chart objects can be tied to other fields. In Figure 1(b), we use a colorscale, a primitive for building linearly interpolated color scales. A legend for the color scale is positioned on the canvas as well. Some scaling of the data can also be useful. We define a 2-D Cartesian frame to provide this scaling, using it instead of the canvas for placing the points.

```
let frame:twodcart with
    frame.map(a,b) = Canvas(a, (3*b)/4+ 10)
in
  let color:colorscale with
      color.min = ColorMap("red"),
      color.max = ColorMap("black"),
      color.minval = 0,
      color.maxval = 10
  in
    {make p:point with
        p.center = frame.map(rec.f, rec.g),
        p.color = color.scale(rec.h)
    |rec in SQL("select f, g, h from Table1")},
    make a:axis with
        a.scale = frame.map,
        a.aorigin = (50,50),
        a.ll = (0,0),
        a.ur = (140,170),
        a.tick = (50,50),
    make c:legend with
        c.scale = color,
        c.location = frame.map(150, 180);
```

A line chart can be generated using a line object instead of a point object. Suppose we take the records in Table 1 to be ordered, so that the lines should connect the points from the table in that order. Then a simple self-join provides the start points (x_1, y_1) and end points (x_2, y_2) for the lines. By specifying the appropriate query in SQL, we can build a line plot (see Figure 1(c)).

```
let frame:twodcart with
    frame.map(a,b) = Canvas(a,(3*b)/4+ 10)
in
```

```
{make l:line with
    l.start = frame.map(record.x1, record.y1),
    l.end = frame.map(record.x2, record.y2)
 | record in SQL("select tab1.f as x1, tab1.g as y1,
                         tab2.f as x2, tab2.g as y2
                  from Table1 as tab1, Table1 as tab2
                  where tab2.recno = tab1.recno+1")},
 make a:axis with
    a.scale = frame.map,
    a.aorigin = (50,50),
    a.ll = (10, 10),
    a.ur = (140,170),
    a.tick = (40,40);
```

The specification language also allows definitions of more abstract notions such as complex objects or groupings. We can use this facility to define a chart as an object that can be instantiated with specified data. This allows generated visualizations themselves, such as scatter plots, to be manipulated as graphical objects. For instance, it is possible to form a scatter plot of scatter plots. Figure 2 depicts a visualization of the data sets for Anscombe's quartet [1], generated by the following specification:

```
define s:splot with
    let frame:twodcart with
            frame.map = s.map
    in
            { make o:oval with
                o.center ~ frame.map(rec.x, rec.y),
                o.width = 8,
                o.height = 8,
                o.fill = true
              | rec in s.recs },
              make a:axis with
                a.scale = frame.map,
                a.aorigin = (0,0),
                a.ll = (0,0),
                a.ur = (20,15),
                a.tick = (10,5)
in
    let outer:twodcart with
        outer.map(x,y) = Canvas(10*x, 10*y)
    in
        let FrameRecs = SQL("select distinct a, b from anscombe")
        in
            {make sp:splot with
                sp.map(x,y) = outer.map(x + 25*frec.a, y + 20*frec.b),
                sp.recs = SQL("select x, y from anscombe where a=frec.a
                              and b=frec.b")
              | frec in FrameRecs};
```

3   Specifying Constraints

All of the examples in the prior section made use of constraints of type equality only. The ability to add other types of constraints dramatically increases the flexibility of the language. For instance, stating that two values are approximately equal (~), instead of strictly equal (=), allows for approximate satisfaction of equality constraints. Further, constraints of non-overlapping (NO) force values apart. Together, these constraints allow dither to be added to graphs.

Suppose we have another table Table with fields id, f, and g, as in Table 2. We again specify a scatter plot of fields f and g but with two differences from our earlier examples: the center of each point is approximately equal to a data value from the table, and none of the point objects can overlap. The resulting graphic is shown in Figure 3(a).

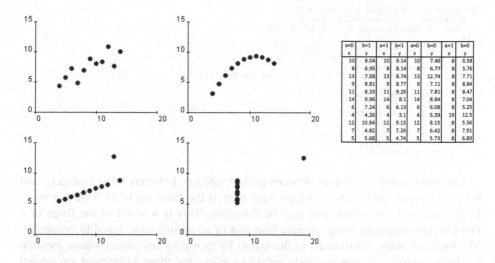

a=0 b=1		a=1 b=1		a=0 b=0		a=1 b=0	
x	y	x	y	x	y	x	y
10	8.04	10	9.14	10	7.46	8	6.58
8	6.95	8	8.14	8	6.77	8	5.76
13	7.58	13	8.74	13	12.74	8	7.71
9	8.81	9	8.77	9	7.11	8	8.84
11	8.33	11	9.26	11	7.81	8	8.47
14	9.96	14	8.1	14	8.84	8	7.04
6	7.24	6	6.13	6	6.08	8	5.25
4	4.26	4	3.1	4	5.39	19	12.5
12	10.84	12	9.13	12	8.15	8	5.56
7	4.82	7	7.26	7	6.42	8	7.91
5	5.68	5	4.74	5	5.73	8	6.89

Fig. 2. Graphic depicting Anscombe's quartet

Table 2.

id	f	g
1	80	80
2	60	120
3	90	140
4	90	140
5	120	60
6	140	135

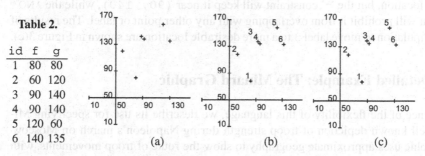

(a)   (b)   (c)

Fig. 3. Data table and scatter plots with positional constraints

```
NO({make p:point with
        p.center ~ Canvas(record.f, record.g)
    | record in SQL("select f, g from Table1")});
make a:axis with
      a.aorigin = (50,50),
      a.ll = (10,10),
      a.ur = (160,170),
      a.tick = (40,40),
      a.color = RGB(0, 0, 0);
```

We would like to add labels at each data point and specify three conditions: (1) no point can overlap with another point, (2) no label can overlap with another label, and (3) no point can overlap with a label. We need the ability to associate a variable name with an object or a set of objects for later reference. This is accomplished with the let statement, which provides access to that variable within the body of the statement. The following code contains the required specifications, with its output shown in Figure 3(b).

```
let points = NO({make p:point with
    p.center ~ Canvas(rec.f, rec.g)
  | rec in SQL("select f, g from Table2")})
  in
    let labels = NO({make l:label with
        l.center ~ Canvas(rec.f, rec.g),
        l.label = rec.id
      | rec in SQL("select f, g, id from Table2")})
    in
        NO(points, labels);
make a:axis with
    a.aorigin = (50,50),
    a.ll = (10,10),
    a.ur = (160,170),
    a.tick = (40,40),
    a.color = RGB(0, 0, 0);
```

The non-overlap constraints between `point` objects, between `label` objects, and between `point` and `label` objects have moved the label on point 3 farther away from the actual data point than may be desirable. This is a result of the force of a non-overlap constraint being stronger than that of an approximate equality constraint. The user can make adjustments to the layout by dragging any object whose location has been specified as approximately equal to a value, but these adjustment are subject to any constraints placed upon that object. Thus, the user may drag the label 3 to a different location, but the '~' constraint will keep it near (90, 140), while the 'NO' constraint will prohibit it from overlapping with any other point or label. The results of user manipulation to move label 3 to a more desirable location are shown in Figure 3(c).

4   A Detailed Example: The Minard Graphic

As evidence of the flexibility of this language, we describe its use for specifying Minard's well known depiction of troop strength during Napoleon's march on Moscow. This graphic uses approximate geography to show the route of troop movements, with line segments for the legs of the journey. Width of the lines is used for troop strength and

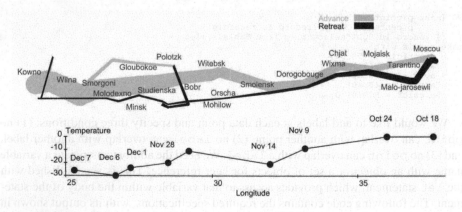

Fig. 4. The graphic generated by a specification extended from that of Figure 5, depicting the troop strength during Napoleon's march on Moscow

```
 1   let scalefactor = 45 in
     let weight = .000002 in

     % Define the depiction of one (multi-leg) branch of the march
 5   % identified by a distinct direction and division
     define m:march with
       let frame:twodcart with
           frame.map = m.map
       in
10      {make l:line with
             l.start = frame.map(rec.cx1, rec.cy1),
             l.end = frame.map(rec.cx2, rec.cy2),
             l.startWidth = scalefactor * weight * rec.r1,
             l.endWidth = scalefactor * weight * rec.r2,
15           l.color = ColorMap(rec.color)
           | rec in m.recs}
       in
       % Extract the set of branches that make up the full march
       let FrameRecs = SQL("select distinct direct,
20                                           division from marchNap")
       in
        % For each branch, depict it
        {make mp:march with
            mp.map(x,y) = Canvas(scalefactor*x - 1075, scalefactor*y - 2200),
25          mp.recs = SQL("select marchNap1.lonp as cx1, marchNap1.latp as cy1,
                           marchNap2.lonp as cx2, marchNap2.latp as cy2,
                           marchNap1.surviv as r1, marchNap2.surviv as r2,
                           marchNap1.color as color
                           from marchNap as marchNap1, marchNap as marchNap2
30                         where marchNap1.direct = framerec.direct
                           and marchNap2.direct = marchNap1.direct
                           and marchNap1.division = framerec.division
                           and marchNap1.division = marchNap2.division
                           and marchNap2.recno = marchNap1.recno+1")
35        | framerec in FrameRecs};

        % Label the cities along the march and do not allow overlap
        NO({make e d2:label with
            d2.center ~ frameTemp.map(recc.lonc, recc.latc),
40          d2.label = recc.city,
            d2.color = ColorMap("blue")
          | recc in SQL("select lonc, latc, city from marchCity")})};
```

Fig. 5. Partial specification of the Minard graphic depicted in Figure 4. This specification does not include the temperature plot.

color depicts direction. The locations of the labels on cities are of approximate equality, as they are not allowed to overlap, and in some cases have been adjusted by the user for better clarity. A parallel graphic shows temperature during the inbound portion of the route, again as a line chart. Our version of the graph is provided in Figure 4.

To generate this graphic, we require appropriate data tables: `marchNap` includes latitude and longitude at each way point, along with location name, direction, and troop strength; `marchTemp` includes latitude, longitude, and temperature for a subset of the inbound journey points, and `marchCity` provides latitude, longitude, and name for the traversed cities.

The main portion of the graphic is essentially a set of line plots, one for each branch of the march, where a branch is defined as a route taken in a single direction by a single division. Additional graphical properties (width and color) are tied to appropriate data fields. Textual labels for the cities are added using a text graphic object. A longitudinally aligned graph presents temperature on the main return branch.

The specification for this graphic (sans the temperature portion) is provided in Figure 5. After specifying some constants (lines 1–2), we define the depiction of a single branch of the march (6–16): a mapping (m.map) specifies a Cartesian frame (7–8) in which a line plot composed of a set of line segments (10–16) is placed, with one segment for each set of records (16). These records provide start and end points, along with widths at start and end (to be interpolated between), and color (11–15).

Thus, to depict a march leg, all that must be provided is the mapping and the set of records. These are constructed in lines 19–35. For each distinct direction and division (19–20), a separate march leg depiction is constructed (23–35). The mapping is a scaling of the Canvas frame (24), with records for the appropriate division and direction extracted from the marchNap database (25–35). Finally, the cities are labeled using the information in the marchCity database by setting the coordinates of the text labels to be approximately equal to latitude and longitude, setting a fixed color, and specifying that labels cannot overlap (38–42).

5   The Language

The examples from this paper were all generated from an implementation of our language. The implementation techniques are outlined in Section 6. The underlying ideas could, however, be implemented in other ways, for instance as a library of functions in a suitable functional language such as Haskell or ML.

The language is built around the specification of objects. The main construct is *objspec*, with a program defined as one or more objspecs. *Objspecs* are used for instantiating a graphical object of a predefined type, specifying relationships between the set of actual data values and their graphical representations, and defining new graphic types.

The make statement is used for instantiating a new instance of an existing object type (either built-in, such as point or line, or user-defined, such as march) with one or more conditions. There are two types of predefined objects: *generic* and *scale*. Generic objects provide visual representations of data values and include the primitive types of point, line, oval, rectangle, polygon, polar segment, and labels. Scales are used for mapping data values to their visual representations and are graphically represented by axes and legends. A scale is associated with a coordinate system object and defines a transformation from the default layout canvas to a frame of type twodcart or twodpolar (at this time).

In addition to a unique identifier, each object type has predefined attributes, with conditions expressed as constraints on these attributes. Constraints can be of type equality ('=') or approximate equality ('~'). Constraints enforcing visual organization features [3] such as non-overlap, alignment, or symmetry, can be applied to a set of graphical objects. These types of constraints are particularly useful in specifying network diagrams [7]. While such constraints can be specified in the language, our current implementation supports constraints in the forms of equality, approximate equality, and non-overlap for a subset of object types.

All of the generic objects have a color attribute and at least one location attribute, represented by a coordinate. For 2-D objects, a Boolean fill attribute defaults to true, indicating that the interior of that object be filled in with the specified color.

This attribute also applies to line objects, which have start and end `widths`. When widths are specified, lines are rendered as four-sided polygons with rounded ends.

Each type of scale object has its own set of attributes. A coordinate system object's `parent` attribute can reference another coordinate system object or the canvas itself. The mapping from data values to graphical values can be specified by conditions on its `origin` and `unit` attributes. Alternatively, a condition can be applied to its `map` attribute in the form of a user-defined mapping function that denotes both the parent object and the scale. Thus, a `twodcart` object whose `parent` is Canvas, `origin` is (5, 10), and `unit` is (30, 40) can be defined by the mapping function: `Canvas.map` `(30x + 5, 40y + 10)`.

Axes are defined in terms of their `origin`, `ll` (lower left) and `ur` (upper right) co-ordinates, and `tick` marks. Legends can be used to graphically associate a color gradient with a range of data values by assigning a color scale object to the `scale` attribute and a coordinate to the `location` attribute. Discrete colors can also be associated with data values, as in the Minard Graph example, where tan represents "Advance" and black represents "Retreat."

Attributes can also be defined by the user within a `make` statement. For example, it is often helpful to declare temporary variables for storing data retrieved from a database. These user-defined attributes are ignored when the specified visualization is rendered.

Another construct for an *objspec* is the `type` statement. This is used for defining a new object *type* that satisfies a set of conditions. These conditions are either constraints on attribute properties or other objspecs. An example of the former would be a condition requiring that the color attribute for a new chart type be "blue." An example of the latter is to require that a chart include a 2-D Cartesian coordinate frame for use in rendering its display and that this frame contain a set of lines and points corresponding to data retrieved from a database.

A third type of *objspec* is a set of objects, a *setspec*. This takes the form of either a query string or a set of objspecs to be instantiated for each record retrieved by a query string. These two constructs are often used in conjunction with one another to associate data with generic graph objects. For example, a setspec may define a query that retrieves two values, x and y, from a database table. A second setspec can then specify a `make` command for rendering a set of points located at the x and y values retrieved by the query. Alternatively, the query can be contained within the `make` statement itself, as shown in the code for the scatter plots in Section 2.

Lastly, the `let` statement associates a variable name with an *objspec* so that it can be referenced later within the body of the statement. Because this construct is commonly used in conjunction with a `make` or `type` statement, two shorthand expressions have been provided. The first of these is an abbreviated form of the `let` statement in which the make clause is not explicitly stated, with: `let var: type` with `conditions` in *body* corresponding to `let var=make var: type` with `conditions` in *body*. Similarly, the `define` statement associates a new object definition with a variable name, where: `define var: type` with `conditions` in *body* expands to `let var = type var: type` with `conditions` in *body*.

Figure 6 demonstrates the usage of the above *objspec* constructs for defining a new type of object called a `netplot`. The code specifies the creation of a set of ovals, referred to as nodes, with width and height of 10, color of "black," and center locations corresponding to x and y values queried from a database and mapped to a frame. (Data values were estimated from a depiction of the ARPAnet circa 1971 available at http://www.cybergeography.org/atlas/historical.html.) A set of blue lines are specified, with the endpoint locations of the lines equal to the centers of the circles. Labels are also specified and are subject to constraints, with the center of each label approximately equal to the center of a corresponding node and no overlap allowed between labels or between labels and nodes.

```
define n:netplot with
  let frame:twodcart with
          frame.map = n.map
  in
    let nodes = {make o:oval with
          o.center = frame.map(rec.x, rec.y),
          o.width = n.radius * 2,
          o.height = n.radius * 2,
          o.color = n.ccolor,
          o.fill = true
        | rec in n.netRecs}
    in
      let lines =  {make l:line with
            l.start = nodes[rec.v1].center,
            l.end = nodes[rec.v2].center,
            l.color = n.lcolor
          | rec in n.edgeRecs }
      in
        let labels =  NO({ make d:label with
          d.center ~ frame.map(rec.x, rec.y),
          d.label = rec.node,
          d.color = n.ccolor
            | rec in n.netRecs })
          in NO(labels, nodes)
in
  make net:netplot with
        net.map(x,y) = Canvas(7*x, 7*y),
        net.netRecs = SQL("select node, x, y from ArpaNodes"),
        net.radius = 5,
        net.edgeRecs = SQL("select v1, v2 from ArpaLinks"),
        net.lcolor = ColorMap("blue"),
        net.ccolor = ColorMap("black");
```

Fig. 6. Specification of the ARPAnet graphic depicted in Figure 7

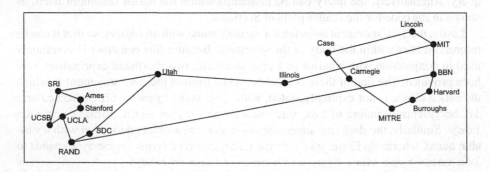

Fig. 7. A network diagram depicting the ARPAnet circa 1971

The output generated from this specification is shown in Figure 7, with the user having made similar adjustments as the one shown in Figure 3(c) to the placement of some labels. This chart and the Minard graphic from the prior section demonstrate the flexibility and control for specifying visualizations that result from working with the right set of first principles. Further, the intricacies of the code required to generate these charts are hidden behind a relatively simple to use but powerful grammar.

6   Implementation

The output generation process that renders visualizations from code written in our language, such as those shown in the prior sections, involves four stages.

1. The code is parsed into an abstract syntax tree representation.
2. The tree is traversed to set variable bindings and to evaluate *objspec*s to the objects and sets of objects that they specify. These objects are collected, and the constraints that are imposed on them are accumulated as well.
3. The constraints are solved as in related work on the GLIDE system [7] by reduction to mass-spring systems and iterative relaxation. Equality constraints between objects are strictly enforced, in that neither the user nor the system can change the position of those objects. Non-overlap constraints between objects are enforced by the placement of a spring between the centers of the objects (at the current time, the system does not solve for non-overlap constraints involving lines). The spring's rest length is the required minimum distance between those objects. Approximate equality, or near constraints, are enforced by the placement of springs with rest lengths of zero between specified points on two objects. The latter set of springs have a smaller spring constant than those used for non-overlap, thereby exerting less force. Solving one constraint may invalidate another, so an iterative process is followed until the total kinetic energy on all nodes falls below an arbitrarily set value.
4. The objects are rendered. Any primitive graphic object contained in the collection is drawn to the canvas using the graphical values determined as a result of the constraint satisfaction process. The user can make positional adjustments to any objects not subject to equality constraints. At the same time, the system continues to reevaluate its solution and update the position of the objects based on the solving of the constraints.

7   Related Work

Standard charting software packages, such as Microsoft Chart or DeltaGraph, enable the generation of predefined graphic layouts selected from a "gallery" of options. As argued above, by providing pre-specified graph types, they provide simplicity at the cost of the expressivity that motivates our work. More flexibility can be achieved by embedding such gallery-based systems inside of programming languages to enable program-specified data manipulations and complex graphical composites. Many systems have this capability: Mathematica, Matlab, Igor, and so forth. Another method for expanding expressiveness is to embed the graph generation inside of a full object-drawing package. Then, arbitrary additions and modifications can be made to the generated graphics by manual direct manipulation. Neither of these methods extends expressivity by deconstructing the graphics into their abstract information-bearing parts, that is, by allowing specification from first principles.

Towards this latter goal, discovery of the first principles of informational graphics was pioneered by Bertin, whose work on the semiology of graphics [2] provides a deep analysis of the principles. It is not, however, formalized in such a way that computer implementation is possible.

Computer systems for automated generation of graphics necessarily require those graphics to be built from a set of components whose function can be formally reasoned about. The seminal work in this area was by Mackinlay [4], whose system could generate a variety of standard informational graphics. The range of generable graphics was extended by Roth and Mattis [6] in the SAGE system. These systems were designed for automated generation of appropriate informational graphics from raw data, rather than for user-specified visualization of the data. The emphasis is thus on the functional appropriateness of the generated graphic rather than the expressiveness of the range of generable graphics.

The SAGE system serves as the basis for a user-manipulable set of tools for generating informational graphics, SageTools [5]. This system shares with the present one the tying of graphical properties of objects to data values. Unlike SageTools, the present system relies solely on this idea, which is made possible by the embedding of this primitive principle in a specification language and the broadening of the set of object types to which the principle can be applied.

The effort most similar to the one described here is Wilkinson's work on a specification language for informational graphics from first principles, a "grammar" of graphics [10]. Wilkinson's system differs from the one proposed here in three ways. First, the level of the language primitives are considerably higher; notions such as Voronoi tesselation or vane glyphs serve as primitives in the system. Second, the goal of his language is to explicate the semantics of graphics, not to serve as a command language for generating the graphics. Thus, many of the details of rendering can be glossed over in the system. Lastly, and most importantly, the ability to embed constraints beyond those of equality provides us the capacity to generate a range of informational graphics that use positional information in a much looser and more nuanced way.

8 Conclusions

We have presented a specification language for describing informational graphics from first principles, founded on the simple idea of instantiating the graphical properties of generic graphical objects from constraints over the scaled interpretation of data values. This idea serves as a foundation for a wide variety of graphics, well beyond the typical sorts found in systems based on fixed galleries of charts or graphs. By making graphical first principles available to users, our approach provides flexibility and expressiveness for specifying innovative visualizations.

References

1. Anscombe, F.J.: Graphs in statistical analysis. American Statistician 27, 17–21 (1973)
2. Bertin, J.: Semiology of Graphics. University of Wisconsin Press (1983)
3. Kosak, C., Marks, J., Shieber, S.: Automating the layout of network diagrams with specified visual organization. Transactions on Systems, Man and Cybernetics 24(3), 440–454 (1994)
4. Mackinlay, J.: Automating the design of graphical presentations of relational information. ACM Transactions on Graphics 5(2), 110–141 (1986)
5. Roth, S.F., Kolojejchick, J., Mattis, J., Chuah, M.C.: Sagetools: An intelligent environment for sketching, browsing, and customizing data-graphics. In: CHI 1995: Conference companion on Human factors in computing systems, pp. 409–410. ACM Press, New York (1995)
6. Roth, S.F., Mattis, J.: Data characterization for graphic presentation. In: Proceedings of the Computer-Human Interaction Conference (CHI 1990) (1990)
7. Ryall, K., Marks, J., Shieber, S.M.: An interactive constraint-based system for drawing graphs. In: Proceedings of the 10th Annual Symposium on User Interface Software and Technology (UIST) (1997)
8. Shneiderman, B.: Creativity support tools: accelerating discovery and innovation. Communications of the ACM 50(12), 20–32 (2007)
9. Thomas, J.J., Cook, K.A.: A visual analytics agenda. IEEE Computer Graphics and Applications 26(1), 10–13 (2006)
10. Wilkinson, L.: The Grammar of Graphics. Springer, Heidelberg (1999)

Generic Components for Static Operations
at Object Level

Andreas P. Priesnitz and Sibylle Schupp

Chalmers University of Technology, 41296 Gothenburg, Sweden
`{priesnit,schupp}@cs.chalmers.se`
`http://www.cs.chalmers.se/~{priesnit,schupp}`

Abstract. Reflection allows defining generic operations in terms of object constituents. A performance penalty accrues if reflection is effectuated at run time,
which is usually the case. If performance matters, some compile-time means of
reflection is desired to obviate that penalty. Furthermore, static meta-information
may be utilized for class creation, e.g., in optimizations. We provide such means
in generic components, employing static meta-programming. Essentially, object
structure is encoded in a generic container that models a statically indexed family.
Optimizations benefit any object defined in terms of that container.

Keywords: Reflection, Serialization, Class Generation, Container.

1 Introduction

Reflection provides access to internal representation details of the class of some given
object. We consider reflection on the fields of objects, which is essential for generic
implementations of low-level tasks. For instance, we are interested in tasks like serializations and optimizing transformations. Many applications of that kind are performance-
critical, whereas reflection usually implies some overhead when it is performed at run
time. Our goal is to be able to implement those tasks both generically and efficiently.

To avoid performance penalties, we want to apply reflection statically. Common languages do not provide corresponding means. But as some languages offer static meta-
programming facilities, we can specify generic components that achieve the desired
effect, thus serve as a portable language extension.

Our design relies on sharing the provision of fields for class implementations in
generic components. The core component is supposed to manage a collection of objects of different types, thus it forms a heterogeneous container. We want to access
those objects in terms of some label, thus the container must model the concept of an
indexed family. By separating the aspects of internal class representation, the container
becomes a stand-alone generic component, and the remaining task-specific components
in our design turn out rather lightweight. Existing implementations of indexed families
do not provide all features necessary to use them as internal representations of class
fields. Thus, we present an enhanced solution that generalizes the applicability of the
container.

The outline of this article is as follows: In Sect. 2, we discuss and specify the applications of reflection that we are particularly interested in. In Sect. 3, we lay out the

J. Filipe et al. (Eds.): ICSOFT/ENASE 2007, CCIS 22, pp. 46–59, 2008.

principal components that constitute our solution. Their heart is the statically indexed family container, whose implementation is in the center of the further discussion. In Sect. 4, we present in general terms the conventional implementation of such containers. In Sect. 5, we motivate and describe in more detail necessary extensions to that approach. In Sect. 6, we put the parts together, we show how to ease this step, and we discuss some consequences of our design. In Sect. 7, we compare our approach with previous work in related domains, and in Sect. 8, we give an evaluation of our results and some concluding remarks.

We chose C++ as language of our implementation and of our examples and discuss this choice in Sect. 5.

2 Why Static Reflection?

The term *reflection* encompasses all kinds of access to software meta-information from within the software itself. We restrict our considerations to accessing the instance fields of a given object. Furthermore, we exclude *alteration*, i.e., the modification of that set of fields. In this section, we exemplify two classes of applications that justify to perform reflection statically.

2.1 Example: Serialization

Algorithms whose effect on an object reduces to inspecting its instance fields limit the role of the object to that of a mere record. In other words, these algorithms do not involve knowledge of the particular semantics of the inspected object. A popular example of such applications is *serialization*, i.e., a mapping of objects to some linear representation. Common cases of such representations are binary encoding, XML objects, or formatted output. In Fig. 1, we depict an object that contains

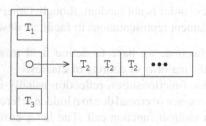

Fig. 1. Exemplary internal representation

– a reference to a collection of objects of type T_2,
– information on the number of objects in the collection as an object of type T_1, and
– some associated information as an object of type T_3.

Consider, for instance, the collection to be a sub-range of a text of characters, not owned by the object, and the associated information to be the most frequent character in that text snippet.

We might think of a serialization of that object as resulting in a representation as in Fig. 2. Each field is transformed to its corresponding representation, as symbolized by dashed lines, and these individual representations are concatenated. The meaning of concatenation depends on the representation. But a second thought reveals that this

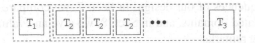

Fig. 2. Schematic serial representation

example is rather a counter-example: In order to serialize the text snippet, we need to know its length, thus we have to inspect the first field to deal with the second, and we rely on a semantic linkage between both. Thus, a "true" serialization could either create a representation of the reference as such, or treat it as being transparent and represent the (first) referenced object. To produce the outcome in Fig. 2, though, we have to specialize the serialization for this type of object.

Some languages provide implementations of certain serialization methods. To define other methods, we need to be able to traverse the fields of objects. Considering common application cases and our previous observation on the example, we restrict these traversals to be

Linear. Each field is mapped at most once. The output of the mapping is valid input to a stream.

Reversible. Each field is mapped at least once. It is possible to create representations that can be *deserialized* to an object that is equal to the original.

Order-independent. The operation does not depend on a particular order of field access. The contrary would require to consider composition information, which we want to exclude.

Deterministic. The access order is not random, though. Otherwise, we were forced to add annotations to element representations to facilitate deserialization.

If we lack means of reflection, we have to define field traversal for each class of interest—in practice, to the majority or entirety of classes.

Like other polymorphic functionalities, reflection usually is a dynamic feature. It causes constant per-object space overhead due to a hidden reference, and a per-reflection time overhead due to an indirect function call. The latter costs are dominated by the impossibility of inlining such calls, and of then applying further optimizations. These costs are insignificant if objects are large and if reflection occurs seldom. Otherwise, the costs of algorithms that make heavy use of reflection may turn out prohibitive, forcing developers again to provide per-class specializations.

This dilemma is avoided if reflection is performed at compile time, abolishing run-time costs in time and space. Run-time reflection is provided by many object-oriented programming languages, e.g., SmallTalk, CLOS, or Java, whereas compile-time reflection is supported only for particular tasks like *aspect-oriented programming*, if at all.

2.2 Example: Memory Layout

Another kind of operations on the fields of a class are optimizing transformations. Usually, only the compiler is responsible for performing optimizations of the internal representation of objects, in terms of space, place, alignment, or accessibility of their allocation. Given adequate means of static reflection and of static meta-programming, one can perform similar transformations within the code. In contrast to the compiler, the developer usually has some high-level knowledge about objects beyond the information expressible in terms of the language. Optimizations that depend on such knowledge complement and support optimizing facilities of the compiler.

As an example, consider the rule-of-thumb of ordering the fields of a class in memory by decreasing size, given that our language allows us to influence that order. The effect of this rule is depicted in Fig. 3 for a 2-byte representation of the number of characters. Fields of less than word size may share a word instead of being aligned at word addresses. If we order the fields in such a way that small fields are adjacent, the compiler is more likely to perform that optimization, reducing the object size in the example from 3 to 2 words. Note that we do not attempt to be smarter than the compiler:

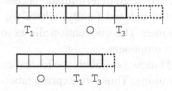

Fig. 3. Differences in memory layout

We can not and do not force it to perform that reduction. The compiler may still decide that it is beneficial to stick to word alignment, e.g., in favor of speed of field access, or even to reorder the fields, if allowed. But vice versa, the compiler might not be allowed to perform that optimization itself unless the proposed order of fields offers or suggests to do so.[1] Thus, our strategy is to supply the compiler with as much useful information as possible.

Given static reflection, we can inspect the fields of a class and create a new class that contains the fields ordered by size. This effect could not be achieved by a hand-coded alternative, as the size of fields may differ on different hardware. A fixed order that suits some hardware can turn out inappropriate on another.

3 Component-Based Static Reflection

We have learned that it is desirable to have static reflection on class fields at our disposal. Thus, we would like to provide this feature ourselves if the language we use does not offer it. In this section, we develop the general outline of our approach to achieve this goal. The individual steps will be discussed in more detail in Sects. 4–6.

---

[1] E.g., the C++ standard introduces several restrictions on actual field order.

3.1   Sharing Internal Representation

In order to share means and applications of static reflection by all class implementations, we provide them within common generic components. As these components shall be in charge of all matters of (here, static) object composition, they have to be shared by inheritance. An application class is privately derived from a component that defines the internal class representation. Thus, unauthorized access by slicing an object to this component type is inhibited. For simplicity, we call the component `Private`. Correspondingly, reflection-based functionalities that are supposed to be available in the signature of the application class are provided in a component `Public` from that the application class is derived publicly.

3.2   Field Access

Access to fields of an object corresponds to an injective mapping from identifying labels to the fields. To allow for evaluating that mapping at compile-time, reflection support has to be provided by a component that models a *statically indexed family* of fields of different types. Because we required order-independence in Sect. 2, that container does not model a tuple. Due to the expected deterministic access order, though, we require that the container is an ordered indexed family, where we adopt the order of labels from the definition of the container. This convention allows to express (e.g., optimizing) transformations in the family component.

In common statically typed languages, labels are lexical entities, on which we cannot operate by type meta-programming. Thus, we express labels by empty classes that do not serve any other purpose than static identification.

3.3   Field Specification

In order to delegate field definition, creation, deletion, access, and further fundamental methods to the family component, we have to statically parametrize it by a collection of field specifications that supply the necessary information. Besides the obviously needed labels and types, one has to specify:

– how to *initialize* fields, depending on the kind of constructor called for the component, and
– how to *delete* field contents in the destructor or when overwriting fields by assignment.

The family has to be self-contained, i.e., able to effect those functionalities for any kind of field types without further support. At the same time, it may not be restrictive; the class developer has to be able to make a choice wherever alternatives of execution exist.

3.4   Separation of Concerns

Not all aspects of reflection on fields can be covered by a general implementation of an indexed family. In order to allow for stand-alone usage of the container, classes are not derived from the family directly. Instead, the component `Private` is a wrapper, which complements the family by non-public features that apply to all application classes but not to a family. In other words, the role of `Private` is to promote the container to a data record, the role of `Public` is to garnish the record with generic, publicly accessible features.

Figure 4 depicts the inheritance relationship between the essential components in our design.

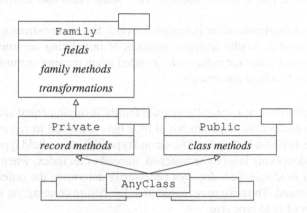

Fig. 4. Principal component hierarchy

4 Implementing Heterogeneous Families

Statically polymorphic containers store objects of different types. They are common in generic models and provided by several, mostly functional languages. We describe an established approach to defining such containers by user-defined constructs. Our language has to allow expressing *generative* constructs, as different container types must be created for different sets of element types. Java, for example, is not suitable: It performs *type erasure* on instances of generic types, instead of creating new types. More generally, our language has to allow conditional type generation, i.e., *static metaprogramming* (*SMP*). Candidates are MetaML, MetaOCaml, Template Haskell, or C++.

Compile-Time Lists: Given SMP, we can define lists of types by a binarily parametrized type `Cons` and a type `Nil` in the same style as lists are defined in LISP [1]. Similarly, a list algorithm is implemented as a type that is parametrized by a type list and that provides the result as an associated type.

Mixins: In order to influence class composition, our language has to allow *parametrized inheritance*, i.e., deriving from a base class that is given as a type parameter. The subclass is then called a *mixin* [2] and generically extends the base class. We are interested in mixins that add a single field and corresponding functionalities to the base class. A particular problem is to provide practicable mixin constructors that call the base class constructors properly [3].

Tuples: Combining the previous idioms, tuples of objects of different types can be defined recursively [4,5]: Cons becomes a mixin that is derived from the list tail and that provides a field of the head type. Nil is an empty class.

Elements are accessed by a static natural index. If the number is nonzero, the mixin forwards the access request with the index' predecessor statically to the base. If it is zero, the field in the mixin is returned. This static recursion causes no run-time overhead.

The tuple class implementation is straightforward. But its constructors are tricky to define, as they need to handle arbitrary numbers of initializing arguments.[2] Furthermore, the creation of temporaries has to be avoided when passing arguments on from a mixin constructor to a base constructor.

Families: An arbitrarily, yet statically indexed family is implemented analogously to a tuple [6,7]: The mixin class is parametrized by a list—assumed to represent a set—of type pairs Field<Label, Type>. The second type indicates a field type, whereas the first type is an identifying label to be matched, instead of an index, when accessing the field. The list of field specifiers does not necessarily determine the order in which the mixins are composed. Thus, we may reorder that list before creating the family, e.g., in decreasing order of field type size.

Constructor implementations differ from those for a tuple, but are similarly intricate. We are not tied anymore to a certain order of arguments nor forced to provide all of them. Instead, constructors take as initializers a set of label-value pairs that need to be searched for a matching label.

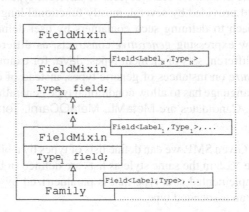

Fig. 5. Family implementation principle

Figure 5 illustrates the general layout of our proposed family implementation. In the following, we will justify the individual components and discuss their relevant features.

---

[2] At least, one has to allow for any argument number up to a practical limit.

5 Increasing Genericity

The semantics of statically indexed families have to be refined in order to serve the internal representation of objects. The principal reason is that any possible field type has to be supported, even types that hardly occur as container element types. Moreover, for each of the fields individual behavior, e.g., that of constructors, may have to be specified.

In this section, we discuss how to deal with different aspects of those problems. We avoid discussing language-specific details and focus on conceptual issues. Our actual implementation, though, was effected in C++. Of the languages that allow static meta-programming (see Sect. 4), C++ is the widest spread and the most suited for modeling low-level performance-critical details.

5.1 Arbitrary Kinds of Element

To allow for any type of elements in the family, we have to provide support for even those types that are not or only partially covered by common implementations of generic containers, e.g., arrays or references in C++. Therefore, each mixin in the original design is replaced by a pair of mixins: We extract the actual field definition from the mixin class to another mixin class that serves as its base. In the remainder of the original mixin we dispatch upon the field type to choose that base mixin appropriately.

5.2 Initialization

In constructors of a generic container, proper element initialization is tedious. Not only does the initialization depend on the element type, but it may also be necessary to treat particular elements in non-"standard" ways. Furthermore, we need a means to initialize in place, to avoid the creation of temporary objects when delegating initialization by an expression. Therefore, we refer for each family element to:

- a nullary functor returning the initializer used by the default constructor,
- a generic unary functor taking a family argument and returning the initializer used by (generic) copy constructors, and
- a unary functor taking the element as argument and returning no result, used by the destructor.

These functors are provided and possibly overwritten in the following order:

1. Type-dependent default implementations are given by the type-specific mixins that provide the fields.
2. For each family element, the user may specify such functors in the corresponding parameter list entry, in addition to element label and type.
3. In a constructor call, functors to use for particular fields can be specified optionally, see below.

A family has not only the usual default and copy constructor:

- A generic copy constructor takes any family argument. Fields are initialized by the result of applying the corresponding functor to that family.

- An initializing constructor takes as arguments an arbitrary number of pairs of labels and nullary functors. Fields are initialized by the result of the functor paired with the corresponding label.
- A hybrid alternative takes as arguments a family and an arbitrary number of initializing pairs. Here, the second entries of the pairs are unary functors that are applicable to the family.

5.3 Element Traversal

Access to family elements is provided in the conventional way (see Sect. 4). For traversals over the elements, we employ combinators like `fold`. But our implementation of `fold` differs from similar approaches [8] in that we fold over the labels rather than the fields, that are then accessed in terms of the labels. This way, we can express role-specific behavior, e.g., serialization to XML.

According to the requirements in Sect. 2, the traversal order has to be deterministic. Thus, `fold` follows the order defined by the user in the specification list parameter of the family. The actual order of mixin composition is not regarded.

Given `fold`, generic serialization is elegantly expressed along the lines of the *Scrap your Boilerplate* pattern [9]. Recursive traversal over an object is expressed by applying `fold` to the object, performing on each of its fields the same kind of traversal. Only for primitive types the serialization has to be specialized appropriately.

5.4 Reflective Operations

Transformations like the memory optimization in Fig. 3 need to be performed once per family type. Therefore, the actual family is represented by a wrapper `Family` that is derived from the recursive implementation, see Fig. 5. The wrapper may pre-process the parameter list, e.g., reorder it, before passing it on to the mixin hierarchy. According to the previous discussion, traversing algorithms are not influenced by this transformation.

6   Component Assembly

We suggested the principle of implementing classes in terms of statically indexed families. Then, we discussed aspects of implementing such families generically. In this section, we put those parts together.

6.1 Sharing Internal Representation

The following example illustrates how to define the text snippet class from Sect. 2. As discussed in Sect. 3, we add fields to the class by deriving it privately from an instance of `Private`, which again wraps the corresponding family container. The fields it provides are specified by its template argument, a `Cons`-created list of label-type pairs `Field<...>`. In this simple example, it is not necessary to provide initializing functors within those field descriptors.

```
class Snippet
  : private Private<Cons<Field<Length,short>,
                    Cons<Field<Text,char const*>,
                    Cons<Field<Most_Frequent,char>, Nil> > > >,
    public Public<Snippet>
{
    friend class Public<Snippet>;
    typedef Private</*...*/> Private_;
public:
    Snippet(char const* text, unsigned from, unsigned to)
      : Private_(init(Length(),to-from),
                 init(Text(),text+from),
                 init(Most_Frequent(), most_frequent(text+from,to-from)))
    {}
    // ...
    void shift(int offset)
    {
        get(Text()) += offset;
        get(Most_Frequent()) = most_frequent(get(Text()),get(Length()));
    }
};
```

The labels, like `Length`, are types without particular semantics which merely serve as tags. The constructor takes a string and start/end positions. To initialize the fields, the constructor delegates label-value pairs created by `init` to the constructor of the family base object. The function `most_frequent` detects the most frequent value in the given C string of the given length. Function `shift` moves the snippet within the text by `offset` characters. It illustrates field access by invoking the function `get` that is inherited from `Private` and the family.

In our family implementation, we perform size-optimizing field reordering as suggested in Sect. 2. In fact, we observe that the size of `Snippet` objects remains minimal, regardless of the order of field specifications. Furthermore, the family provides combinators as known from the library fusion [8].

`Private` wraps the family and provides features of rather technical nature. These features can be characterized as transforming the container into a record, i.e., making it more than the sum of its parts. For instance, `Private` might introduce an object identifier or an instance counter. But more importantly, it allows for providing generic solutions to repeated initialization or serialization problems: As explained in Sect. 2, proper use of a reference to dynamically allocated storage requires knowledge of the size of the latter. This information is likely to be dependent on some other fields' values. Given a representation of that dependency, operations like initialization, assignment, or serialization can be expressed generically. Being out of the scope of the family but common to many class implementations, such generic operations belong into `Private`.

The implementation of some `Private` features may require that class definitions provide additional arguments to each `Field` instance. When introducing such features, we therefore need to upgrade existing class definitions. Our explicit composition idiom supports performing this task by re-factoring tools.

`Public` provides generic methods that become part of the signature of any derived class. As these methods require access to the internal representation, we make `Public` a friend of the derived class. More elegant alternatives to this rather straightforward solution require further structural abstractions, which we omit from this discussion.

Some of the methods that `Public` provides rely on the actual type `Snippet`. For instance, assignment should return the assigned object of this type. Therefore, the derived class type is passed to `Public` as a template argument. `Public` has access to the internal representation of the derived class, including combinators over its fields and possible extensions provided in `Private`. Thus, serialization and other common methods can be implemented rather easily.

6.2 Class Composition

In its current form, our approach relies on a fixed inheritance structure and on a closed specification of all fields of a class. We run into trouble when we define a class not in a monolithic way, but by deriving it from some other class. That definition bases on (at least) two families whose features are not automatically combined. A straightforward solution is to avoid class composition and to rely exclusively on object composition. In fact, inheritance is hardly used in performance-critical software where the need for static reflection arises.

But class composition from components is feasible, as long as these do not contain field definitions. We see the mission of our approach in role-based and generative design, where classes are assembled from collections of components. The extraction of the respective fields into a family serving as base class is an initial step of class generation. A final step would be the provision of the common class front-end mentioned before. The application to component-based class design was a major motivation behind our requirement of independence on field order in Sect. 2.

7 Related Work

To our knowledge, no previous attempt has been undertaken to provide static reflection on object fields or corresponding effects by means of appropriately designed generic containers.

Several publications under the label *SYB/Scrap your Boilerplate* [9] aim at generically expressing recursive traversals of data structures in Haskell. They rely either on constructs that need to be specialized per data structure, or on appropriate language extensions. The *spine view* approach [10] generalizes SYB by mapping algebraic data types to recursively defined types that expose the fields one by one. A fundamental difference to our approach is that Haskell is functional, which makes the exposition of fields uncritical. To ensure encapsulation, our internal representation has to be kept private, and functions like serialization have to be part of the class interface.

A previous study on expressing SYB in C++ [11] applied a `fold` definition in terms of a heterogeneous container. The container served itself as object, similar in style to Haskell, causing conflicts between encapsulation and applicability. Our approach avoids such problems by separating an object from its field container and encapsulating both the container and field traversal functions. Class definitions only expose a high-level interface to methods like serialization.

The fusion library [8] provides a rich selection of sophisticated implementations of heterogeneous containers of different kind, including families (called maps) as well as the corresponding algorithms and adapters in the style of the C++ STL/Standard Template Library. As for now, there is no sufficient support for the issues addressed in

Sect. 5. Hence, these containers can not serve to provide the fields of classes. But the differences are slight, and the combination of fusion constructs with our approach is both feasible and promising.

As discussed in Sect. 6, our work aims at solving common problems in component-based class design. Research in that domain focused on splitting classes horizontally by functionality [1,12] rather than on vertical splits by class structure. The question of forwarding constructor calls through mixin hierarchies has been addressed before [3]. The solution relies on wrapping constructor arguments in a tuple, obeying the composition order.

Another article [13] proposes to add more extensive meta-information to C++ classes by means of pre-processor instructions and meta-programming. On the one hand, field and method definitions have to be given separately and are not created by the same instructions. On the other hand, the meta-information is provided at compile time, but not used for class generation.

Finally, *multi-stage programming* languages like MetaOCaml [14] incorporate meta-information and code creation in the language itself. Unfortunately, they are not very spread yet. Our approach provides a subset of such features in reusable components. In contrast to language-based features, library-based language extensions are portable, applicable in a popular performance-oriented language like C++, and open to high-level extensions beyond the language's expressiveness. We believe the combination of both approaches the key to mastering complex and computationally expensive software.

8 Conclusions

We discussed how to provide generic algorithms that rely on meta-information about their arguments' internal representation, but to avoid the performance penalties of dynamic reflection. Static reflection is a rare language feature, but statically generic programming allows to provide such features in reusable components. We restricted our discussion to linear, order-independent, but order-preserving algorithms, like serializations. Our solution allows to implement generically the serialization and the optimization described in Sect. 2, without using run-time constructs. Similarly, other type-dependent operations of the named kind can be implemented.

We extract field information from class definitions and represent it in containers that model statically indexed families. It is known how to implement such containers, but shortcomings of conventional implementations have to be resolved for our purpose. In particular, support of unusual element types and proper initialization need to be ensured. Linear algorithms are expressed elegantly in terms of higher-order combinators. Optimizations and other features are beneficial for either use of the construct as generic container or as internal class representation.

Class definitions that inherit their fields from a container involve boilerplate code. We share such code in a wrapper that makes the container a record. A more general solution is left to future work on component-based class generation.

As our solution relies on static meta-programming, it pays run-time efficiency by increased compilation efforts. We consider these costs tolerable for performance-critical systems, and for objects with rather few fields. In other cases, dynamic reflection may be more appropriate.

Our approach attempts to deal with as many aspects and uses of field definition and reflection as possible within the underlying container. But we do not claim or require their complete coverage. A class definition in the proposed style allows to influence or override any behavior of the underlying family. Only essential and invariable properties are encapsulated.

The solution is a flexibly applicable implementation pattern, sufficiently abstract to be expressed entirely in terms of portable library components, and does not rely on compiler extensions. We can conveniently add extensions to the components that automatically enrich the semantics of all classes defined by that pattern. We consider these aspects significant advantages of our solution over alternative approaches that rely on language extensions and compiler facilities. These solutions are safer to use, but harder to propagate and extend.

Acknowledgements. We are grateful to Gustav Munkby and Marcin Zalewski for inspiring discussions, to the anonymous ICSOFT reviewers for useful comments, and to the Swedish Research Council (Vetenskapsrådet) for supporting this work in part.

References

1. Czarnecki, K., Eisenecker, U.W.: Generative Programming: Methods, Tools, and Applications. Addison-Wesley, Reading (2000)
2. Bracha, G., Cook, W.: Mixin-Based Inheritance. In: SIGPLAN Not., vol. 25(10), pp. 303–311. ACM Press, New York (1990)
3. Eisenecker, U.W., Blinn, F., Czarnecki, K.: A Solution to the Constructor-Problem of Mixin-Based Programming in C++. In: Meyers, S. (ed.) First Workshop on C++ Template Programming (2000), http://www.oonumerics.org/tmpw00
4. Järvi, J.: Tuples and Multiple Return Values in C++. Technical Report 249, Turku Centre for Computer Science (1999)
5. Järvi, J.: Boost Tuple Library Homepage (2001), http://www.boost.org/libs/tuple
6. Winch, E.: Heterogeneous Lists of Named Objects. In: Josuttis, N., Smaragdakis, Y. (eds.) Second Workshop on C++ Template Programming (2001), http://www.oonumerics.org/tmpw01
7. Weiss, R., Simonis, V.: Storing Properties in Grouped Tagged Tuples. In: Broy, M., Zamulin, A.V. (eds.) PSI 2003. LNCS, vol. 2890, pp. 22–29. Springer, Heidelberg (2004)
8. de Guzman, J., Marsden, D., Schwinger, T.: Fusion Library 2.0 Homepage (2007), http://spirit.sourceforge.net/dl_more/fusion_v2/libs/fusion
9. Lämmel, R., Peyton Jones, S.: Scrap Your Boilerplate: A Practical Design Pattern for Generic Programming. In: SIGPLAN Not., vol. 38(3), pp. 26–37. ACM Press, New York (2003)
10. Hinze, R., Löh, A.: Scrap Your Boilerplate Revolutions. In: Uustalu, T. (ed.) MPC 2006. LNCS, vol. 4014, pp. 180–208. Springer, Heidelberg (2006)
11. Munkby, G., Priesnitz, A.P., Schupp, S., Zalewski, M.: Scrap++: Scrap Your Boilerplate in C++. In: Hinze, R. (ed.) 2006 SIGPLAN Workshop on Generic Programming, pp. 66–75. ACM Press, New York (2006)

12. Smaragdakis, Y., Batory, D.: Mixin-Based Programming in C++. In: Butler, G., Jarzabek, S. (eds.) GCSE 2000. LNCS, vol. 2177, pp. 163–177. Springer, Heidelberg (2001)
13. Attardi, G., Cisternino, A.: Reflection Support by Means of Template Metaprogramming. In: Bosch, J. (ed.) GCSE 2001. LNCS, vol. 2186, pp. 118–127. Springer, Heidelberg (2001)
14. Calcagno, C., Taha, W., Huang, L., Leroy, X.: Implementing Multi-Stage Languages using ASTs, Gensym, and Reflection. In: Pfenning, F., Smaragdakis, Y. (eds.) GPCE 2003. LNCS, vol. 2830, pp. 57–76. Springer, Heidelberg (2003)

A Visual Dataflow Language for Image Segmentation and Registration

Hoang D.K. Le[1], Rongxin Li[1], Sébastien Ourselin[1], and John Potter[2]

[1] BioMedIA, Autonomous Systems Laboratory, CSIRO, Sydney, Australia
{hoang.le,ron.li,sebastien.ourselin}@csiro.au
[2] Computer Science and Engineering, University of New South Wales, Sydney, Australia
potter@cse.unsw.edu.au

Abstract. Experimenters in biomedical image processing rely on software libraries to provide a large number of standard filtering and image handling algorithms. The Insight Toolkit (ITK) is an open-source library that provides a complete framework for a range of image processing tasks, and is specifically aimed at segmentation and registration tasks for both two and three dimensional images.

This paper describes a visual dataflow language, ITKBoard, designed to simplify building, and more significantly, experimenting with ITK applications. The ease with which image processing experiments can be interactively modified and controlled is an important aspect of the design. The experimenter can focus on the image processing task at hand, rather than worry about the underlying software. ITKBoard incorporates composite and parameterised components, and control constructs, and relies on a novel hybrid dataflow model, combining aspects of both demand and data-driven execution.

1 Introduction

Segmentation and registration are two common tasks that are conducted in processing biomedical images, such as those produced by Computed Tomography Imaging (CT scanners) and Magnetic Resonance Imaging (MRI scanners). *Segmentation* involves identifying and classifying features of interest within an image, and *registration* involves the alignment of corresponding parts of different images or underlying grids and meshes. A biomedical application being explored in our BioMedIA laboratory at CSIRO performs non-invasive modelling of the internal 3D structure of an artery. Potentially this can be used to customise the design of a good-fitting stent as used in angioplasty procedures. Segmentation is used to identify the arterial region of interest, and registration used to align the 3D model of the stent.

Researchers working in this area typically experiment with different image filtering algorithms, and modify parameters of the algorithms in attempting to achieve good images for diagnostic or other purposes. Their concern is with the kind of algorithm and the tuning process involved, in order to achieve accurate reconstructions from the images. They do not, in general, want to be concerned with the development of the underlying software.

J. Filipe et al. (Eds.): ICSOFT/ENASE 2007, CCIS 22, pp. 60–72, 2008.
© Springer-Verlag Berlin Heidelberg 2008

The *Insight Toolkit* (ITK) is an open-source software toolkit for performing registration and segmentation. ITK is based on a demand-driven dataflow architecture, and is implemented in C++ with a heavy reliance on C++ templates to achieve efficient generic components. Only competent C++ developers can build image processing applications using ITK in the raw. Although the toolkit does allow Python, Java or Tcl scripting of ITK components, the level of programming expertise and time required is still too high for convenient experimentation with ITK. In particular, experimenters working with scripted models, still have a cognitive mismatch between the image processing problems they want to solve, and the task of writing scripts to run different experiments.

Our aim with *ITKBoard* is to provide a simpler platform for biomedical researchers to experiment on, leveraging the underlying strengths of ITK, but requiring little or no programming knowledge to use. More specifically, we aim to overcome some existing limitations such as a lack of efficiency in dataflow model construction, difficulty in customising filter's properties for running different experiments, and lack of high-level constructs for controlling execution.

To this end, we have designed a visual dataflow language which provides extra features above and beyond ITK. The *novel contributions* of ITKBoard are: *visual construction* of ITK applications, by graphical manipulation of filtering models as reported elsewhere [1]); a *hybrid dataflow* model, combining both *demand-driven* execution for images, and *data-driven* execution for filter parameters; explicit *visual parameterisation* of filters, with graphical input/output parameter dependencies between filters; *visual composition* of filters that can be saved and re-deployed in other applications; explicit *visual control flow* with selection and repetition constructs; and explicit construction of expressions combining parameters through the visual interface.

The combination of features in ITKBoard is unique, and designed to suit experimentation in biomedical image processing. In particular, our hybrid dataflow model, incorporating both data-driven and demand-driven computation, is novel. The control constructs in our language are also interesting, being tailored specifically to cope with the underlying ITK execution model, combined with our hybrid model.

The paper is organised as follows. Section 2 provides a summary of the software architecture of ITK. In particular we discuss how the uniformity in the ITK design allows dataflow models to be "wired up", and filters to be executed on demand. The user-based perspective of ITKBoard, presented in Section 3, gives an indication of the ease-of-use that we achieve with the interactive visual layer that we place on top of the ITK model. In Section 4 we focus on the hybrid dataflow model that we employ in ITKBoard. Here we extend the demand-driven dataflow model of ITK, in which image data is cached at filter outputs, with a data-driven model for handling filter parameters. This adds a uniformity to handling parameters which is not directly provided in the ITK model. Further details of the ITKBoard architecture are outlined in Section 5: we consider wrapper and plug-in mechanisms for ITKBoard, composite filters, a simple way of expressing combinations of parameters, and explicit visual control flow constructs for selection and repetition. In Section 6 we discuss the contribution of our work, and compare it with other related work, before a brief conclusion in Section 7 where we also point to possible future work.

2 ITK: The Insight Toolkit

We provide a brief summary of the features of ITK which are relevant for the design of ITKBoard. There is a comprehensive software guide [2] which should be consulted for further information.

ITK is based on a simple demand-driven dataflow model [3] for image processing. In such a model, the key idea is that computation is described within process elements, or *filters*, which transform input data into output data. Data processing *pipelines* form a directed graph of such elements in which the output data of one filter becomes the input for another. In a demand-driven model, computation is performed on request for output data. By caching output data it is possible to avoid re-computation when there have been no changes to the inputs. Image processing is computationally intensive, and so redundant computations are best avoided. ITK has therefore adopted this lazy, memo-ised evaluation strategy as its default computational mechanism.

ITK is implemented in C++ and adopts standardised interfaces for the constituent elements of the dataflow model. In particular it provides a generic interface for *process objects* (filters). In ITK images and geometric meshes are the only types of data transformed by filters. They are modelled as generic data objects, which can be specialised for two and three dimensional image processing. Typically output data objects for a filter are cached, and only updated as necessary.

When inputs are changed, downstream data becomes out-of-date. Subsequent requests will cause re-computation of only those parts of the model which are out-of-date. As well as tracking the validity of the cached output images, ITK provides a general event notification mechanism, based on the *Observer* design pattern [4] for allowing arbitrary customisation of behaviours that depend on the state of computation. This mechanism makes it feasible for us to trap particular events in the standard ITK execution cycle, and intersperse the standard behaviours with further updating as required by our ITKBoard model.

ITK uses data other than the objects (image data) that are transformed by filters. This other data is constrained to be local to a particular filter. Such data is usually used to parameterise the behaviour of the filters, or in some cases, provide output information about a computation which is distinct from the image transformation typical of most filters. Such parametric data are treated as properties of a filter, and there is a standard convention in ITK for providing access to them with get and set methods. There is however, no standard mechanism in ITK to model dependencies between the parameters of different filters in ITK. This is one of our contributions with ITKBoard.

To build an ITK application, the basic approach requires all elements of the model to be instantiated as objects from the C++ classes provided by ITK. The toolkit does encourage a particular idiomatic style of programming for creating filter instances and constructing filter pipelines by hard-wiring the input-output connections. Nevertheless, programming in any form in C++ is not for the faint-hearted.

Consequently, to escape from the complexities of C++, there is a wrapper mechanism which allows ITK filters to be accessed from simpler, interpreted languages. Currently those supported are Java, Python and Tcl. Image processing applications can be written in any of these languages, using standard ITK filters as the primitive components. However, even with such high-level languages, the text-based programming task still slows down the experimenter who simply wants to investigate the effect of different image processing algorithms on a given image set.

3 A User's Perspective of ITKBoard

ITKBoard is a system that implements a visual dataflow language, providing a simple means of interactively building a dataflow model for two- and three-dimensional image processing tasks. Unlike the underlying ITK system, outlined in Section 2, ITKBoard requires no knowledge of programming. It provides a much less intimidating approach to the image processing tasks that ITK is designed to support. To execute an image processing task simply involves clicking on any image viewer element in the model. By offering a single user interface for both model construction and execution, ITKBoard encourages experimenters to try out different filtering algorithms and different parameter settings.

We summarise the key features that are evident from the screenshot displayed in Figure 1. This shows a small example of an image processing task to illustrate the key concepts behind our design of ITKBoard. The simplicity and ease-of-use of the interface and the intuitive nature of the way in which tasks are modelled should be evident from the figure. The panel on the left shows the palette of available elements, which are typically wrapped versions of ITK process objects. This provides the interface with the library of components that ITK provides—no other exposure of ITK functionality is required in ITKBoard.

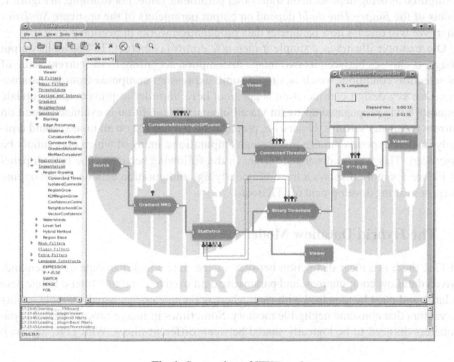

Fig. 1. Screenshot of ITKBoard

The main panel displays a complete filtering application mid-way through execution, as suggested by the execution progress bar. To construct such an application, the user selects elements from the left panel, placing them in the application panel, and connecting outputs to inputs as required. We visually distinguish between image data, the thick dataflow edges, and parametric data, the thin edges. Image inputs are provided on the left of filter elements, and outputs on the right. Parameter inputs are via pins at the top of filters, outputs at the bottom. ITKBoard uses colour coding of image dataflows to indicate when image data is up-to-date. In the colour version of Figure 1, red image dataflows indicate that the data is out-of-date, yellow flows indicate that data is being updated, and green ones indicate up-to-date data.

Viewer elements act as probes, and provide the mechanism for execution. The user may request any particular view to be updated. This demand is propagated upstream, just as in ITK, until the cached image data is found to be up-to-date. Viewers cache their image data. This allows experimenters to attach multiple viewers to a given filter output, and update them independently. This is an easy-to-use device for simultaneously observing images produced before and after changes to the upstream model which is particularly helpful for experimenting with filtering models. Figure 2 illustrates the display for a viewer, in this case one that compares two-dimensional cross sections of three-dimensional images.

Parameters of a filter are interactively accessible via property tables, which can be displayed by clicking on the filter icon. Input parameters can be set in one of three ways: at initialisation, with some default value, by explicit user interaction, or by being configured as being dependent on some other parameter value. For example, in Figure 1, inputs of the *Binary Threshold* depend on output parameters of the upstream *Statistics* filter. We will discuss how parameter data propagation works in Section 4.

Our example illustrates a simple *if-then-else* control construct. To update its output image, it chooses between its two image data inputs, according to the current value of its selection condition, which is simply some expression computed from its parameters. One key design consideration is apparent here: in order to prevent image update requests being propagated upstream via all inputs, the condition is evaluated based on current parameter settings. Otherwise, propagating the request on all inputs would typically involve expensive image processing computations, many of which may simply be discarded. This will be discussed further in the following sections. The bottom panel of Figure 1 simply displays a trace of underlying configuration actions for debugging purposes during development of ITKBoard.

4   The Hybrid Dataflow Model

In ITK there is a clear distinction between image data which participates in demand-driven dataflow computations, and parametric data used to configure filters. Image data is large and must be managed with care, whereas parameters are usually simple scalars or vectors that consume negligible memory. Sometimes in image processing tasks, parameters do need to be logically shared between different filters. With ITK, any such

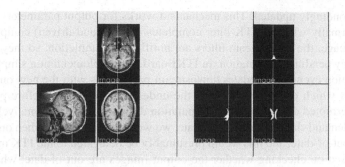

Fig. 2. Comparing images with a Viewer

sharing must be configured explicitly by the user when they set the parameters for individual filters. For ITKBoard we introduced a design requirement that sharing of such data between filters must be made explicit in the model that a user builds.

A couple of design alternatives presented themselves to us. First, instead of treating parameters as local to individual filters, we could simply take them to be global variables for the whole model. This solution is certainly implementable, but requires a global namespace for the parameters, and binding of the global parameter names to the local names used within each of the individual filters. On the down-side, this approach requires the user to invent a global name and keep track of its local variants, in order to understand what parameters are shared where. With the ability to form composite models, we must also deal with the added complexity of nested namespaces. Furthermore, we want parameters of some filters to be able to determine those of other filters. This would require the user to distinguish between those parameters which had to be set externally by the user, and those which would be the responsibility of a filter to set.

Instead of using global variables for handling parametric data, we have opted for a dataflow solution. In our ITKBoard model, we deem this appropriate because we are already using a dataflow model for image data, and because it avoids some of the problems mentioned above in treating parameters as global variables. First, with a visual dataflow language, there is no need to provide global names for the dataflow parameters; second, the dataflow connections make apparent any dependencies between parameters, and who is responsible for setting the data values, by distinguishing between input and output parameters—more on this soon; third, with a composite model, it is a relatively simple matter to choose which internal parameters to expose at the composite level, with external inputs mapping to internal inputs, and internal outputs mapping to external outputs.

Perhaps the most interesting design choice in ITKBoard has been to opt for a *data-driven propagation scheme for data parameters*. Given that ITK already supports a demand-driven model of computation for image data, why not instead just go with the flow, and make the update of parameters demand-driven as well? There are three main reasons for making parameter updates data-driven.

The first reason is implementation biased—it is simpler to extend ITK with a data-driven propagation scheme for parameter updates. Whenever a parameter is updated, we simply ensure that any downstream dependants are immediately notified of the change, and correspondingly updated. This mechanism works for output parameters of a filter as well. Normally when an ITK filter completes its (demand-driven) computation of its output image, the downstream filters are notified of completion, so they may proceed with any pending computation. In ITKBoard, our implementation simply catches this notification event, and updates downstream parameters with the new output parameter values which can be gotten from the underlying ITK filter, before proceeding with the interrupted demand-driven computation downstream. If, alternatively, we implemented demand-driven parameter update, we would have to track when output parameters are out-of-date, and this would presumably be delegated to the ITK mechanism which is based on checking whether the output images are out-of-date; whatever the implementation trick used, the demand-driven approach implies that requests for parameter updates will trigger (potentially expensive) image processing computations. With the data-driven approach, (cheap) parameter updates are propagated immediately, and never trigger image processing without a separate request for an up-to-date image.

The second reason for adopting the data-driven scheme for parameter propagation is more user focused. Our scheme implies that all image processing computations are driven by requests for up-to-date images, and nothing else. These requests are propagated upstream, and the state of computation is solely determined by the current validity of image data and the current thread of incomplete requests. In particular, a dependency of an input parameter on an upstream filter's output parameter will not trigger computation for the upstream filter, even if that filter is not up-to-date.

The third reason for not applying the demand-driven scheme to parametric data is to allow a richer class of models, while still avoiding recursion caused by cycles between image data and parameter dependencies. In a standard demand-driven model, downstream output parameters cannot connect to upstream filters without causing a feedback loop. By separating the image demands from the data-driven parameter updating mechanism, ITKBoard can handle such dependencies. So, ITKBoard allows cyclic dependencies that involve at least one image and one parameter data dependency. ITKBoard does not allow cycles of image dataflow, or of parameter dataflow; only mixed cycles are allowed. ITKBoard provides built-in support for iterative filters, so image feedback is not needed for expressing repetitive processing.

Because the user interface gives visual cues about which data is not up-to-date, and where parameter dependencies lie, it is a simple matter for a user to explicitly request that an output parameter for a particular filter be brought up-to-date by requesting an updated view of the output image for that filter. The rationale for this is to give the user better control over which parts of the model are actually re-computed after changes are made, thereby avoiding potentially expensive but unnecessary image processing. For biomedical experimenters, we decided this finer granularity of control over where image processing occurs was a worthwhile feature.

Input parameters can be set in a number of ways. Unconnected parameters have a default value. Input parameter pins are positioned along the top of the filter icon, and outputs at the bottom. Connections can be made from either input or output pins to other input pins. When a connection is made between two input parameters, the downstream input parameter is overridden by the value of the upstream input. For all parameter connections, whenever the upstream value is updated, the connected downstream value is correspondingly updated. Users may directly override the current value of an input parameter associated with a filter; in this case, any downstream dependants will be updated, but upstream parameters will retain their old value—again, this gives experimenters some extra flexibility. They have the opportunity to test local parameter changes in a model, without restructuring the dependencies in the model. This is similar to the way some program debuggers allow the values of variables to be set externally. Note that the value of a user-modified input parameter will be retained until there is some other upstream variation causing the parameter to be modified via some inward parameter dataflow, if such a connection exists.

Our design choice in opting for a hybrid dataflow model does have an impact on the behaviour of control constructs as described later in Section 5.

5 More Details of ITKBoard

We detail some of the more useful features of ITKBoard: auto-wrappers for ITK filters and a plug-in mechanism, ITKBoard's take on composite filters, support for expressing parameter-based computations, and finally control constructs.

5.1 Wrappers and Plug-Ins for ITKBoard

The main goal of ITKBoard has been to present an easy-to-use interface for experimenters who want to use ITK without worrying about programming details. To this end it is critical that ITKBoard can easily access all of the ITK infrastructure and that new pre-compiled components can be easily added to ITKBoard.

The first mechanism is an auto-wrapper for ITK filters so that ITKBoard is easily able to leverage all of the filtering functionality provided by ITK, or by any other C++ code implementing similar interfaces. Our auto-wrapper parses the C++ code for a filter, and generates C++ code that wraps the filter so that it can be used within ITKBoard. In particular, the auto-wrapping process is able to extract input and output parameters for an ITK filter, assuming the convention that input parameters are those defined with get and set methods (or macros), and output parameters those just with get methods. The auto-wrapper allows a user to intervene to adapt the auto-wrapper's translation as required.

The second mechanism provides support for plug-ins. This allows us to include newly developed filters into the ITKBoard system without recompiling the source code for the system. This is important for effective sharing and distribution of extensions. We rely on a shared library format (.so or .dll) to achieve this. Every shared library can provide a collection of filters, and must provide a creation routine for instantiating the filters implemented by the library.

Details of both the auto-wrapper and plug-in mechanisms for ITKBoard have been presented elsewhere [1].

5.2 Composite Filters

Although ITK itself can support the construction of composite filters made up of other filters, we provide a separate mechanism in ITKBoard. The reason for doing this is that

Fig. 3. A Composite Filter: collapsed form

Fig. 4. A Composite Filter: expanded form

we wish to distinguish between primitive filters (typically ITK filters), and those which can be composed of sub-filters.

So we simply provide an ITKBoard implementation of the *Composite* design pattern [4]. The *Component* interface is the standard ITKBoard abstraction for representing a filter. *Leaf* components are ITK filters with their standard ITKBoard wrapper. What perhaps is interesting here, is that the composite structure is not hard-coded, as it is in ITK. The definition of the composite is simply an XML-based description of the structure of the composite filter which describes the individual components of the composite, together with their data dependencies (the "wiring"). When a composite filter is instantiated by the user, the actual internal structure of the composite is dynamically configured by interpreting the XML description.

At the user interface level, composites can either be collapsed, appearing as a single filter icon, or expanded, showing the internal structure of the composite. For example Figure 3 displays a composite filter in collapsed form, which can be used just like any other filter. The expanded form of the same composite filter, Figure 4 shows both the internal data dependencies and the distribution of parameters amongst the internal components.

5.3 Parameter Expressions

Most filter parameters are simple scalar or vector values used to customise filters, such as the threshold level for a simple filter. In some cases we may wish to combine parameters in simple ways. To that end we define a simple expression interpreter for defining arithmetic and comparison operations on data values. These expressions can be entered into the property tables that define the parameters for a component. Although a more elaborate mechanism could be built in, we think that a simple expression interpreter is likely to suffice for most kinds of applications; anything more complex can easily be implemented

5.4   Control Constructs

Structured control flow has three aspects: sequencing, choice and repetition. The dataflow model naturally supports sequencing of computations through its directed graph of data dependencies. Given that filters may have multiple inputs, it is apparent that we can encode choice and repetition within the implementation of a filter. In fact, some of the standard ITK components have optimising filters which may already implement repeated filtering behaviour until some criterion is met. However, our goal with ITK-Board is to provide a visual language to allow users to specify selective and repetitive behaviour.

Selection. Conditional selection of inputs is achieved with the *If-*-Else* construct, as depicted in Figure 5.

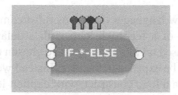

Fig. 5. Conditional construct

Each image data input is guarded by a boolean expression. The first of these to be true, starting from the top, is selected for input. Given inputs In_i for $i = 0 \ldots n$, with boolean guards $Cond_i$ for $i = 0 \ldots n - 1$, we can define the output Out by:

$$Out = In_0, \quad \text{if } Cond_0$$
$$= \ldots$$
$$= In_{n-1}, \text{ if } Cond_{n-1}$$
$$= In_n, \quad \text{otherwise}$$

The guard expressions are defined in terms of the input parameters, as properties of the *If-*-Else* construct. With the data-driven policy for parameters, it is clear that the guards do not directly depend on the input image data. So, with demand-driven computation, we find that each request for output is propagated to just one input, at most. In other words, the selection of the input dataflow In_i only depends on i, which, in turn only depends on the input parameters.

This deterministic behaviour for conditional input selection, which may seem somewhat unorthodox on first sight, is in fact, one of the main reasons for opting for the hybrid dataflow model, as already discussed in Section 4. If instead, we had opted for a demand-driven model for input parameters, then this conditional expression could generate multiple upstream requests until a true condition was found. In Figure 1, for example, the parameters for the conditional block are only dependent on the input parameters of the preceding *Connected Threshold* filter. So, while those parameters remain unaltered, the *If-*-Else* will always select the same input when the right-most viewer is updated—the selection does not depend on the input image data. If we want downstream choices to depend on upstream images, then we must make it explicit in the model, by arranging for some filter to process the upstream image, generating output parameters which can then be used as inputs to an *If-*-Else*. When such a dependency is modelled, the actual behaviour for the conditional selection can depend on where the user makes update requests. If the user *guides* the computation downstream, by ensuring that all up-stream image data is up-to-date, then any conditional selections made downstream will be up-to-date. However, if the user only requests updates from the downstream side, it is possible that, even with repeated requests, that some of the upstream data remains out-of-date.

Although it may seem undesirable to allow different results to be computed, depend-ing on the order of update requests, we claim this to be a benefit of our hybrid model. As mentioned in Section 4, the user is always given visual cues to indicate which image data is not up-to-date. This gives the experimenter control over which parts of the model are brought up-to-date, and which parts may be ignored while attention is focused on another part of the model.

Repetition. In some dataflow languages that focus on stream-based dataflow, iteration is often presented in terms of a feedback loop with a delay element. Although our implementation is indeed based on this idea, we have chosen to keep this as implemen-tation detail which need not be exposed to the user. Instead we provide a simple way of wrapping an existing filter in a loop construct, with explicit loop control managed as a property of the loop construct.

In Figure 6 we illustrate the manner in which a *while loop* can be expressed. Observe how the initial, test and update expressions are written in terms of properties of the loop construct. Although logically redundant, we also provide a counting *for loop*.

6   Related Work

A recent survey [3] provides a good account of dataflow programming, including data-flow visual programming languages (DFVPLs). The current state of play for DFVPLs with iteration constructs is reviewed by [5].

Prograph [6] is a general purpose DFVPL that uses iconic symbols to represent ac-tions to be taken on data. Its general-purpose features are not ideal for supporting com-putationally intensive domains such as signal processing and image analysis. However, control flow in Prograph is implicitly defined, forcing users to implement their own control constructs rather than directly use available constructs.

The Laboratory Virtual Instrumentation Engineering Workbench(LabVIEW) [7] is a platform and development environment for a VPL from National Instruments. Lab-VIEW aims to provide design tests, measurement and control systems for research and industry. It is not geared towards experimental image processing like ITKBoard is. Two iterative forms are supported in LabVIEW namely the for-loop and the while-loop, with a square-shaped block to represent the loop's body.

Cantata [8] is a language originally designed for image processing within the Khoros system [9], a visual programming environment for image processing. It is also useful for signal processing and control systems. In Cantata, icons (called glyphs) represent programs in Khoros system. Unlike ITKBoard, Cantata provides a coarse-grained com-putation model more suited to distributed computation, where each glyph corresponds to an entire process rather than to a process component. There are two iterative forms

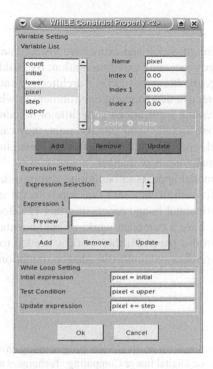

Fig. 6. While Loop construct

provided in Cantata namely the count-loop and the while-loop, similar to our approach. VisiQuest [10] is a commercial scientific data and image analysis application that builds on the foundation laid by Khoros and Cantata.

VIPERS [11] is a DFVPL based on the standard scripting language Tcl, which is used to define the elementary functional blocks (the nodes of the data-flow graph) which are similar to Khoros programs in Cantata. Interestingly, each block corresponds to a Tcl command so VIPERS is less coarse-grained than Cantata. VIPERS programs are graphs that have data tokens travelling along arcs between nodes (modules) which transform the data tokens themselves. VIPERS has some basic representations to allow node connections, ports to link blocks into a network. However, the use of enabling signals in each block introduces a new boolean data-flow channel on top of the main input-output data-flow. VIPERS relies on enabling signals to support language control constructs. See [5] for examples and screen-shots.

MeVisLab [12] is a development environment for medical image processing and visualisation. MeVisLab provides for integration and testing of new algorithms and the development of application prototypes that can be used in clinical environments. Recently, ITK algorithms have been integrated into MeVisLab as an "add-on" module. However, in ITKBoard, ITK provides the core image processing functionality; with its extensive user base, ITK library code reliable and well validated. Moreover, ITKBoard provides a full DFVPL with extra expressiveness in comparison to MeVisLab.

7 Conclusions

We have presented ITKBoard, a rich and expressive visual dataflow language tailored towards biomedical image processing experimentation. By building on top of ITK, we have been able to leverage the strengths of ITK's architecture and libraries. At the same time, we provide an interactive development environment in which experimenters can develop new image processing applications with little or no concern for programming issues. Furthermore, in one and the same environment, they can interactively execute all or part of their model, and investigate the effect of changing model structure and parameterisation in a visually intuitive manner.

The hybrid dataflow model appears to be unique amongst dataflow languages, and matches well with the division of data into two kinds in the underlying ITK framework, and with the experimental image processing that ITKBoard has been designed to support. Currently we have a complete working implementation of the features of ITKBoard as discussed in this paper. Further work is needed to develop ITKBoard to support streaming of image data and multi-threaded computation that ITK already goes some way to supporting.

References

1. Le, H.D.K., Li, R., Ourselin, S.: Towards a visual programming environment based on itk for medical image analysis. In: Digital Image Computing: Techniques and Applications (DICTA 2005), p. 80. IEEE Computer Society, Los Alamitos (2005)
2. Ibáñez, L., Schroeder, W., Ng, L., Cates, J.: The Insight Software Consortium: The ITK Software Guide, 2nd edn. (November 2005), http://www.itk.org
3. Johnston, W.M., Hanna, J.R.P., Millar, R.: Advances in dataflow programming languages. ACM Computing Surveys 36(1), 1–34 (2004)
4. Gamma, E., Helm, R., Johnson, R., Vlissides, J.: Design Patterns: Elements of Reusable Object-Oriented Software. Addison-Wesley, Reading (1995)
5. Mosconi, M., Porta, M.: Iteration constructs in data-flow visual programming languages. Computer Langages 22, 67–104 (2000)
6. Cox, P.T., Giles, F.R., Pietrzykowski, T.: Prograph: A step towards liberating programming from textual conditioning. In: Proceedings of the IEEE Workshop on Visual LanguagesVL 1989, Rome, Italy, pp. 150–156 (1989)
7. National Instruments Corporation: LabVIEW. User Manual (2003)
8. Young, M., Argiro, D., Kubica, S.: Cantata: visual programming environment for the khoros system. SIGGRAPH Computer Graphics 29(2), 22–24 (1995)
9. Konstantinides, K., Rasure, J.R.: The Khoros software development environment for image and signal processing. IEEE Transactions on Image Processing 3(3), 243–252 (1994)
10. VisiQuest: Visual Proramming Guide (2006)
11. Bernini, M., Mosconi, M.: Vipers: a data flow visual programming environment based on the tcl language. In: AVI 1994: Proceedings of the workshop on Advanced visual interfaces, pp. 243–245. ACM Press, New York (1994)
12. Rexilius, J., Spindler, W., Jomier, J., Link, F., Peitgen, H.: Efficient algorithm evaluation and rapid prototyping of clinical applications using itk. In: Proceedings of RSNA 2005, Chicago (December 2005)

A Debugger for the Interpreter Design Pattern

Jan Vraný[1] and Alexandre Bergel[2]

[1] Department of Computer Science, Faculty of Electrical Engineering
Technical University in Prague, Czech
`vranyj1@fel.cvut.cz`
[2] INRIA Futurs, Lille, France
`alexandre.bergel@inria.fr`

Abstract. Using Interpreter and Visitor design patterns is a widely adopted approach to implement programming language interpreters. The popularity of these patterns stems from their expressive and simple design. However, no general approach to conceive a debugger has been commonly adopted.

This paper presents the debuggable interpreter design pattern, a general approach to extend a language interpreter with debugging facilities such as step-over and step-into. Moreover, it enables multiple debuggers to coexists. It extends the Interpreter and Visitor design patterns with few hooks and a debugging service. SmallJS, an interpreter for Javascript-like language, serves as illustration.

Keywords: Program interpretation, debugger, design pattern, programming environment.

1 Introduction

A *design pattern* is a general repeatable solution to a commonly occurring problem in software design. It is a description or template for how to solve a problem that can be used in many different situations. Design patterns gained popularity after Gamma, Helm, Johnson, and Vlissides compiled and classified what were recognized as common patterns [5,4].

The interpreter and visitor design patterns are usually described in terms of interpreting grammars. Given a language, they defines a representation for its grammar along with an interpreter sentences in the language [4]. Whereas the ability of the visitor and interpreter patterns to define programming language interpreters is widely recognized [2,1,7], no approaches to facilitate the realization of a debugger is currently available, as far as we are aware of.

The *Debuggable Interpreter Design Pattern* describes a programming language interpreter that offers debugging facilities. It augments the Interpreter pattern [4] with some hooks in the "visiting" methods and employs a debugging service to model operations (*i.e.,* step-in, step-over, ...).

The contribution of this paper are the following: (i) description of the debuggable interpreter pattern, and (ii) illustration with SmallJS, a subset of Javascript.

Section 2 illustrates the challenges in implementing a debugger. Section 3 presents the debuggable interpreter pattern and its illustration with SmallJS, a minimal procedural language. Section 4 discusses several points and shows some properties of the

J. Filipe et al. (Eds.): ICSOFT/ENASE 2007, CCIS 22, pp. 73–85, 2008.

pattern. Section 5 provides a brief overview of related work. Section 6 concludes by summarizing the presented work.

2   Interpreting and Debugging Languages

2.1   The SmallJS Interpreter

SmallJS is a JavaScript-subset interpreter written in Smalltalk/X, a dynamically-typed object-oriented programming language[1]. SmallJS contains usual language constructions to define variables and functions. As an illustration, the following code describes the factorial function:

```
function fact(i) {
  if ( i > 0 ) {
    return i * fact(i - 1);
  } else {
    return 1;
  };
}
var a;
a = 6;
fact(a);
```

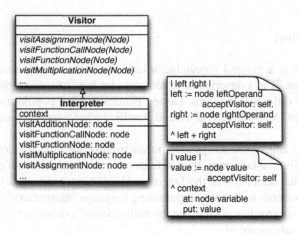

Fig. 1. The SmallJS interpreter

Figure 1 provides an excerpt of a visitor-based interpreter for SmallJS and presents the body of the visitAdditionNode: aNode and visitAssignmentNode: aNode methods[2]. An addition is realized by running the visitor on the left operand, then on the right operand, and to finally return the sum of these two values. An assignment is realized

---

[1] http://www.exept.de/exept/english/Smalltalk/frame_uebersicht.html

[2] We adopted the Smalltalk/X syntax in UML diagrams to use a homogeneous notation.

by running the visitor on the value of the assignment, and then storing the value in the context of the interpreter.

The interpretation of the SmallJS programming language is realized through a direct application of the Interpreter and Visitor design pattern [4].

2.2 Realizing a Debugger

A debugger is a tool that is used to test and debug programs. Typically, debuggers offer sophisticated functionalities such as running a program step by step, stopping (pausing the program to examine the current state) at some kind of event by means of breakpoint, and tracking the values of some variables.

The interpreter maintains several registers such as the program counter (instruction pointer) and the stack pointer. These registers represent the state of execution of process, which defines the *interpreter state*.

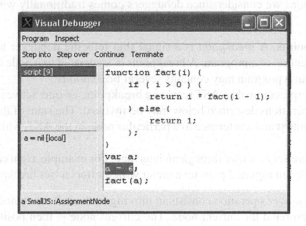

Fig. 2. A graphical user interface of a debugger

Figure 2 shows a debugging session involving the piece of code given above. Advanced debugging environments (*e.g.,* Smalltalk/X, VisualWorks[3], Dolphin Smalltalk[4]) enable several debuggers and interpreters for the same program to coexist. Operations such as opening a new debugger from a debugger, debugging two different pieces of the same program, or debugging a multi-threaded program may be performed.

Whereas the state of an instruction-based interpreters is contained in a set of registers, recursive function invocations define the state of a visitor-based interpreter. State of the interpreter is determined by the set of function activation records, contained in the method calls stacks. Local context visualization is achieved by sophisticated reflective feature, such as stack reification, which might result in a lack of performances or raise technical issues that are difficult to address.

---

[3] http://www.cincomsmalltalk.com/
[4] http://www.object-arts.com

A visitor-based interpreter allow for breakpoints to be set and offers a mechanism to perform debugging operations such as step by step instruction execution.

3 The Debuggable Interpreter Pattern

This section describes a general approach to realize and implement a debugger for a visitor-based interpreter. It augments the visitor interpreter with a set of hooks inserted in the visit* methods. As part of the Interpreter design pattern, the dynamic information needed for a program interpretation is stored in a context. Debugging operations such as step-into, step-by, and continue are offered by a debugging service.

3.1 Debugging Operations

Before describing the Debuggable Interpreter design pattern, it is important to list the different operations we consider since debuggers comes traditionally with its own set of definitions.

Setting Breakpoints. A *breakpoint* is a signal that tells the debugger to temporarily suspend the execution of a program. A breakpoint is associated to a node in the abstract syntax tree. A same program may contains several breakpoints.

When the interpretation of an AST reaches a breakpoint, an interactive session begins for which the operations described below may be invoked. The state of the debugger is modeled by a context and a reference to a particular node in the AST, which we call the *current node*.

Breakpoints are set by a user through an interface. For example, right clicking on the interface presented in Figure 2 pops up a menu which offers a 'set breakpoint' entry.

Step-over. A *step-over* operation consists in moving to the following node in the AST after having interpreted the current node. The current node is then positioned on this new node.

Figure 3 illustrates a step-over operation. The current node is fact(a). By performing a step-over operation, the current node for which the user has the hand is print(fact(a)) node.

Step-into. A *step-into* operation consists in moving to the next node in the AST according to the application control flow. This operation differs from *step-over* by entering recursion.

A step-into operation is illustrated in Figure 4. In this situation, the interpreter halts at the first node in the recursion which is i > 0.

Continue. The execution of an application may be resumed by *continuing* it.

Terminate. The program execution might be prematurely ended with the *terminate* operation. As a consequence, no subsequent nodes are evaluated and allocated resources such as context stack and the interpreter are freed.

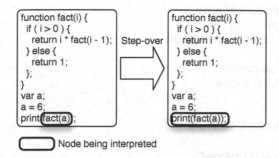

Fig. 3. A step-over operation does not go into recursion

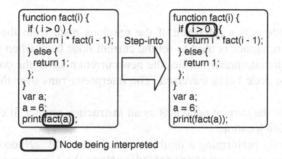

Fig. 4. A step-into operation goes into recursion

3.2 Hooks in the Visitor

The flow of a program interpretation stems from the invocation order of the visiting methods. Suspending and resuming the interpretation flow and capturing recursion are the primitives for the debugging operations.

In the remaining of this paper, the interpreter class is called Interpreter. As described in Figure 1, it implements the visiting methods.

Hooks need to be inserted in the Interpreter class to enable step-over and step-into operations. These hooks inform a debugger service which part of code is being executed. As it will be described further (Section 3.5), a service is a placeholder for debugging operations. The method Interpreter.onTracepoint: aNode enables interpretation of an AST to be driven by a service:

```
Interpreter. onTracepoint: aNode {
    debuggerService onTracepoint: aNode
}
```

These two methods have to be invoked when a node is visited. A visit: method maintains the current node reference in a context:

```
Interpreter.visit: aNode {
    | value previousNode |
```

```
"A reference of the current node is
temporarily stored"
previousNode := context currentNode.

"The node is set as current"
context currentNode: aNode.

"Visit the node"
value := aNode acceptVisitor: self.

"The previous node is restored"
context currentNode: previousNode.
^value
}
```

First, visit: aNode gets a reference of the previous node from the current activation context. This reference is used to set the current node back when visit: aNode has completed. Then the interpreter notifies the new current node to the context. This new current node is the node being traversed. The interpreter runs over this node using a double dispatch.

The reference of the current node acts as an instruction pointer. It clearly identifies the current execution location.

Instead of directly performing a double dispatch, the visit: aNode has to be used. For example, in the method visitAdditionNode: aNode the recursion is obtained from invoking visit: aNode:

```
Interpreter.visitAdditionNode: aNode {
    | left right |
    left := self visit: aNode left.
    right := self visit: aNode right.
    self onTracepoint: aNode.
    ^left+right
}
```

Each visit* method must call onTracepoint: aNode after traversing all branches and before synthesizing the result.

Compared with the code shown in Figure 1, this new version of visitAdditionNode: aNode makes the interpreter aware of breakpoints. When a breakpoint is reached, the execution of the interpreter is suspended. Subsequent subsections illustrates how breakpoints and debugging modes are modeled.

3.3 Context Definition

Associations between variables and values are stored within a context object [4]. The debuggable interpreter pattern augments this context with dynamic information related to the parent context and the current node under execution. Each function invocation creates a new context.

The class InterpreterContext contains three variables: sender referring to the parent context, currentNode for the node in the abstract syntax tree, and returnReached indicating if a *return* node has been reached or not. The interpreter should not evaluate subsequent nodes when a return node has been reached. Typically, this occurs when a return statement is interpreter. The method visitReturnNode: aNode is therefore defined as follows:

```
Interpreter.visitReturnNode: aNode {
    | value |
    value := self visit: expression.
    self onTracepoint: aNode.
    self returnReached: true.
    context returnReached: true.
    ^value
}
```

3.4 Separate Control Flow

An interpreter has to be embedded in a thread. This is necessary for several reasons:

- Multiple execution of the same program enables a debugger to launch another debugger. Although not essential, this feature leads to a better comfort when debugging.
- If the executed program fails, it should not impact the enclosing application environment. The interpreter cannot run in the same control flow that the one of the programming environment. The program under execution and the debugger cannot be executed in the same thread.

The Interpreter class defines a variable process and an evaluate: aNode method to trigger an execution:

```
Interpreter.evaluate: aNode {
    | value semaphore |
    semaphore := Semaphore new: 0.
    context := Context new.
    process := ([value := self visit: aNode] newProcess)
        addExitAction: [semaphore signal];
        resume.
    semaphore wait.
    ^value
}
```

The evaluate: aNode method creates a semaphore. This is necessary to block the execution of the interpreter. A new context is created. Then a new thread (called process in the ST/X terminology) is created, intended to execute the code value := self visit: aNode. Independently if an exception is raised or not, once completed, the exit action is triggered which releases the semaphore.

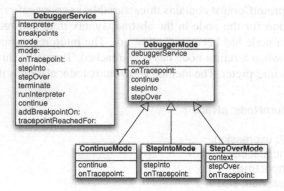

Fig. 5. Debugger services definitions

3.5 Debugging Service

In addition to keep a reference to a context, an interpreter must refer to a *debugger service*. This service implements the debugging operations such as step-into, step-over and continue. Figure 5 provides an overview.

The control flow of an application is halted when it reaches a breakpoint, which identifies a location in the program[5]. With the debuggable interpreter pattern, a breakpoint is identified as a node in the abstract syntax tree.

When a breakpoint is reached, the interpreter enters the interaction mode, which allows the user to perform further operations. The mode in which the service is set reflects the operation that is currently performed. Each mode represents a debugging operation. We consider only 3 modes: continue, step-into, and step-over.

A service maintains a list of breakpoints, accessible by the modes. A breakpoint is added to a service by the user through an user interface.

The methods continue, stepInto, stepOver defined on DebuggerService are delegated to the current mode:

```
DebuggerService.continue {
    mode continue }

DebuggerService.stepInto {
    mode stepInto }

DebuggerService. stepOver {
    mode stepOver }
```

The TracepointReachedFor: and run methods are used to steer the process associated to an interpreter. TracepointReachedFor: is invoked by the mode when a breakpoint is reached. This method simply suspends the process associated to the interpreter:

---

[5] Note that advanced definition of breakpoint such as conditional are out of the scope of this paper.

```
DebuggerService.runInterpreter { interpreter run }
Interpreter.run { process resume }
```

Setting Breakpoints. A debugger service maintains a list of nodes that represent breakpoints in the program.

```
DebuggerService.addBreakpointOn: aNode {
    breakpoints ifNil: [
        breakpoints := OrderedCollection new ].
    breakpoints add: aNode.
}
```

The method addBreakpointOn: is invoked by the debugger user interface. The aNode parameter corresponds to the node in the abstract syntax tree that should halt the program interpretation.

Continue Mode. The continue mode is the initial mode of the debugger service. When the debugger service is in the continue mode, the program is executed until a breakpoint is reached. In that case, the interpreter thread is suspended, and a debugger opened. The service can either switch to a step-into or step-over mode, or a continue mode. The two methods that define this mode are:

```
ContinueMode.onTracepoint: aNode {
    (debuggerService isBreakpoint: aNode) ifTrue: [
        debuggerService TracepointReachedFor: aNode
    ].
}
```

```
ContinueMode.continue {
    debuggerService runInterpreter
}
```

Step-into Mode. When the debugger service is in the step-into mode, the program interpretation is stopped (and a debugger is opened) when onTracepoint: is invoked.

```
StepIntoMode.onTracepoint: aNode {
    debuggerService TracepointReachedFor: aNode
}
```

```
StepIntoMode.stepInto {
    debuggerService runInterpreter
}
```

Step-over Mode. When the debugger service is in the step-over mode, stepping does not follow recursions and method calls. The StepOverMode has a context variables. This variable captures the state of the current interpretation. It is initialized when the debugger service switches to this mode.

The two methods that define this mode are:

```
StepOverMode.onTracepoint: aNode {
  ((context = debuggerService context) ifTrue: [
    debuggerService TracepointReachedFor: aNode]
}

StepOverMode.stepOver {
  debuggerService runInterpreter
}
```

If the current context and the current node match the ones referenced by the step-mode, then the debugger switches for a step-into mode, so that the execution will be halted on the node that follows the "step-overed" one.

4 Discussion

Coexisting Debuggers. Since multiple interpreters can interpret the same program, several debuggers may be active at the same time. Although this feature does not figure as a priority for the Java debugger[6]java.sun.com/j2se/1.4.2/docs/jguide/jpda/-jarchitecture.html., it greatly enhances the debugging activity.

Breakpoints. The debuggable interpreter pattern emits a breakpoint signal when the control flow reaches a particular node in the abstract syntax tree. Note that this definition of breakpoint might slightly diverge from widely spread debugger such as Gdb[7] where a breakpoint signal may be triggered when the control flow reaches a particular line of the source code.

New Operations. New debugging operations can easily be added by subclassing DebuggerMode and adding the corresponding methods in DebuggerService. For example, a mode that holds a condition to enable the interpretation can be implemented in a class ConditionalInterpretationMode in which the method onTracepoint: aNode checks for the conditional expression.

Speed and Memory Overhead. A natural question to be raised is about the cost in terms of speed and memory consumption. The table below shows the time in millisecond taken to execute the factorial function with the debuggable interpreter design pattern (DIDP) and the classical visitor pattern (VDP). Those figures are obtained while disabling the just-in-time compiler (JIT).

Iteration	DIDP (ms)	VDP (ms)	ratio
fac(100)	5.0	2.0	2.50
fac(200)	8.0	5.0	1.60
fac(400)	17.0	12.0	1.41
fac(900)	38.0	20.0	1.90
fac(10000)	973.0	540.0	1.80
fac(70000)	22774.0	19722.0	1.15

---

[6] http://java.sun.com/j2se/1.4.2/docs/jguide/jpda/jarchitecture.html
[7] http://sourceware.org/gdb/

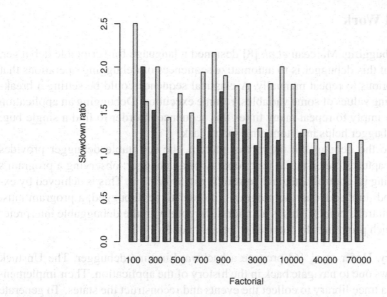

Fig. 6. Ratio between a non-debuggable and a debuggable interpreter. Use of a just-in-time compiler (JIT) is denoted in black.

Figure 6 shows for each factorial expression the overhead of the DIDP with the VDP. It also illustrates the benefit of having a JIT. The black bar indicates a measurement obtained with the JIT enabled, whereas the value denoted by a white bar is obtained with the JIT disabled.

As a result, we see that the ratio is asymptotic to 1. This means that the cost of the debuggable interpreter design pattern compared the classical visitor one is negligible for deep recursion.

The table below compares the memory consumption of DIDP with VDP. The total number of created objects is shown for each factorial expression.

Iteration	DIDP	VDP	ratio
fac(100)	102	102	1
fac(200)	202	202	1
fac(400)	402	402	1
fac(900)	902	902	1
fac(10000)	10002	10002	1
fac(70000)	70002	70002	1

During the evaluation of factorial, contexts are created for each recursion, plus a mode and a service. This table shows that the DIDP does not occurs any memory overhead.

The computer used for this experiment is an Intel Pentium M 1,6GHz, Linux (kernel 2.6.20.1), 512MB RAM, ST/X 5.2.8.

5 Related Work

Scripting Debugging. Marceau *et al.* [8] designed a language for scriptable debugger. The purpose of this debugger is to automatize sequences of debugging operations that might be laborious to repeat manually. A classical sequence could be setting a breakpoint, examining values of some variables, resume execution. Debugging an application generally may imply to repeat many times this sequence in order to find a single bug. Scripting a debugger helps in automating such a task.

Similarly to the debuggable interpreter pattern, the scriptable debugger provides primitives to captures the essential functionality of a debugger: observing a program's state, monitoring its control path, and controlling its execution. This is achieved by explicit commands embedded in the program. In order to be debugged, a program must contain explicit trace points. This is a major difference with the debuggable interpreter pattern for which programs do not need to be annotated.

Trace Library. Hofer *et al.* [6] propose a backward-in-time debugger. The Unstuck debugger allows one to navigate back in the history of the application. Their implementation uses of a trace library to collect the events and reconstruct the states. To generate events (method invocation, variable access and method return), the methods are instrumented using ByteSurgeon [3], a high-level library to manipulate method bytecode.

Unstuck assumes that a program is interpreted by a virtual machine, whereas the debuggable interpreter design pattern relies on an interpretation driven by a visitor.

AST Instrumentation. The Relational Meta-Language (RML) [9] is a language for writing executable Natural Semantics specifications. It is used to formally specify programming languages such as Java, Pascal, and MiniML. The RML debugger is based on an abstract syntax tree instrumentation to capture and records particular events. A postmorten analysis tool is then provided to walk back and forth in time, display variable values, and execution points.

The AST is instrumented with debugging annotation related to trace generation. From its design, the programming environment of RML is limited to one single debugger per session. Our approach allows several debugger to coexist.

Grammar Weaving. Wu *et al.* [10] claims that the debugging is a concern that crosscuts a domain specific language specification. They propose to use AspectJ[8] to weave the debugging semantics into the code created by a parser generator.

Their work is restricted to programming languages that are translated into a general purpose language. Our approach is different since it assumes a program interpretation through a visitor and interpreter design pattern.

6 Conclusion and Future Work

This paper presents a general approach for implementing and realizing a debugger for visitor-like interpreter. It extends a visitor with a set of hooks embedded in the visiting

---

[8] http://eclipse.org/aspectj/

methods. The context, primarily used to hold variables bindings, has been extended with a reference to a parent context and keeps a reference to the node currently being interpreted. A debugger service models different operation available by means of a set of mode.

Easy to implement, coexistence of multiple debuggers, and open to new debugging operations are the benefits of the Debuggable Interpreter Design Pattern.

As future work, we plan to assess the scalability of the debuggable interpreter design pattern by implementing a larger language such as Java, which involves dozens of AST nodes.

Acknowledgements. We gratefully thank the financial support of Science Foundation Ireland and Lero – the Irish Software Engineering Research Centre. We also would like to thank Marcus Denker for his precious comments.

References

1. Acebal, C.F., Castanedo, R.I., Lovelle, J.M.C.: Good design principles in a compiler university course. SIGPLAN Not. 37(4), 62–73 (2002)
2. Cheong, Y.C., Jarzabek, S.: Frame-based method for customizing generic software architectures. In: SSR 1999: Proceedings of the 1999 symposium on Software reusability, pp. 103–112. ACM Press, New York (1999)
3. Denker, M., Ducasse, S., Tanter, É.: Runtime bytecode transformation for Smalltalk. Journal of Computer Languages, Systems and Structures 32(2-3), 125–139 (2006)
4. Gamma, E., Helm, R., Johnson, R., Vlissides, J.: Design Patterns: Elements of Reusable Object-Oriented Software. Addison Wesley, Reading (1995)
5. Gamma, E., Helm, R., Vlissides, J., Johnson, R.E.: Design patterns: Abstraction and reuse of object-oriented design. In: Nierstrasz, O. (ed.) ECOOP 1993. LNCS, vol. 707, pp. 406–431. Springer, Heidelberg (1993)
6. Hofer, C., Denker, M., Ducasse, S.: Design and implementation of a backward-in-time debugger. In: Proceedings of NODE 2006. Lecture Notes in Informatics, vol. P-88, pp. 17–32. Gesellschaft für Informatik (GI) (2006)
7. Lorenz, D.H.: Tiling design patterns a case study using the interpreter pattern. In: OOPSLA 1997: Proceedings of the 12th ACM SIGPLAN conference on Object-oriented programming, systems, languages, and applications, pp. 206–217. ACM Press, New York (1997)
8. Marceau, G., Cooper, G.H., Spiro, J.P., Krishnamurthi, S., Reiss, S.P.: The design and implementation of a dataflow language for scriptable debugging. Automated Software Engineering Journal (2006)
9. Pop, A., Fritzson, P.: Debugging natural semantics specifications. In: AADEBUG 2005: Proceedings of the sixth international symposium on Automated analysis-driven debugging, pp. 77–82. ACM Press, New York (2005)
10. Wu, H., Gray, J., Roychoudhury, S., Mernik, M.: Weaving a debugging aspect into domain-specific language grammars. In: SAC 2005: Proceedings of the 2005 ACM symposium on Applied computing, pp. 1370–1374. ACM Press, New York (2005)

Part II

Software Engineering

Concepts for High-Perfomance Scientific Computing

René Heinzl, Philipp Schwaha, Franz Stimpfl, and Siegfried Selberherr

Institute for Microelectronics, TU Wien, Gusshausstrasse 27-29, Vienna, Austria

Abstract. During the last decades various high-performance libraries were developed written in fairly low level languages, like FORTRAN, carefully specializing codes to achieve the best performance. However, the objective to achieve reusable components has regularly eluded the software community ever since. The fundamental goal of our approach is to create a high-performance mathematical framework with reusable domain-specific abstractions which are close to the mathematical notations to describe many problems in scientific computing. Interoperability driven by strong theoretical derivations of mathematical concepts is another important goal of our approach.

1 Introduction

This work reviews common concepts for scientific computing and introduces new ones for a timely approach to library centric application design. Based on concepts for generic programming, e.g. in C++, we have investigated and developed data structures for scientific computing. The Boost Graph Library [11] was one of the first generic libraries, which introduced concept based programming for a more complex data structure, a graph. The actual implementation of the Boost Graph Library (BGL) is for our work of secondary importance, however, we value the consistent interfaces for graph operations. We have extended this type of concept based programming and library development to the field of scientific computing. To give a brief introduction we use an example resulting from a self-adjoint partial differential equation (PDE), namely the Poisson equation:

$$\operatorname{div}(\varepsilon \operatorname{grad}(\Psi)) = \rho$$

Several discretization schemes are available to project this PDE into a finite space. We use the method of finite volumes. The resulting equations are given next, where A_{ij} and d_{ij} represents geometrical properties of the discretized space, ρ the space charge, Ψ the potential, and ε the permittivity of the medium.

$$\sum_{j} D_{ij} A_{ij} = \rho \tag{1}$$

$$D_{ij} = \frac{\Psi_j - \Psi_i}{d_{ij}} \frac{\varepsilon_i + \varepsilon_j}{2} \tag{2}$$

J. Filipe et al. (Eds.): ICSOFT/ENASE 2007, CCIS 22, pp. 89–100, 2008.
© Springer-Verlag Berlin Heidelberg 2008

An example of our domain specific notation is given in the following code snippet and explained in Section 4:

```
value =
(
  sum<vertex_edge>
  [
      diff<edge_vertex>
      [
        Psi(_1)
      ] * A(_1)/d(_1) *
      sum<edge_vertex>[eps(_1)]/2
  ] - rho(_1)
)(vertex);
```

Generic Poisson Equation

As can be seen, the actual notation does not depend on any dimension or topological type of the cell complex (mesh) and is therefore dimensionally and topologically indepent. Only the relevant concepts, in this case, the existence of edges incident to a vertex and several quantity accessors, have to be met. In other words, we have extended the concept programming of the standard template library (STL) and the generic programming of C++ to higher dimensional data structures and automatic quantity access mechanisms.

Compared to the reviewed related work given in Section 2, our approach implements a domain specific embedded language. The related topological concepts are given in Section 3, whereas Section 4 briefly overviews the used programming paradigms. In Section 5 several application examples are presented. The first example introduces a problem of a biological system with a simple PDE. The second example shows a non-linear system of coupled PDEs, which makes use of the linearization framework introduced in Section 4.1, where derivatives are calculated automatically.

For a successful treatment of a domain specific embedded notation several programming paradigms are used. By object-oriented programming the appropriate iterators are generated, hidden in this example in the expression vertex_edge and edge_vertex. Functional programming supplies the higher order function expression between the [and] and the unnamed function object _1. And finally the generic programming paradigm (in C++ realized with parametric polymorphism or templates) connects the various data types of the iterators and quantity accessors.

A significant target of this work is the separation of data access and traversal by means of the mathematical concept of fiber bundles [5]. The related formal introduction enables a clean separation of the internal combinatorial properties of data structures and the mechanisms of data access. A high degree of interoperability can be achieved with this formal interface. Due to space constraints the performance analysis is omitted and we refer to a recent work [6] where the overall high performance is presented in more detail.

2 Related Work

In the following several related works are presented. All of these software libraries are a great achievement in the various fields of scientific computing. The **FEniCS** project [9], which is a unified framework for several tasks in the area of scientific computing, is a great step towards generic modules for scientific computing.

The **Boost Graph Library** is a generic interface which enables access to a graph's structure, but hides the details of the actual implementation. All libraries which implement this type of interface are interoperable with the BGL generic algorithms. This approach was one of the first in the field of non-trivial data structures with respect to interoperability. The property map concept [11] was introduced and heavily used. The **Grid Algorithms Library, GrAL** [4] was one of the first contributions to the unification of data structures of arbitrary dimension for the field of scientific computing. A common interface for grids with a dimensionally and topologically independent way of access and traversal was designed.

Our Developed Approach, the Generic Scientific Simulation Environment (GSSE [8]) deals with various modules for different discretization schemes such as finite elements and finite differences. In comparison, our approach focuses more on providing building blocks for scientific computing, especially an embedded domain language to express mathematical dependencies directly, not only for finite elements. To achieve interoperability between different library approaches we use concepts of fiber bundle theory to separate the base space and fiber space properties. With this separation we can use several other libraries (see Section 3) for different tasks. The theory of fiber bundles separates the data structural components from data access (fibers).

Based on this interface specification we can use several libraries, such as STL, BGL, GrAL, and accomplish high interoperability and code reuse.

3 Concepts

Our approach extends the concept based programming of the STL to arbitrary dimensions similar to GrAL. The main difference to GrAL is the introduction of the concept of fiber bundles, which separates the base mechanism of application design into base and fiber space properties. The base space is modeled by a CW-complex and algebraic topology, whereas the fiber space is modeled by a generic data accessor mechanism, similar to the cursor and property map concept [3].

Table 1. Comparison of the cursor/property map and the fiber bundle concept

	cursor and property map	fiber bundles
isomorphic base space	no	yes
traversal possibilities	STL iteration	cell complex
traversal base space	yes	yes
traversal fiber space	no	yes
data access	single data	topological space
fiber space slices	no	yes

3.1   Theory of Fiber Bundles

We introduce concepts of fiber bundles as a description for data structures of various dimensions and topological properties.
- Base space: topology and partially ordered sets
- Fiber space: matrix and tensor handling

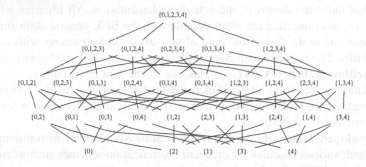

Fig. 1. Cell topology of a simplex cell in four dimensions

Based on these examples, we introduce a common theory for the separation of the topological structure and the attached data. The original contribution of this theory was given in Butler's vector bundle model [5], which we compactly review here:

Definition 1   (Fiber Bundle). *Let E, B be topological spaces and $f : E \to B$ a continuous map. Then (E, B, f) is called a **fiber bundle**, if there exists a space F such that the union of the inverse images of the projection map f (the fibers) of a neighborhood $U_b \subset B$ of each point $b \in B$ are homeomorphic to $U_b \times F$, whereby this homeomorphism has to be such that the projection pr_1 of $U_b \times F$ yields U_b again.*

E is called the *total space*, B is called the *base space*, and F is called the *fiber space*. This definition requires that a total space E can locally be written as the product of a base space B and a fiber space F. The decomposition of the *base space* is modeled by an identification of data structures by a CW-complex [7]. As an example Figure 2 depicts an array data structure based on the concept of a fiber bundle. We have a simple fiber space attached to each cell (marked with a dot in the figure), which carries the data of our array.

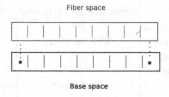

Fig. 2. A fiber bundle with a fiber space over a 0-cell complex. A simple array is an example of this type of fiber space.

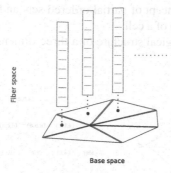

Fig. 3. A fiber space over a 2-simplex cell complex base space. An example of this type of fiber space is a triangle mesh with an array over each triangle.

Figure 3 depicts a fiber bundle with a 2-cell complex as base space. For the base space of lower dimensional data structures, such as an array or single linked list, the only relevant information is the number of elements determined by the index space. Therefore most of the data structures do not separate these two spaces. For backward compatibility with common data structures the concept of an index space depth is used [7].The advantages of this approach are similar to those of the cursor and property map [3], but they differ in several details. The similarity is that both properties can be implemented independently. However, the fiber bundle approach equips the fiber space with more structure, e.g., storing more than one value corresponding to the traversal position as well as preservation of neighborhoods. This feature is especially useful in the area of scientific computing, where different data sets have to be managed, e.g., multiple scalar or vector values on vertices, faces, or cells. Another important property of the fiber bundle approach is that an equal (isomorphic) base space can be exchanged with another cell complex of the same dimension. An overview of the common topological features and differences for various data structures are presented in Table 1.

Table 2. Classification scheme based on the dimension of cells, the cell topology, and the complex topology

data structure	cell dimension	cell topology	complex topology
array/vector	0	simplex	global
SLL/stream	0	simplex	local(2)
DLL/binary tree	0	simplex	local(3)
arbitrary tree	0	simplex	local(4)
graph	1	simplex	local
grid	2,3,4,..	cuboid	global
mesh	2,3,4,..	simplex	local

3.2 Topological Interface

We briefly introduce parts of the interface specification for data structures and their corresponding iteration mechanism based on algebraic topology. A full reference can be found in [7]. With the concept of partial ordered sets and a Hasse diagram we can order and depict the structure of a cell.

As an example the topological structure of a three-dimensional simplex is given in Figure 4.

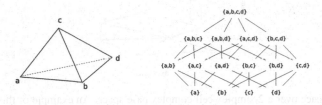

Fig. 4. Cell topology of a 3-simplex cell

Inter-dimensional objects such as edges and facets and their relations within the cell can thereby be identified. The complete traversal of all different objects is determined by this structure. We can derive the vertex on edge, vertex on cell, as well as edge on cell traversal up to the dimension of the cell in this way. Based on our topological specification arbitrary dimensional cells can be easily used and traversed in the same way as all other cell types, e.g., a 4-dimensional simplex shown in Figure 1.

Next to the cell topology a separate complex topology is derived to enable an efficient implementation of these two concepts. A significant amount of code can be reduced with this separation. Figure 5 depicts the complex topology of a 2-simplex cell complex where the bottom sets on the right-hand sides are now the cells. The rectangle in the figure marks the relevant cell number.

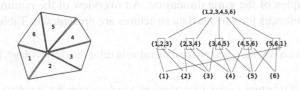

Fig. 5. Complex topology of a simplex cell complex

The topology of the cell complex is only available locally because of the fact that subsets can have an arbitrary number of elements. In other words, there can be an arbitrary number of triangles attached to the innermost vertex. Our final classification scheme uses the term local to represent this fact.

A formal concise definition of data structures can therewith be derived and is presented in Figure 2. The complex topology uses the number of elements of the corresponding subsets.

```
complex_t<cells_t,global   > cx;  //{1}
complex_t<cells_t,local<2>> cx;  //{2}
complex_t<cells_t,local<3>> cx;  //{3}
complex_t<cells_t,local<4>> cx;  //{4}
```

STL Data Structure Definitions

Here {1} describes an array, {2} a stream or a singly linked list, {3} a doubly linked list or a binary tree, and finally {4} an arbitrary tree. To demonstrate the equivalence of the STL data structures and our approach we present a simple code snippet (the typedefs are omitted due to space constraints):

```
cell_t<0, simplex>                  cells_t;
complex_t<cells_t, global>          complex_t;
container_t<complex_t, long>        container;
// is equivalent to
std::vector<data_t>                 container;
```

Equivalence of Data Structures

3.3 Data Access

In the following code snippet a simple example of the generic use of a data accessor similar to the property map concept is given, where a scalar value is assigned to each vertex. The data accessor implementation also takes care of accessing data sets with different data locality, e.g., data on vertices, edges, facets, or cells. The data accessor is extracted from the container to enable a functional access mechanism with a key value which can be modeled by arbitrary comparable data types.

```
da_type da(container, key_d);
for_each(container.vertex_begin(),
         container.vertex_end(), da = 1.0 );
```

Data Accessor

Several programming paradigms are used in this example which are presented in detail in the next section, especially the functional programming, given in this example with the da = 1.0.

4   Programming Paradigms

Various areas of scientific computing encourage different programming techniques, even demands for several programming paradigms can be observed:

- Object-oriented programming: the close interaction of content and functions is one of the most important advantages of the object-oriented programming
- Functional programming: offers a clear notation, is side-effect free and inherently parallel

– Generic programming: can be seen as the glue between object-oriented and functional programming

Our implementation language of choice is C++ due to the fact, that it is one of the few programming languages where high performance can be achieved with various paradigms.

To give an example of this multi-paradigm approach, a simple C++ source snippet is given next.

```
std::for_each(v.begin(),v.end(),
    if_(arg1 > 5)
    [
        std::cout << arg1 << std::cout
    ]
);
```

Multiple Paradigms

The object-oriented programming paradigm is used to create the iterator capabilities of the container structures of the STL. This paradigm is not used anywhere else in our approach. Functional programming is used to create function objects which are passed into the generic for_each algorithm. In this example the notation of the Boost Phoenix 2 [2] library is used to create a functional object context, marked by the [and]. The generic paradigm uses the template mechanism of C++ to bind these two paradigms together efficiently. A more complex example is given in the following expression. Here a cell complex of arbitrary dimension is used and the vertex to vertex iteration is expressed.

```
gsse::for_each_vertex(domain
    result=gsse::add<vertex_vertex> [  quan  ]
);
```

Complex Functor

The same paradigms as in the example before can be seen, but in this case, a complex topological traversal is used instead of simple container traversal. The topological properties of the GSSE are demonstrated twofold: on the one side, the topological concept programming allows the implementation of a dimensionally independend algorithm. On the other side, different data structures of library approaches can be used, which fullfill the basic requirements of the required topological concept.

This GSSE algorithm sums up the potential values of all vertices adjacent to a vertex. The data accessor quan handles the storage mechanism for the value attached to a vertex. Here the interaction of programming paradigms related to the base space and fiber space can be seen clearly. The base space traversal is built with the generic programming paradigm, whereas the fiber space operation is implemented by means of functional programming. A lot of difficulties with conventional programming can be circumvented by this approach. Functional programming enables great extensibility due to the modular nature of function objects. Generic programming and the corresponding template mechanisms of C++ offer an overall high performance. In addition arbitrary data structures of arbitrary dimensions can be used. The only requirement is that the data structure models the required concept, in this case a vertex to vertex information.

4.1 Automatic Linearization

A calculation framework is used where derivatives are implicitly available and do not have to be specified explicitly. This enables the specification of nonlinear differential equations in a convenient way. The elements of the framework are truncated Taylor series of the following form $f_0 + \sum_i c_i \cdot \Delta x_i$. To use a quantity x_i within a formula we have to specify its value f_0 and the linear dependence $c_i = 1$ on the vector $\mathbf{x}$ of quantities. This non-trivial and highly complex scenario yields itself exceptionally well to the application of the functional programming paradigm. In general, all discretization schemes which use line-wise assembly based on finite differences as well as finite volumes can be handled with the described formalism. Basic operations on Taylor series can handle truncated polynomial expansions. In the following we specify our nonlinear functionals using linearized functions in upper case letters. All necessary numerical operations on these data structures can be performed in a straight forward manner, e.g. multiplication:

$$F = f_0 + \sum_i c_i \cdot \Delta x_i, \qquad G = g_0 + \sum_i d_i \cdot \Delta x_i \qquad (3)$$

$$F \otimes G = (f_0 \cdot g_0) + \sum_i (g_0 \cdot d_i + f_0 \cdot c_i) \cdot \Delta x_i \qquad (4)$$

Having implemented these schemes we are able to derive all required functions on these mathematical structures. This means that we have a consistent framework for formulas in the following sense: if A is the linearized version of function $\mathcal{A}$ at x_0, we obtain $\partial A / \partial x_i = \partial \mathcal{A} / \partial x_i \mid_{x_0}$ around the point of linearization. Figure 6 shows the multiplication of two truncated expansions $F = 3 + \Delta x_1$, and $G = 3 + \Delta x_3$. As a result we obtain $F \otimes G = 12 + 4\Delta x_1 + 3\Delta x_3$.

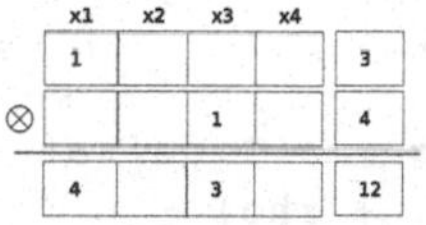

Fig. 6. The multiplication of two Taylor series

By implementing only the linear (or higher order polynomial) functional dependence of equations on variables around $\mathbf{x}$ we reduce the external specification effort to a minimum. Thus, it is possible to ease the specification with the functional programming approach, while also providing the functional dependence of formulas.

5 Application Design

In the following we briefly review a few applications based on the introduced concepts with their respective paradigms.

5.1 Biological System

Electric phenomena are common in biological organisms such as the discharges within the nervous system, but usually remain within a small scale. In some organisms, however, the electric phenomena take a more prominent role. Some fish species, such as *Gnathonemus petersii* from the family of Mormyridae [12], use them for detection of their prey. The up to 30 cm long fish actively generates electric pulses with an organ located near its tail fin (also marked in Figure 7). More information can be found in [10].

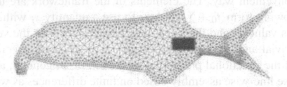

Fig. 7. Discretized domain of a fish with a red marked electrically active organ

For this case we derive the equation system based on a quasi-electro-statical system directly from the corresponding Maxwell equations. The charge separation of the electrically active organ which is actively taking place within parts of the simulation domain also has to be taken into account. We use the conservation law of charge and the divergence theorem (Gauss's law) and finally get:

$$\partial_t \left[\operatorname{div}(\varepsilon \operatorname{grad}(\Psi)) \right] + \operatorname{div}(\gamma \operatorname{grad}(\Psi)) = P \tag{5}$$

Equation 5 is discretized using the finite volume discretization scheme. The high semantic level of the specification is illustrated by the following snippet of code:

```
linearized_equ = sum<vertex_edge>(_v)
[
    orient(_v,_1) * sum<edge_vertex>(_e)
    [
    equ_pot * orient(_e,_1)
    ]
    * (area  / dist ) * (gamma * deltat + eps)
] + vol * ( (P * deltat) + rho)
```

This source snippets presents most of the application code which has to be developed. In addition, only a simple preprocessing step which creates the necessary quantity accessors is required.

The simulation domain is divided into several parts including the fish itself, its skin, that serves as insulation, the water the fish lives in, and an object, that represents either an inanimate object or prey. The parameters of each part can be adjusted separately. A result of the simulation is depicted in the following figure:

5.2 Drift-Diffusion Equation

Semiconductor devices have become an ubiquitous commodity and people expect a constant increase of device performance at higher integration densities and falling prices.

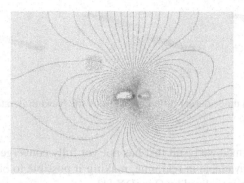

Fig. 8. Result of a simulation with a complete domain, the fish, and a ideally conductor as response object

To demonstrate the importance of a method for device simulation that is both easy and efficient we review the drift diffusion model that can be derived from Boltzmann's equation for electron transport by applying the method of moments [8]. Note that this problem is a nonlinear coupled system of partial differential equations where our linearization framework is used. This results in current relations as shown in Equation 6. These equations are solved self consistently with Poisson's equation, given in Equation 7.

$$\mathbf{J}_n = q\,n\,\mu_n\,\mathrm{grad}\,\Psi + q\,D_n\,\mathrm{grad}\,n \qquad (6)$$

$$\mathrm{div}(\mathrm{grad}(\varepsilon\,\Psi)) = -\rho \qquad (7)$$

The following source code results from an application of the finite volume discretization scheme:

```
linearized_eq_t equ_pot, equ_n;
equ_pot = (sum<vertex_edge>
              [
                diff<edge_vertex>[pot_quan]
              ] + ( n_quan - p_quan + nA - nD ) *
              vol * q / (eps0 * epsr)
            )(vertex);
equ_n = (sum<vertex_edge>
            [
              diff<edge_vertex>
              ( -n_quan*Bern( diff<edge_vertex>[ pot_quan / U_th] ),
                -n_quan*Bern( diff<edge_vertex>[-pot_quan / U_th] )
              )* (q * mu_h * U_th)
            ])(vertex);
```

Drift-Diffusion Equation

To briefly present a simulation result we provide Figure 9 which shows the potential in a pn diode at different stages of a nonlinear solving procedure. The leftmost figure shows the initial solution, while the rightmost depicts the final solution. The center

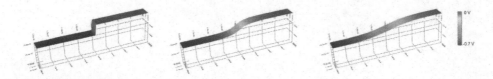

Fig. 9. Potential in a pn diode during different stages of the Newton iteration. From initial (left) to the final result(right).

image shows an intermediate result that has not yet fully converged. The visualization of the calculation is available in real time, making it possible to observe the evolution of the solution, which is realized by OpenDX [1].

References

1. DX, IBM Visualization Data Explorer. IBM Corporation, Yorktown Heights, 3rd edn., NY, USA (1993)
2. Phoenix2, Boost Phoenix 2. Boost C++ Libraries (2006), http://spirit.sourceforge.net/
3. Abrahams, D., Siek, J., Witt, T.: New Iterator Concepts. Technical Report N1477 03-0060, ISO/IEC JTC 1, Information Technology, Subcommittee SC 22, Programming Language C++ (2003)
4. Berti, G.: Generic Software Components for Scientific Computing. Doctoral thesis, Technische Universität Cottbus (2000)
5. Butler, D.M., Bryson, S.: Vector Bundle Classes From Powerful Tool for Scientific Visualization. Computers in Physics 6, 576–584 (1992)
6. Heinzl, R., Schwaha, P., Spevak, M., Grasser, T.: Performance Aspects of a DSEL for Scientific Computing with C++. In: Proc. of the POOSC Conf., Nantes, France, pp. 37–41 (2006a)
7. Heinzl, R., Spevak, M., Schwaha, P., Selberherr, S.: A Generic Topology Library. In: Library Centric Sofware Design, OOPSLA, Portland, OR, USA, pp. 85–93 (2006c)
8. Heinzl, R.: Concepts for Scientific Computing. Dissertation, Technische Universität Wien (2007)
9. Logg, A., Dupont, T., Hoffman, J., Johnson, C., Kirby, R.C., Larson, M.G., Scott, L.R.: The FEniCS Project. Technical Report 2003-21, Chalmers Finite Element Center (2003)
10. Schwaha, P., Heinzl, R., Mach, G., Pogoreutz, C., Fister, S., Selberherr, S.: A High Performance Webapplication for an Electro-Biological Problem. In: Proc. of the 21th ECMS 2007, Prague, Czech Rep. (2007)
11. Siek, J., Lee, L.-Q., Lumsdaine, A.: The Boost Graph Library: User Guide and Reference Manual. Addison-Wesley, Reading (2002)
12. Westheide, W., Rieger, R.: Spezielle Zoologie. Teil 2: Wirbel- oder Schädeltiere. Elsevier, Amsterdam (2003)

A Model Driven Architecture Approach to Web Development

Alejandro Gómez Cuesta[1], Juan Carlos Granja[1], and Rory V. O'Connor[2]

[1] Software Engineering Department, University of Granada, Spain
elales@gmail.com, jcgranja@ugr.es
[2] School of Computing, Dublin City University, Ireland
roconnor@computing.dcu.ie

Abstract. The rise of the number and complexity of web applications is ever increasing. Web engineers need advanced development methods to build better systems and to maintain them in an easy way. Model-Driven Architecture (MDA) is an important trend in the software engineering field based on both models and its transformations to automatically generate code. This paper describes a a methodology for web application development, providing a process based on MDA which provides an effective engineering approach to reduce effort. It consists of defining models from metamodels at platform-independent and platform-specific levels, from which source code is automatically generated.

Keywords: Model Driven Development, Web Engineering, Software Engineering, Agile Development.

1 Introduction

The requirements for web applications involve a great diversity of different services; multimedia, communication and automation, which reside in multiple heterogeneous platforms. The development of such systems is a difficult task due to the large number of complexities involved. Accordingly there is a need for solid engineering methods for developing robust web applications.

The development of most systems is based on models as abstractions of the real world. In software engineering the Unified Modelling Language (UML) is becoming the standard for Object-Oriented (OO) modelling. While OO models traditionally serve as blueprints for systems implementation, the Model-Driven Architecture (MDA) [11], [10], which is a software design approach, promotes the usage of models throughout the entire development process. Such an approach provides a set of guidelines for specifying, designing and implementing models.

Currently there are many methodologies to define web applications which are based on models such as: OO-H [6], UWE [9], ADM [5] or WebML [3]. These methodologies have different levels of abstraction, but all require spending much time on defining conceptual and usually do not take into account tangible elements such as pages. However, other approaches such as agile development are not so abstract.

J. Filipe et al. (Eds.): ICSOFT/ENASE 2007, CCIS 22, pp. 101–113, 2008.

MIDAS [12] is an approach to create web applications which merges models and agile development. The MDA approach defines three levels of model abstraction and a way to transform them:

- **CIM** (Computational Independent Model): A model of a system from a computation independent viewpoint where details of the structure of systems are not shown. It is used to represent the specific domain and the vocabulary for the specification of a system.
- **PIM** (Platform Independent Model): A model of a software or business system that is independent of the specific technological platform used to implement it.
- **PSM** (Platform Specific Model): A model of a software or business system that is linked to a specific technological platform (programming language, operating system or database).
- **QVT** (Query, View, Transformations): A way to transform models (e.g., PIM can be translated into PSM).

The application of both MDA and agile methods to web applications development can help to build better and fast systems in an easier way than applying traditional methods [2]. MDA supports agile development, providing an easy way to create prototypes through automatic code generation.

We have chosen the MIDAS [12], approach as our starting point, as it is both a development process based on an agile process and an architecture for web applications based on models. Using an agile approach prioritises the client-developer relationship and a using model-based approach improves communication between client and developer models, which allows the system to be seen from a higher level of abstraction.

The basic structure of web applications consists of three tiers: the Client, the Server and the Data store. In MIDAS, there are: graphical user interface (GUI), business logic and persistence, where each tier can be implemented with a different programming language, but utilising the same data structures. Thus MDA can be applied to this kind of software: joining all the data structures into a PIM, we can define the problem as only one and afterwards, splitting it into each tier. Thus we can define one problem and obtain code for different platforms / programming languages.

The work presented in this paper proposes a methodology for agile web applications development based on models. We provide a model-driven development method for the specification and implementation of web applications through our own metamodels. Our approach establishes a methodological framework by automating the construction of high-quality web applications (PIM) in a productive way, starting from a UML based representation for the, then an automated transformation method for the refinement of engineering models into technology models (PSM), and finally an automated method for the generation of platform specific code. Our approach has several advantages, e.g. the diagrams to use are very known and the metamodels are very simple so developers has not to learn new techniques. In addition, the automatic code generation provides boosts development time.

In this paper we use the case study of a web bookshop which allows the purchaser to search books by author, subject or title from a database. Once the user has a listing with the result, he can select several books which will be added to a cart. The user can manage this cart adding or deleting his selected books as times as necessary. Finally, the user can buy those books which are in the cart, and in this moment is when he has to insert his personal details in the system. We construct a web system for this example using own described methodology.

2 MDA-Based Approach for Web Applications

The development of a web application system implies the use of many different technologies in order to satisfy all user requirements. Usually these technologies provide low abstraction level constructs to the developer. Therefore, applying a MDA approach to web applications involves bridging an abstraction gap that must deal with the technology heterogeneity. Thus, MDA approach raises the abstraction level of the target platform and, therefore, the amount of code is clearly reduced. We reach a balance between model based and agile development. Although we are not completely fulfilling the requirements of MDA (that of abstracting as much as possible) we are adding some level of work schema to an agile development, even though this kind of development is not based on restrictions when writing code.

Accordingly our proposed approach for web applications development is based on:

1. Web application PIM.
2. Web application PSMs.
3. Transformations from PIM into PSM.
4. Code generation from PSMs.
5. Validation of this approach.

We can see this methodology is purely based on MDA. As we will see the metamodels used are not very complex and therefore we are also taking into account a certain level of agile development.

2.1 The Web Application PIM

Our PIM is intended to represent web application which consists of services. A service can be defined as functionalities offered by the system and whose results satisfy some specific needs from a user.

Our approach consists in defining at an early stage the services we want our application to fulfil. The second step is to refine those services into others more detailed. Finally, for each detailed service a set of messages is defined. These messages specify the relationship between the previous defined services and the final system. Moreover, we need to set up a class model collecting the elements that have to be modelled into the web application and can be used into the previous models.

So, our methodology define four stages - or from an MDA perspective - four different submodels, which all together shape our web application platform-independent model:

- **Class model:** a typical definition of the problem by means of classes.
- **Conceptual user service model:** defines the actors and services for a web application.
- **Extended service model:** details the functionality of a conceptual user service dividing it into extended services.
- **System sequence model:** establishes how each actor acts on the web application through messages or callings to other extended services within a extended service.

The conceptual user and extended services are definitions which are taken from MIDAS, but which have simplified the conceptual user service model and added new features to the extended service model to support our system sequence model as well as our class model.

2.1.1 Class PIM

This is a typical class model, but using many simplifications. There is neither encapsulation for attributes (all are public), complex associations between classes nor interfaces. It is a simple metamodel of a class model which is useful for the agile development and keeping the MDA process.

2.1.2 Conceptual User Service Model

This model is similar to a use case model where the system services as well as who uses them are represented. System services are called Conceptual User Services (CUS) and are depicted like a use case, where an actor is somebody or something who executes a CUS. Actors and services connect themselves the former to the latter.

Figure 1 shows an example CUS. The execution of the *BuyBooks* service consists of: searching some books, adding them into a cart. The execution of a CUS is related to the execution of a web by their actors. This concept is a new feature on conceptual user service that MIDAS does not include.

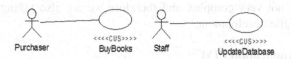

Fig. 1. Conceptual user service model

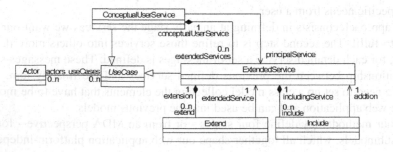

Fig. 2. Metamodel for CUS and ES models

In Figure 2 we can see the metamodel of this model. Actors are instances of the class called Actor. CUSs are instances of the *ConceptualUserService* class. As the *ConceptualUserService* class is a subclass of the abstract class *UseCase*, an unspecified number of actors can be associated to CUS's.

2.1.3 Extended Service Model

Each CUS is related to an extended service model. This model breaks up each CUS into simpler elements called Extended Services (ES) which are usually either a data input or output. Figure 3 shows the ES model associated to the *Purchaser* actor depicted in Figure. We can see the functions such actor performs within the CUS called *BuyBook*; the ESs describe how the actor buys a book, managing them by means of a cart. The *AddBookToCart* ES is called by both *UpdateCart* and *OrderBooks* due to the <<*include*>> associations which go out from them, and *AddBookToCart* ES calls to *SearchBook*. On other hand, we have that *SearchBook* ES could be replaced with *SearchForAuthor*, *SearchForSubject* or *SearchForTitle* due to the <<*extends*>> associations.

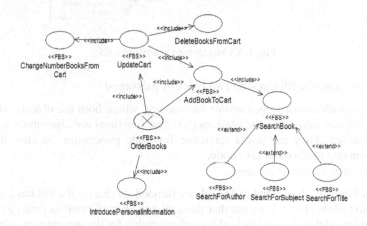

Fig. 3. ES model for the purchaser

The ES model execution is similar to that of the CUS. It starts from the principal service and it continues through the <<*include*>> relations to other ESs depending on the associated system sequence. If the user executes an ES where <<*extend*>> associations are coming in, as *SearchBook* ES, means that any of those ESs can be executed instead; the final selection depends on the user. As ESs are similar to functions, when one is ended, the execution returns to the ES which made the call.

Using the example in Figure 3, a user executes the first extended service, *OrderBook*. Later he/she can select either *UpdateCart* or *AddBookToCart* extended services. From *UpdateCart* can change the number of books selected for a specific one and he/she can delete one, as well as to add other book selecting the appropriate extended service. From *AddBookToCart* the user can makes a search, and this search can be done for: author, subject o title. This model extends the requirements specified in the CUS model to have a second level of abstraction. Using this model, it is easy to check how many parts have a specific CUS. In Figure 2 we have the metamodel of the

ES model. We can see the relation between CUSs and ESs, where the latter owns several elements of the former, besides the relation *principalService*.

2.1.4 System Sequence Model

A System Sequence (SS) model describes the functions performed by an ES. This model is simpler than the legacy sequence models which exist in UML. In the diagrams of these models there are two vertical lines: first one for actors and the second one for objects, but in our case we use only one: the *System*. Figure 4 shows an example of a system sequence model, in this case for the *OrderBook* extended service Figure 3.

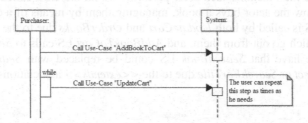

Fig. 4. SS Model for OrderBooks ES

Actors from a specific ES can send two types of messages:

- Functions with return values and/or parameters, where both are objects which belong to classes defined in the class model. The functions are algorithms which are not modelled. They are just a reference for code generation, because they will transform into methods in Java code.
- Calls to other extended services.

Note in Figure 4, the second call to other extended service on the left has a little box with the text while above. It means that this call can be performed as times as the user wants. We have defined three kinds of way of execution for the messages on a box:

- Sequential: is the normal order.
- Repetition: messages are always repeated while the user wants.
- Choose one: from all messages the user selects one.

It is possible to have multiple levels of execution, so for example, we could have several messages upon a *choose one* execution and upon that one, other messages as well as other *repetition* executions. Therefore we define a specific order for the execution. MIDAS defines an activity diagram for this purpose, but we have the same idea using the previous execution concept. Using our approach we define a number of messages which are closer to implementation without getting into this low level favoring the agile development.

Figure 5 shows the system sequence metamodel. We can note that every ES has only a message sequence. An object of a *MessageSequence* class can have just one type of elements: Objects from Message class (functions) or *MessageSequence* objects (calls to other Ess). The attribute *MessageSequenceDefault* indicates what order of execution owns such sequence: sequential, repetition or choose one.

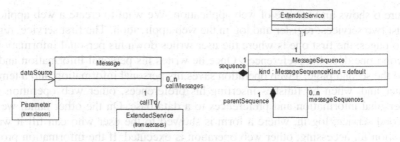

Fig. 5. SS meta-model

2.2 PSM

MIDAS defines web applications with three-tiers: the GUI, the business logic and the content. In our case we have one PSM for each part, as each part is a platform-specific model. We have chosen SQL for the content, Java servlets for the business logic and a Java servlet application for the GUI.

2.2.1 The Logic Business and Content PSMs

The logic business PSM is a class model, but in this case is the same Class-PIM but adding interfaces. The content model is a database model which has been taken out from [8]. Summarizing, we have tables which consist of columns, where each column owns a SQL data type. Tables have both primary keys and foreign keys and both are related to a column.

2.2.2 The Web Application PSM

A web is the place where users or actors connect to access the web services which are related to CUSs. A service consists of a number of web pages which are executed in a certain order. Each page is related to a ES. Every page has associations to others which are next (*outgoings*) and previous (*incomings*) ones; then, crawling among these pages a concrete service is executed. As each service has a first page we know from where that listing of services has to start. If we need to define a page which has the same execution than other one already defined, we can associate the first one to the second one, e.g., if we need to request certain information twice in different points, we create one page and the other just *calls to* this one.

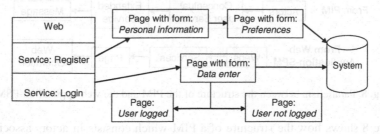

Fig. 6. Web application example

Figure 6 shows an example of web application. We want to create a web application with just two services: register and log in the web application. The first service, register, has two pages: the first one is where the user writes down his personal information and the second one does his preferences. Once he writes his personal information and tries to go to the second page, a web operation saves his personal information into a temporal container and, when he finishes inserting his preferences, other web operation sends both personal information and preferences to a data base. On the other hand, we have the second service, log in, where a form is showed to the user who can fill it with its information for accessing; other web operation is executed. If the information provided is not of any user, he is sent to the register page which belongs to the register service; in other case, he becomes registered. At the end, the web application will show to the user what service he wants to select: register or login. Depending on what he chooses, the specific service will be run. Figure 7 shows our web metamodel. *WebOperation* class is related to the messages defined in the previous system sequence model.

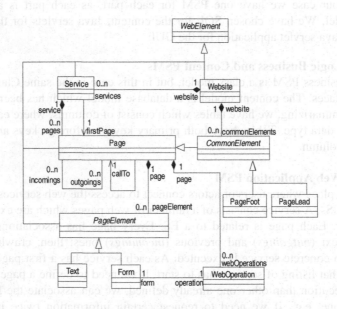

Fig. 7. Web application meta-model

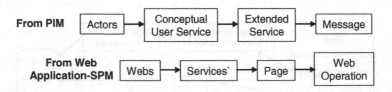

Fig. 8. Parallelism between the structure of the PIM and the web application-PSM

Figure 8 shows how the structure of a PIM which consists in actors associated to CUSs, CUSs associated to a set of ESs, ESs associated to a SS model which has messages, it is related to a structure in the web applications-PSM where webs have services,

services have pages, and pages have web operations. At this level a relationship exists between both structures and that is the reason to create a transformation between each level.

2.3 Transformations

A transformation is defined as an operation that takes as input a model and produces another one [8]. This process is described by a transformation definition, which consists of a number of transformation rules. We have three transformations:

1. Class-PIM to content-PSM
2. Class-PIM to business logic-PSM
3. PIM to web application-PSM.

As transformations for this methodology are many, due to space limitations we just introduce them a little bit. In a future work, we will develop them with more detail.

2.3.1 From Class-PIM to Content-PSM

The first transformation we have to perform is to transform the class platform-independent model into the content model. There are many ways to approach this problem [7] from which we have chosen 'one class one table'. This transformation is made up of the rules we can see in Table 1.

Table 1. Transformation Rules

Class PIM	Content PSM
Class	A table with a column which is a identifier and a principal key
Extended class	A table with a column which is identifier. This identifier is a forcign key pointing to the identifier of the resulting transformed table of the parent class. See Figure 10
Multiple class association	A table with two columns being both identifiers and foreign keys and...
1:1	... each one is also unique and both are the primary key
1:n	.. the column which makes references to '1' in the multiplicity, is not a primary key, while the other one is the primary key
n:m	... both columns are the primary key
An attribute	A column which is inserted into the resulting table of the class which has the attribute

2.3.2 From Class-PIM to Business Logic-PSM

This is a complex transformation. We copy the same model from PIM to PSM but we change each class for an interface and we add a new class which implements such interface. [1] describe how this is done.

Table 2. Transformation Rules

Class PIM	Business logic PSM
Package structure	The same package structure is copied.
Datatype structure	The same data type structure is copied.
Class c	An interface i and a class c'. Both are added to the same transformed package that the initial class c was. This new class c' implements the new interface i.
Attribute a (belonging to a class c)	An attribute a' whose visibility is private. As c is transformed into i and c', the attribute a' is added into i and c'. Besides, two new methods are added to interface i: some get- and set- public methods which let the access a'.
An operation o (belonging to a class c)	An operation o' added to the transformed interface i from the class c. The data types used in o' have to be the copied ones.
Association between classes	An association, with the same features, between the created interfaces from the initial classes.
Association end	An attribute. An association end with multiplicity equals to '1' is transformed into an attribute whose type is the class which points to. If the multiplicity is 'n', the association end is transformed into other attribute but in this case the type is *Vector<class_which_points_to>* (using Java 5 notation)

2.3.3   From PIM to Web Application-PSM

Even though it is quite easy to see this transformation, it has many rules. A summary are the following rules which are explained in Table 3.

Table 3. Transformation Rules

Class PIM	Graphical User Interface PSM
An actor	A web
Conceptual user service associated to actor	A service associated to the transformed web for such actor
An extended service	A page. The same structure defined for a extended service model is built using pages from the web application-PSM
A message	A web operation

2.4   Code Generation

The last stage of our methodology is code generation. Once the PSMs are automatically obtained from the created PIM, they are used to generate source code. There are many ways to generate source code from models. Most of them are based on templates considered as other kind of model transformation: model-to-text transformation [4]. The code generation helps the developer to not start from scratch his development. We have defined these transformations:

1. From the content PSM, SQL code.
2. From the business logic PSM, Java code.
3. A general web application is created for the web application PSM. This application just needs the model to run.

In summary, code generation covers business logic which is the data structures, and data base. Finally instead of creating code for the web application, its model is directly executed.

2.4.1 From Business Logic-PSM to Java

From a given class-PSM for each class or interface we need to create a new file where we have to:

- Write the name of the package and define the class or interface.
- Write every attribute: visibility, type & name.
- Write every method with all its parameters. It is possible that some methods have some associated source code as get and set methods.

[4] provide the following example to illustrate how a template-based model-to-text transformation is.

```
<<DEFINE Root FOR Class>>
public class <<name>>{
  <<FOREACH attrs AS a>>
  private <<a.type.name>> <<a.name>>;
  <<ENDFOREACH>>
  <<EXPAND AccessorMethods FOREACH
    attribute>>  }
<<ENDDEFINE>>
<<DEFINE AccessorMethods FOR
  Attribute>>
public <<type.name>>
  get<<name.toFirstUpper>>() {
    return this.<<name>>;    }
public void
  set<<name.toFirstUpper>>(
  <<type.name>> <<name>> )
{  this.<<name>> = <<name>>;  }
<<ENDDEFINE>>
```

2.4.2 From Content PIM to SQL

From a given relational model and for each table create a new file where we have to:

- Write code which defines a table.
- Look up all its columns and write its name along with its type.
- Usually every table has a primary key which has to be written in the code.
- Finally, if the class has some foreign key.

2.4.3 Web Application-Model to Servlets

Creating the web application has been done by means of other different kind of transformation. Instead of creating directly code for the web application using JSP or similar, we directly execute the created model. The final web application is a web previously constructed which only needs a parameter which is a model, in our case, the web model automatically generated, to work properly. This is a complex transformation. We copy the same model from PIM to PSM but we change each class for an interface and we add a new class which implements such interface. [1] describe how this is done.

2.5 Validation

A plugin for Eclipse has been developed to validate this process. Such a tool comes with a GUI which allows to directly draw the diagrams we have seen before for the PIM. Once this model has been created the next step is to transform it. At this moment the tool does not allow to modify PSM models, but the transformation can be performed. When it is done, a new Eclipse project is created and you can see both the PSM models created and the generated code. Models created with this tool use XMI format, therefore can be exported to other tools. The proposed example was created using this approach. Although this application is not very complicated, we have seen that from CUSs we are defining at the same time requirements and a certain degree of the navigation of the web application which is being created. Such a navegation is completely described by means of ESs. Finally, when defining SSs, we are getting some functions which are very close to code but they are enough abstract to be a part of the PIM. For the future, the tool should add the possibility of modifying the PSM models and to define a customed presentation of the web pages. We are now working on these aspects.

3 Conclusions

In this paper we have shown a methodological approach to web applications development. We have kept to MDA framework, where the development is performed by two different abstraction levels: independent and specific from the platform. The chosen models allow the user who uses this methodology to not lose in any moment the sense of what he is doing: he will generate code. This method help us construct better web applications in a time not very high, because it joins on the one hand advantages of having a set of steps very marked from a model-based methodology and on the other hand the agile development allows to construct prototypes of the final system from very early development stages. The automatic code generation makes MDA promote agile development, since the models from the code which are generated are kept, and they are the documentation of the final web application.

Using this process, we extend the work made by MIDAS adding new models to take into account, as well as a new approach closer to agile development such as our system sequence models. It should be feasible to make a fusion between MIDAS and our approach to build one more complete one.

Using this method, the web engineering industry has a new way to build simple web application in a faster manner. Our contribution is to provide metamodels which are applied. It is not easy to find metamodels which are applied to a specific field. Usually, other proposed model-based methodologies only offer diagrams and one cannot see how the process is really working. Besides, they used to be models for UML class or Java diagrams. We have proposed a set of metamodels for web engineering as well as concrete syntax for those ones, explaining what each one does and how the models are transformed.

There is much possible future work to be done. It would be useful to include new models from UML but considering our goal of simplifies them to keep to agile development. Also, we could leave the agile development to centre our effort in constructing a comprehensive methodology for web engineering using complex models and complex transformations. Independently, our transformations have to be improved

References

1. Bézivin, J., Hammoudi, S., Lopes, D., Jouault, F.: Applying MDA Approach forWeb Service Platform. In: Enterprise Distributed Object Computing Conference (2004)
2. Cáceres, P., Marcos, E.: Procesos ágiles para el desarrollo de aplicaciones Web. In: Taller de Web Engineering de las Jornadas de Ingeniería del Software y Bases de Datos de 2001 (JISBD2001) (2001)
3. Ceri, S., Fraternali, P., Bongio, A.: Web Modeling Language (WebML): a modeling language for designing Web sites. Computer Networks 3(1-6), 137–157 (2000)
4. Czarnecki, K., Helsen, S.: Feature-based survey of model transfomation approaches. IBM Systems Journal 45(3) (2006)
5. Díaz, P., Aedo, I., Montero, S.: Ariadne, a development method for hypermedia. In: Mayr, H.C., Lazanský, J., Quirchmayr, G., Vogel, P. (eds.) DEXA 2001. LNCS, vol. 2113, pp. 764–774. Springer, Heidelberg (2001)
6. Gómez, J., Cachero, C.: OO-H Method: Extending UML to Model Web Interfaces. Idea Group Publishing (2002)
7. Keller, W.: Mapping Objects to Tables (2004), http://www.objectarchitects. de/ObjectArchitects/papers/Published/ZippedPapers/mappings04 .pdf
8. Kleppe, A., Warmer, J., Bast, W.: MDA Explained - The Model-Driven Architecture: Practice and Promise. Addison-Wesley, Reading (2003)
9. Koch, N., Kraus, A.: The Expressive Power of UML-based Web Engineering. In: Second International Workshop on Web-oriented Software Technology (IWWOST 2002) (2002)
10. Mellor, S., Scott, K., Uhl, A., Weise, D.: MDA Distilled, Principles of Model Driven Architecture. Addison-Wesley, Reading (2004)
11. Millar, J., Mukerji, J.: MDA Guide Version 1.0.1 (2003), http://www.omg.org/cgi-bin/doc?omg/03-06-01
12. Vela, B., Cáceres, P., de Castro, V., Marcos, E.: MIDAS: una aproximación dirigida por modelos para el desarrollo ágil de sistemas de información web, Ingeniería de la web y patrones de diseño, Coordinadores: Mª Paloma Díaz, Susana Montero e Ignacio Aedo, ch. 4. Prentice Hall, Pearson (2005)

Reverse-Architecting Legacy Software Based on Roles: An Industrial Experiment

Philippe Dugerdil and Sebastien Jossi

Department of Information Systems, HEG-Univ. of Applied Sciences
7 route de Drize, CH-1227 Geneva, Switzerland
{philippe.dugerdil,sebastien.jossi}@hesge.ch

Abstract. Legacy software systems architecture recovery has been a hot topic for more than a decade. In our approach, we proposed to use the artefacts and activities of the Unified Process to guide this reconstruction. First, we recover the use-cases of the program. Then we instrument the code and "run" the use-cases. Next we analyse the execution trace and rebuild the run-time architecture of the program. This is done by clustering the modules based on the supported use-case and their roles in the software. In this paper we present an industrial validation of this reverse-engineering process. First we give a summary of our methodology. Then we show a step-by-step application of this technique to real-world business software and the result we obtained. Finally we present the workflow of the tools we used to perform this experiment. We conclude by giving the future directions of this research.

Keywords: Reverse-engineering, software process, software clustering, software reengineering, program comprehension, industrial experience.

1 Introduction

To extend the life of a legacy system, to manage its complexity and decrease its maintenance cost, it must be reengineered. However, reengineering initiatives that do not target the architectural level are more likely to fail [3]. Consequently, many reengineering initiatives begin by reverse architecting the legacy software. The trouble is that, usually, the source code does not contain many clues on the high level components of the system [12]. However, it is known that to "understand" a large software system, which is a critical task in reengineering, the structural aspects of the software system i.e. its architecture are more important than any single algorithmic component [4] [19]. A good architecture is one that allows the observer to "understand" the software. To give a precise meaning to the word "understanding" in the context of reverse-architecting, we borrow the definition of Biggerstaff et al. [5]: "A person understands a program when able to explain the program, its structure, its behaviour, its effects on its operational context, and its relationships to its application domain in terms that are qualitatively different from the tokens used to construct the source code of the program".

In other words, the structure of the system should be mappable to the domain concepts (what is usually called the "concept assignment problem"). In the literature,

J. Filipe et al. (Eds.): ICSOFT/ENASE 2007, CCIS 22, pp. 114–127, 2008.
© Springer-Verlag Berlin Heidelberg 2008

many techniques have been proposed to split a system into components. These techniques range from clustering [1] [23], slicing [6] [22] to the more recent concept analysis techniques [8] [20] or even mixed techniques [21]. However the syntactical analysis of the mere source code of a system may produce clusters of program elements that cannot be easily mapped to domain concepts because both the domain knowledge and the program clusters have very different structures. Then, to find a good clustering of the program elements (i.e. one for which the concept assignment is straightforward) one should first understand the program. But to understand a large software system one should know its structure. This resembles the chicken and egg syndrome. To escape from this situation, we proposed to start from an hypothesis on the architecture of the system. Then we proceed with the validation of this architecture using a run time analysis of the system. The theoretical framework of our technique has been presented elsewhere [7]. In this paper we present the result of the reverse engineering of an industrial-size legacy system. This shows that our technique scales well and allows the maintenance engineer to easily map high-level domain concepts to source code elements. It then helps him to "understand" the code, according to the definition of Biggerstaff et al.

2   Short Summary of Our Method

Generally, legacy systems documentation is at best obsolete and at worse nonexistent. Often, its developers are not available anymore to provide information of these systems. In such situations the only people that still have a good perspective on the system are its users. In fact they are usually well aware of the business context and business relevance of the programs. Therefore, our iterative and incremental technique starts from the recovery of the system use-cases from its actual use and proceeds with the following steps [7]:

- Re-document the system use-cases;
- Re-document the corresponding business model;
- Design the robustness diagrams associated to all the use-cases;
- Re-document the high level structure of the code;
- Execute the system according to the use-cases and record of the execution trace;
- Analyze the execution trace and identify the classes involved in the trace;
- Map the classes in the trace to the classes of the robustness diagram with analysis of the roles.
- Re-document the architecture of the system by clustering the modules based on their role in the use-case implementation.

Figure 1 shows a use-case model and the corresponding business analysis model. Then, for each use-case we rebuild the associated robustness diagram (usual name for the Analysis Model of the Unified Process). These robustness diagrams represent our best hypothesis on the actual architecture of the system. Then, in the subsequent steps, we must validate this hypothesis and identify the roles the modules play. Figure 2 presents an example of a robustness diagram with its UML stereotypical classes that represent software roles for the classes [11].

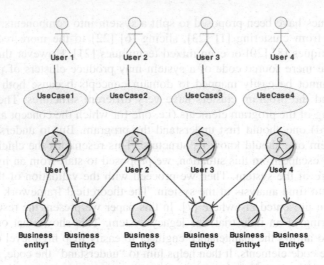

Fig. 1. Use-case model and business model

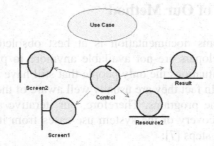

Fig. 2. Use-case and robustness diagram

The next step is to recover the visible high level structure of the system (classes, modules, packages, subsystems) from the analysis of the source code, using the available syntactic information (fig 3).

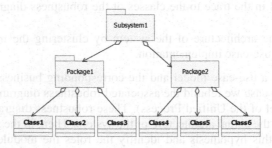

Fig. 3. High-level structure of the code

Now, we must validate our hypothetical architecture (robustness diagrams) against the actual code of the system and find the mapping from the stereotypical classes to the implementation classes. First, we run the system according to each use-case and record the execution trace (fig. 4). Next, the functions listed in the trace are linked to the classes they belong to. These are the classes that actually implement the executed scenario (instance of a use-case). These classes are then highlighted in the high level structure of the code (fig. 5).

Fig. 4. Use-case and the associated execution trace

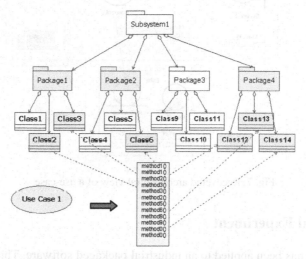

Fig. 5. Execution trace linked to the implementation classes

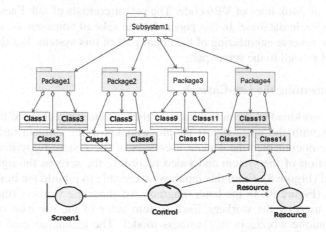

Fig. 6. Map from the software roles to the implementation classes

The classes found are further analysed to find evidences of a database access function or an interface (for example a screen display function). This let us categorize the classes as entities (access to database tables) or boundaries (interface). The remaining classes will be categorized as control classes. Figure 6 presents the result of such a mapping. The last step in our method is to cluster the classes or modules according to the use-case they implement and to their role as defined above. This represents a view of the functional architecture of the system. Figure 7 shows such a view for a use-case that shares some of its classes with another use-cases (UC-common1-2).

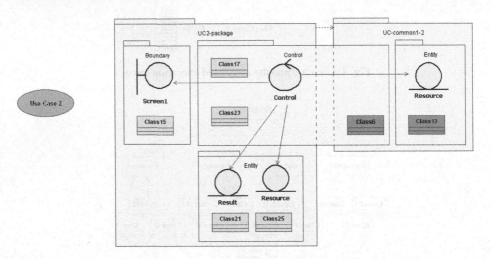

Fig. 7. Recovered architectural view of a use-case

3   Industrial Experiment

This technique has been applied to an industrial packaged software. This system manages the welfare benefit in Geneva. It is a fat-client kind of client-server system. The client is made of 240k lines of VB6 code. The server consists of 80k lines of PL/SQL code with an Oracle database. In this paper, for the sake of conciseness, we will concentrate on the reverse engineering of the client part of this system. But the technique can be applied as well to the server part.

3.1   Re-documenting the Use-Cases

Due to heavy workload of the actual users of this system we recovered the use-cases by interacting with the user-support people who know the domain tasks perfectly well. Then we documented the 4 main use-cases of the system by writing down the user manipulation of the system and video recording the screens through which the user interacted (Figure 8). From this input, we were able to rebuild the business model of the system (Figure 9). In the latter diagram, we show the workers (the tasks) and the resources used by the workers. Each system actor of the use-case model corresponds to a unique worker in the business model. The technique used to build the business model come from the Unified Process.

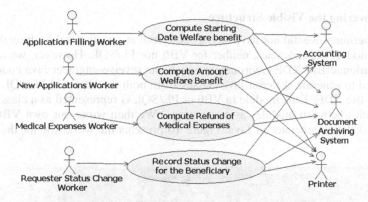

Fig. 8. The main use-cases of the system

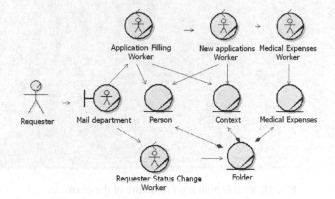

Fig. 9. The recovered high level business model

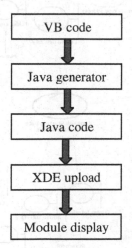

Fig. 10. Visible high level structure extraction workflow

3.2 Recovering the Visible Structure

In our experiment we did not have any specific tool at our disposal to draw the modules and module dependencies, neither for VB6 nor PL/SQL. However, we regularly use the Rational/IBM XDE environment which can reverse-engineer Java code. Then, we decided to generate skeleton Java classes from both the VB6 and PL/SQL code to be able to use XDE. Each module in VB6 or PL/SQL is represented as a class and the dependencies between modules as associations. We then wrote our own VB6 parser and Java skeleton generator in Java. Figure 10 presents the workflow for the display

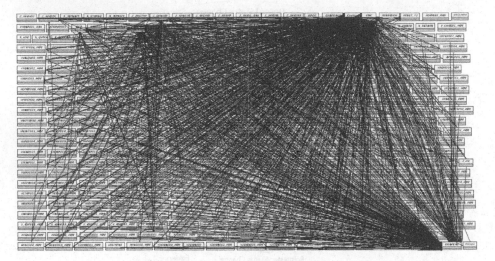

Fig. 11. Visible high level structure of the client tier

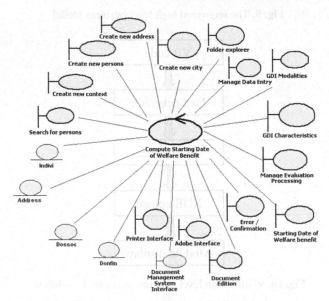

Fig. 12. Robustness diagram of the first use-case

of the visible high-level structure of the VB6 client tier of the system. The resulting high-level structure diagram is presented in figure 11. There are 360 heavily inter-linked modules in this diagram. Looking at this picture, we understand why mainte-nance engineers are afraid each time they have to maintain this system!

3.3 Building the Robustness Diagram

The robustness diagram is built by hand using the heuristics set forth by the Unified Process. Figure 12 presents the robustness diagram of the first use-case called "Com-pute the starting date of welfare benefit". It is the second largest use-case of this system. This diagram has been built from the analysis of the use-case only, without taking the actual code into account.

3.4 Running the Use-Cases

The next step is to "execute" the use-cases i.e. to run the system according to scenar-ios that represent instances of these use-cases. Then the execution trace must be recorded. Again, we have not found any specific environment able to generate an execution trace for the client part written in VB6. Then we decided to instrument the code to generate the trace (i.e. to insert trace generation statement in the source code). Therefore we wrote an ad-hoc VB6 instrumentor in Java. The modified VB6 source code was then recompiled before execution. The format of the trace we generated is:

```
<moduleName><functionSignature><parameterValues>
```

The only parameter values we record in the trace are the one with primitive types, because we are interested in SQL statements passed as parameters. This will help us find the modules playing the role of "entities" (database tables access). For the server part (PL/SQL), the trace can be generated using the system tools of Oracle.

3.5 Execution Trace Analysis

At this stage, we analysed the trace to find the modules involved in the execution. The result for the client part is presented in figure 13. We found that only 44 modules are involved in the processing of this use-case, which is one of the biggest in the applica-tion. But this should not come as a surprise. Since this system is a packaged software, then a lot of the implemented functions are unused.

The last step is to sort out the roles of the modules in the execution of the use case. This will allow us to cluster the modules according to their role. Then we analysed the code of the executed functions to identify screen-related functions (i.e. VB6 functions used to display information). The associated modules then play the role of the bounda-ries in the robustness diagram. Next, we analysed the parameter values of the functions to find SQL statements. The corresponding modules play the role of the entities in the robustness diagram. The remaining modules play the role of the control object. The result of this analysis is presented in figure 14. The modules in the top layer are boundaries (screens), the bottom layer are the entities and the middle layer contains the modules playing the role of the control object.

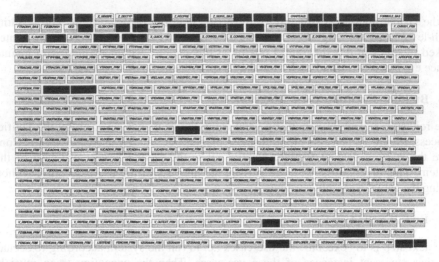

Fig. 13. Modules involved in the first use-case

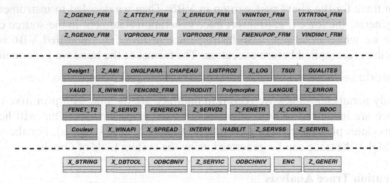

Fig. 14. Software roles of the involved modules

As an alternative view, we mapped the modules of the client part to the robustness diagram we built from the use-case. By correlating the sequence of involvement of the screens in the use-case and the sequence of appearance of the corresponding modules in the execution trace, we could map each boundary object to the corresponding module. However, we found that the "entity" modules were not specific to any database table. In fact, each "entity" module was involved in the processing of many tables. Then, we represented all of them in the robustness diagram without mapping to any specific database table. The result of the mapping is presented in figure 15. Since the number of modules that play the role of the control object is large, these modules are not shown in the diagram.

3.6   Role-Based Clustering of Modules

Figure 16 summarizes the role-based clustering of the client modules identified in our experiment, for three use-cases. First, the modules are grouped in layers according to

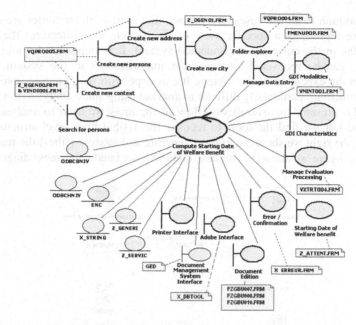

Fig. 15. Mapping of the implementation modules to the robustness diagram

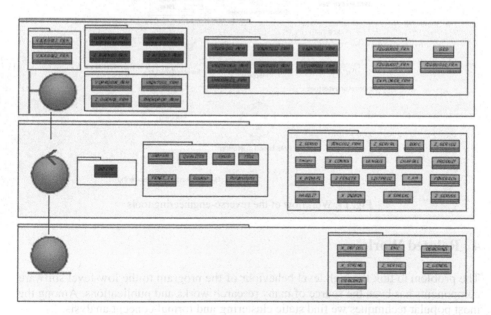

Fig. 16. Recovered architectural view of the system

the robustness-diagram role they play in the implementation. Each role is symbolized by the corresponding UML stereotype. Second, the modules are grouped according to the use-case they implement. The rightmost package in each layer contains the modules common to all use-cases. We can see that most of the modules are common. This picture represents a specific view of the system's architecture: the role the modules play in the currently implemented system. It is important to note that this architecture can be recovered whatever the maintenances to the system. Since it comes from the execution of the system, it can cope with the dynamic invocation of modules, something particularly difficult to analyse using static analysis only.

Figure 17, presents the overall workflow of the tools we used to analyse the system. On the left we find the tools to recover the visible high level structure of the system. On the right we show the tools to generate and analyse (filter) the trace. In the centre of the figure we show the use-case and the associated robustness diagram.

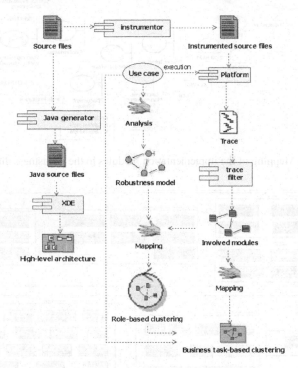

Fig. 17. Workflow of the reverse-engineering tools

4   Related Work

The problem to link the high level behaviour of the program to the low-level software components has been the source of many research works and publications. Among the most popular techniques we find static clustering and formal concept analysis.

- The clustering algorithms group the statements of a program based on the dependencies between the elements at the source level, as well as the analysis of the cohesion and coupling among candidate components [15].
- Formal concept analysis is a data analysis technique based on a mathematical approach to group the «objects» that share some common «attributes». Here

the object and attributes can be any relevant software elements. For example, the objects can be the program functions and the attributes the variables accessed by the functions [13] [18]. For example, this technique has been proposed to identify the program elements associated to the visible features of the programs [17] [8].

In fact, these techniques try to partition the set of source code statements and program elements into subsets that will hopefully help to rebuild the architecture of the system. The key problem is to choose the relevant set of criteria (or similarity metrics [24]) with which the "natural" boundaries of components can be found. In the reverse-engineering literature, the similarity metrics range from the interconnection strength of Rigi [16] to the sophisticated information-theory based measurement of Andritsos [1] [2], the information retrieval technique such as Latent Semantic Indexing [14] or the *kind* of variables accessed in formal concept analysis [18] [20]. Then, based on such a similarity metric, an algorithm decides what element should be part of the same cluster [15]. On the other hand, Gold proposed a concept assignment technique based on a knowledge base of programming concepts and syntactic "indicators" [10]. Then, the indicators are searched in the source code using neural network techniques and, when found, the associated concept is linked to the corresponding code. However he did not use his technique with a knowledge base of domain (business) concepts. In contrast with these techniques, our approach is "business-function-driven" i.e. we cluster the software elements according to the supported business tasks and functions. The domain modelling discipline of our reverse-engineering method presents some similarity with the work of Gall and Weidl [9]. These authors tried to build an object-oriented representation of a procedural legacy system by building two object models. First, with the help of a domain expert, they build an abstract object model from the specification of the legacy system. Second, they reconstruct an object model of the source code, starting from the recovered entity-relationship model to which they append dynamic services. Finally, they try to match both object models to produce the final object oriented model of the procedural system. In contrast, our approach does not try to transform the legacy system into some object-oriented form. The robustness diagram we build is simply a way to document the software roles. Our work bears some resemblance to the work of Eisenbarth and Koschke [8] who used Formal Concept Analysis. However the main differences are:

1. The scenarios we use have a strong business-related meaning rather than being built only to exhibit some features. They represent full use-cases.
2. The software clusters we build are interpretable in the business model. We do group the software elements after their roles in the implementation of business functions.
3. We analyse the full execution trace from a real-use-case to recover the architecture of the system.
4. The elements we cluster are modules or classes identified in the visible high-level structure of the code.

Finally, it is worth noting that the use-cases play, in our work, the same role as the test cases in the execution slicing approach of Wong et al. [25]. However, in our work, the "test cases" are not arbitrary but represent actual use-cases of the system.

5 Conclusions

The reverse-engineering experiment we presented in this article rests on the Unified Process from which we borrowed some activities and artefacts. The techniques are based on the actual working of the code in real business situations. Then, the architectural view we end up with is independent on the number of maintenances to the code. Moreover it can cope with situations like dynamic calls, which are tricky to analyse using static techniques. We actually reverse-engineered all the use-cases of the system and found that their implementation had most of their modules in common. Finally, this experiment seems to show that our technique is scalable and able to cope with industrial size software. As a next step in this research we are developing a semi-automatic robustness diagram mapper that will take a robustness diagram and a trace file as input and produce a possible match as output. This system will use a heuristic-based search engine coupled to a truth maintenance system.

Acknowledgements. We gratefully acknowledge the support of the "Reserve Strategique" of the Swiss Confederation (Grant ISNET 15989). We also thank the people at the "Centre des Technologies de l'Information" of the Canton Geneva (Switzerland) who helped us perform the industrial experiment.

References

1. Andritsos, P., Tzerpos, V.: Software Clustering based on Information Loss Minimization. In: Proc. IEEE Working Conference on Reverse engineering (2003)
2. Andritsos, P., Tzerpos, V.: Information Theoretic Software Clustering. IEEE Trans. on Software Engineering 31(2) (2005)
3. Bergey, J., et al.: Why Reengineering Projects Fail. Software Engineering Institute, Tech Report CMU/SEI-99-TR-010 (April 1999)
4. Bergey, J., Smith, D., Weiderman, N., Woods, S.: Options Analysis for Reengineering (OAR): Issues and Conceptual Approach. Software Engineering Institute, Tech. Note CMU/SEI-99-TN-014 (1999)
5. Biggerstaff, T.J., Mitbander, B.G., Webster, D.E.: Program Understanding and the Concept Assignment Problem. Communications of the ACM, CACM 37(5) (1994)
6. Binkley, D.W., Gallagher, K.B.: Program Slicing. In: Advances in Computers, vol. 43. Academic Press, London (1996)
7. Dugerdil, P.: A Reengineering Process based on the Unified Process. In: IEEE International Conference on Software Maintenance (2006)
8. Eisenbarth, T., Koschke, R.: Locating Features in Source Code. IEEE Trans. On Software Engineering 29(3) (March 2003)
9. Gall, H., Weidl, J.: Object-Model Driven Abstraction to Code Mapping. In: Proc. European Software engineering Conference, Workshop on Object-Oriented Reengineering (1999)
10. Gold, N.E.: Hypothesis-Based Concept Assignment to Support Software Maintenance. PhD Thesis, Univ. of Durham (2000)
11. Jacobson, I., Booch, G., Rumbaugh, J.: The Unified Software Development Process. Addison-Wesley Professional, Reading (1999)

12. Kazman, R., O'Brien, L., Verhoef, C.: Architecture Reconstruction Guidelines, 3rd edition. Software Engineering Institute, Tech. Report CMU/SEI-2002-TR-034 (2003)
13. Linding, C., Snelting, G.: Assessing Modular Structure of Legacy Code Based on Mathematical Concept Analysis. In: Proc IEEE Int. Conference on Software Engineering (1997)
14. Marcus, A.: Semantic Driven Program Analysis. In: Proc IEEE Int. Conference on Software Maintenance (2004)
15. Mitchell, B.S.: A Heuristic Search Approach to Solving the Software Clustering Problem. In: Proc IEEE Conf on Software Maintenance (2003)
16. Müller, H.A., Orgun, M.A., Tilley, S., Uhl, J.S.: A Reverse Engineering Approach To Subsystem Structure Identification. Software Maintenance: Research and Practice, vol. 5(4). John Wiley & Sons, Chichester (1993)
17. Rajlich, V., Wilde, N.: The Role of Concepts in Program Comprehension. In: Proc IEEE Int. Workshop on Program Comprehension (2002)
18. Siff, M., Reps, T.: Identifying Modules via Concept Analysis. IEEE Trans. On Software Engineering 25(6) (1999)
19. Tilley, S.R., Santanu, P., Smith, D.B.: Toward a Framework for Program Understanding. In: Proc. IEEE Int. Workshop on Program Comprehension (1996)
20. Tonella, P.: Concept Analysis for Module Restructuring. IEEE Trans. On Software Engineering 27(4) (2001)
21. Tonella, P.: Using a Concept Lattice of Decomposition Slices for Program Understanding and Impact Analysis. IEEE Trans. On Software Engineering 29(6) (2003)
22. Verbaere, M.: Program Slicing for Refactoring. MS Thesis. Oxford University, Oxford (2003)
23. Wen, Z., Tzerpos, V.: An Effective measure for software clustering algorithms. In: Proc. IEEE Int. Workshop on Program Comprehension (2004)
24. Wiggert, T.A.: Using Clustering Algorithms in Legacy Systems Remodularisation. In: Proc. IEEE Working Conference on Reverse engineering (1997)
25. Wong, W.E., Gokhale, S.S., Horgan, J.R., Trivedi, K.S.: Locating Program Features using Execution Slices. In: Proc. IEEE Conf. on Application-Specific Systems and Software Engineering & Technology (1999)

A Supporting Tool for Requirements Elicitation Using a Domain Ontology

Motohiro Kitamura[1], Ryo Hasegawa[1], Haruhiko Kaiya[2], and Motoshi Saeki[1]

[1] Dept. of Computer Science, Tokyo Institute of Technology
Ookayama 2-12-1, Meguro-ku, Tokyo 152, Japan
[2] Dept. of Computer Science, Shinshu University
Wakasato 4-17-1, Nagano 380-8553, Japan
saeki@se.cs.titech.ac.jp, kaiya@cs.shinshu-u.ac.jp

Abstract. Since requirements analysts do not have sufficient knowledge on a problem domain, i.e. domain knowledge, the technique how to make up for lacks of domain knowledge is a key issue. This paper proposes the usage of a domain ontology as domain knowledge during requirements elicitation processes and a supporting tool based on this technique. In addition, we had several experiments to show the usefulness of our tool.

1 Introduction

Knowledge on a problem domain where software is operated (simply, domain knowledge) plays an important role on eliciting requirements of high quality. For example, to develop e-commerce systems, the knowledge on marketing business processes, supply chain management, commercial laws, etc. is required as well as internet technology. Although requirements analysts have much knowledge of software technology, they may have less domain knowledge. As a result, lack of domain knowledge allows the analysts to produce requirements specification of low quality, e.g. an incomplete requirements specification where mandatory requirements are lacking. In addition, communication gaps between requirements analysts and with domain experts resulted from their knowledge gaps [1]. Thus, the techniques to provide domain knowledge for the analysts during their requirements elicitation and computerized tools based on these techniques to support the analysts are necessary.

We have already proposed how to use domain ontologies for requirements elicitation [2] where domain ontologies are used to supplement domain knowledge to requirements analysts during requirement elicitation processes. However, it mentioned just an idea and a methodology, called ORE (Ontology driven Requirements Elicitation), but did not address the issues on how the analyst can utilize a domain ontology more concretely. This paper presents a more concrete methodology and a tool for supporting the usage of a domain ontology. By using this tool, lacking requirements and inconsistent ones are incrementally suggested to a requirements analyst and she or he evolves a list of the current requirements based on these suggestions. The tool deduces lacking requirements and inconsistency ones by using inference rules on the domain ontology. As mentioned in [3], we consider an ontology as a thesaurus of words and inference

J. Filipe et al. (Eds.): ICSOFT/ENASE 2007, CCIS 22, pp. 128–140, 2008.

rules on it, where the words in the thesaurus represent concepts and the inference rules operate on the relationships on the words. Each concept of a domain ontology can be considered as a semantic atomic element that anyone can have the unique meaning in a problem domain. That is to say, the thesaurus part of the ontology plays a role of a semantic domain in denotational semantics, and the inference rules help a requirements analyst in detecting lacks of requirements and semantically inconsistent requirements during her or his elicitation activities.

Although we can hardly get a complete ontology for our problem domain yet, we can have several (semi) automated techniques to do it. Like DAML Ontology Library [4], researchers and developers in knowledge engineering communities, in particular Semantic Web community are creating many ontologies in wide varieties of domains, and they are represented with standardized OWL based language so as to be exchanged by many persons. Furthermore, some studies to extract ontological concepts and their relationships by applying text-mining techniques to natural-language documents exist [5]. We apply this technique to various kinds of documents related on a problem domain in order to automate partially the creation of a domain ontology. We customize and adapt usual logical structure of ontologies into requirements elicitation. More concretely, as a result, we adopt varieties of types of ontological types and their relationships so that an ontology can express domain knowledge for requirements elicitation. Following this adaptation, we should develop a newly devised text-mining technique fit to our ontological structure to achieve the creation of domain ontologies from documents.

To sum up, we get a computerized tool to support the analysts' activities following the process of Figure 1, which two tools are seamlessly integrated into. The first part of the tool is for supporting requirements elicitation tasks and it suggests to an analyst where the problems, e.g. lacks of mandatory requirements (incompleteness) and inconsistency of requirements are included. A kind of text-mining tool for supporting ontology creation is the second part of the integrated tool. In addition, requirements specification documents that have been produced with the first part of the tool may be able to revise and improve the domain ontology as shown in the figure. On account of space, we discuss nothing but the first part, i.e. the tool for supporting requirements elicitation tasks following our methodology ORE. The readers who have an interest in the ontology creation tool can find its details in [6].

The rest of this paper is organized as follows. In the next section, we explain the basic idea and show the logical structure of domain ontologies. In section 3, we clarify the tool for supporting our methodology ORE, i.e. requirements elicitation by using a domain ontology. Section 4 presents experimental studies to assess our supporting tool. In sections 5 and 6, we discuss related works and our current conclusions together with future work, respectively.

2   Basic Idea

2.1   Using a Domain Ontology

In this section, we present the basic idea how to use a domain ontology to detect lacking requirements and inconsistent requirements. Below, let's consider how a requirements analyst uses a domain ontology for completing requirements elicitation. Suppose that a

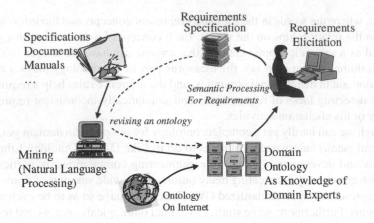

Fig. 1. Requirements and Knowledge Elicitation Cycle

requirements document initially submitted by a customer are itemized as a list. At first, an analyst should map a requirement item (statement) in a requirement document into atomic concepts of the ontology as shown in Figure 2. In the figure, the ontology is written in the form of class diagrams. For example, the item "bbb" is mapped into the concepts A, B, and an aggregation relationship between them. The requirements document may be improved incrementally through the interactions between a requirements analyst and stakeholders. In this process, logical inference on the ontology suggests to the analyst what part she or he should incrementally describe. In the figure, although the document S includes the concept A at the item bbb, it does not have the concept C, which is required by A. The inference resulted from "C is required by A" and "A is included" suggests to the analyst that a statement having C should be added to the document S. In our technique, it is important what kind of relationship like "required by" should be included in a domain ontology for inference, and we will discuss this issue in the next sub section.

2.2 Domain Ontology

As mentioned in section 1, our domain ontology consists of a thesaurus and inference rules. Figure 3 shows the overview of a meta model of the thesaurus part of our ontologies. Thesauruses consist of concepts and relationships among the concepts and they have varies of subclasses of "concept" class and "relationship". In the figure, "object" is a subclass of a concept class and a relationship "apply" can connect two concepts. Concepts and relationships in Figure 3 are introduced so as to easily represent the semantics in software systems. Intuitively speaking, the concepts "object", "function", "environment" and their subclasses are used to represent functional requirements. On the other hand, the concepts "constraint" and "quality" are used to represent non-functional requirements. The concept "constraint" is useful to represent numerical ranges, e.g., speed, distance, time expiration, weight and so on. The concept "ambiguity" is also related to quality of software, but its meaning is ambiguous. Ambiguous words mainly adjectives

A requirements document "S" (consists of req. items.)

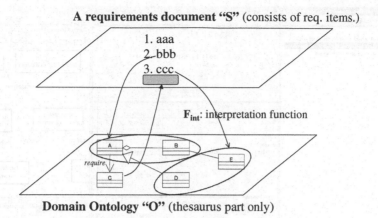

Domain Ontology "O" (thesaurus part only)

Fig. 2. Mapping from Requirements to Ontology

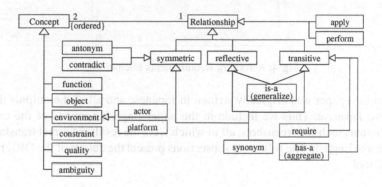

Fig. 3. Ontology Meta model

and adverbs such as "as soon as" and "user-friendly" belong to this concept so as to detect ambiguous requirements.

3 Requirements Elicitation Tool Based on ORE

In ORE method, the following two semi-automated tasks to elicit requirements are iterated collaborating with a requirements analyst and our tool during requirements elicitation.

- Develop mapping between requirements and ontological concepts.
- Analyze requirements by using a domain ontology.

Figure 4 illustrates a screenshot of the tool for supporting ORE method, and we will use it to explain the elicitation process following the ORE method. The example problem used here is a library system and its initial requirements are displayed in the left upper area of the window. In the right upper area, a thesaurus part of the ontology "Library System" is depicted in class diagram form. Note that this example used

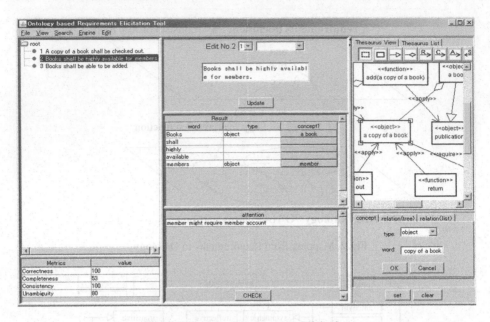

Fig. 4. A Tool for Requirements Elicitation

throughout the paper was originally written in Japanese and all of the outputs that our tools were Japanese. Thus we include in the paper English versions of the example and the outputs such as screenshots, all of which were the results of direct translation of Japanese into English. The following subsections present the details of the ORE method using the tool.

3.1 Mapping between Requirements and Ontological Concepts

To handle with the meaning of each sentence in requirements symbolically, a set of concepts in an ontology, i.e. a set of ontological concepts is related to each sentence in the requirements list. As shown in Figure 4, the initial requirements of the library system consists three itemized sentences. Lexical and morphological analysis is automatically executed on the sentences so as to extract relevant words from them. In the case of Japanese, specific morphological analyzer tools are necessary because Japanese does not have explicit separators such as white spaces between words. We use a morphological analyzer called "Sen" [7] written in Java. The morphological analyzer identifies the part of speech of a morpheme (lexical categories such as nouns, verbs, adjectives, etc.). By using its result, several morphemes that are not so significant can be removed from the input sentences. For example, articles such as "a" or "the" are removed in general. After filtering out insignificant morphemes and identifying words and their parts of speech, the tool finds corresponding ontological concepts to each morpheme using a synonym dictionary.

We illustrate the above procedure by using a sentence item # 2 "Book shall be highly available for members" in Figure 4. After morphological analysis, an analyst can get the five words "books", "shall", "highly", "available", and "members" as shown in the first column "word" of the center table of the figure. And then the tool constructs automatically two mappings from the word "books" to the concept "a book" and from "members" to "member", because the concepts having the same labels as these words are included in the ontology shown in the right window of Figure 4 and in Figure 5. Figure 5 shows the ontology of Library domain, which our tool created, and we represent an ontology with class diagram, where classes and associations denote domain concepts and their relationships respectively. See the columns "word" and "concept1" in the center table "Result" of the figure. However, the analyst finds that the word "Books" semantically implies "a copy of a book" in addition to "book", and she or he tries to modify the mapping from the word "Books" by clicking "a copy of a book" in the right window of the figure showing the ontology. Finally she or he can have the mappings as follows.

- "books" → object(a book), object(a copy of a book)
- "available" → quality(availability)
- "members" → actor(member)

3.2 Analyzing Requirements by Using an Ontology

By using concepts and morphemes corresponding to each requirements item, requirements are improved incrementally in this task. Our tool detects 1) the concepts to be added into the current requirements and 2) the concepts to be removed from the current requirements, by using inference on the ontology, and advises suitable actions, e.g. adding a new requirement or removing the existing one, to the requirements analysts. The analysts collaborate with stakeholders to decide whether the advices can be accepted or not. If they accept the advices, they update the requirements items, e.g. add new items or remove the exiting one following the advices. These steps are iterated and the requirements items are incrementally updated and improved.

By using the mapping mentioned in the previous sub section, we will intuitively explain how the inference rules can find the lacks of mandatory requirements, i.e. the requirements to be added. See Figure 5 together with Figure 4. In our example, a requirements item #1 is mapped to function(check out). However, no items are mapped to object(member account) even though there is a require-relationship between function(check out) and object(member account) as shown in Figure 5 . Our tool detects the missing of "member account" by the inference on this require-relationship, and gives an advice to add object(member account) automatically. Following this advice, an analyst adds new requirements item about member account, e.g., "Each member has his/her own member account". The rule A1 shown in Figure 6 is defined based on this idea. In the figure, we illustrate these rules for producing advices to requirements analysts, and they can be recursively applied to produce the advices. The rules are represented with Horn clauses of Prolog.

They can be categorized with respect to their contribution to characteristics of a good Software Requirements Specification of IEEE 830 [8] as follows.

- Completeness:
  When a concept x in current requirements requires another concept y but y is not included in the current requirements, requirements about y can be considered as

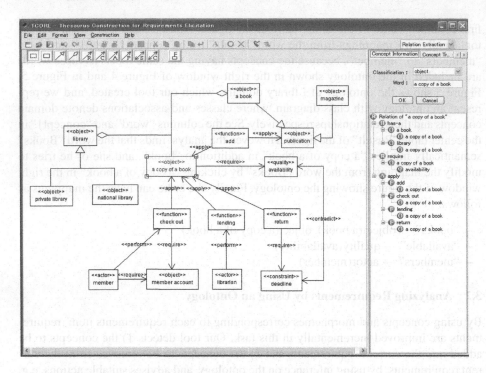

Fig. 5. Ontology of Library Domain

missing ones. Thus the current requirements are incomplete. The rules of this category suggest such missing concepts to improve the completeness of requirements.

– Correctness:
  If a requirement cannot be related to any elements of a domain ontology, the requirement could specify a requirement irrelevant to the domain. These rules suggest such irrelevant requirements to improve the correctness.

– Consistency:
  Suppose that requirements S1 and S2 are related to concepts C1 and C2 respectively. If it is inferred on an ontology that C1 contradicts to or excludes C2, S1 and S2 can be inconsistent. These rules give suggestions to exclude mutually inconsistent requirements.

– Unambiguity:
  If a requirement is related to several concepts and these concepts are not semantically related with each other on an ontology, the requirement can be interpreted in several ways. These rules suggest such requirements to improve unambiguity.

The rules A1, A2, A3, A4 and A5 listed in Figure 6 are for checking completeness in the above category.

Let's turn back to our example, Figure 4. When the analyst clicks a CHECK button located in the bottom of the center area in the figure, the tool starts inference to detect missing mandatory requirements, inconsistent ones and ambiguous ones. In this example, according to the ontology "member" requires the concept "member account"

Rule A1: If concepts(all) includes concept(A), concepts(all) does not include concept(B) and Ontology() includes require(concept(A), concept(B)), then our tool gives an advice to add or modify requirements item(s) so that concepts(all) includes concept(B).

Rule A2: If concepts(all) includes object(A), concepts(all) does not include function(B), and Ontology() includes apply(function(B), object(A)), then our tool gives an advice to add or modify requirements item(s) so that concepts(all) includes function(B).

Rule A3: If concepts(all) does not include object(A), concepts(all) includes function(B), and Ontology() includes apply(function(B), object(A)), then our tool gives an advice to add or modify requirements item(s) so that concepts(all) includes object(A).

Rule A4: If concepts(all) includes environment(A), concepts(all) does not include function(B), and Ontology() includes perform(environment(A), function(B)), then our tool gives an advice to add or modify requirements item(s) so that concepts(all) includes function(B).

Rule A5: If concepts(all) does not include environment(A), concepts(all) includes function(B), and Ontology() includes perform(environment(A), function(B)), then our tool gives an advice to add or modify requirements item(s) so that concepts(all) includes environment(A).

Abbreviations:

- concepts(item x): a set of concepts related to a requirements item x.
- concepts(all): the union of the sets of concepts related to all requirements items in a requirements list. ($\cup_x$ concepts(item x))
- morphemes(item x): a set of morphemes related to an item x.
- morphemes(all): the union of the sets of morphemes related to all items in a list. ($\cup_x$ morphemes(item x))
- mapc(morpheme x): a set of concepts such that the mapping between x and the concept exists.
- mapm(concept x): a set of morphemes such that the mapping between the morpheme and x exists.
- Ontology(): a set of all relationships and concepts of a domain ontology.

Fig. 6. Inference Rules for Completeness

to distinguish members of the library from non-members, and the rule A1 mentioned above suggests that "member account" should be added. This inference result is shown in the bottom area "attention" in the figure, and the analyst can add the sentence related to "member account" at the top area "Edit".

Our tool also tells us the measures of correctness, consistency, unambiguity and completeness with the ratio of the detected faults to the total number of requirements items [2]. The reader can find them in the left bottom window of Figure 4. They help an analyst to decide when she or he may terminate her or his elicitation activities.

4 Case Studies for Assessing Our Approach

We expect our tool to help requirements analysts without domain knowledge in eliciting requirements. In this section, we show experimental results to validate this assumption.

We pick up the problem domain of feed readers in these experiments. A feed reader is also called as "News Aggregator", and it is client software to retrieve frequently updated Web contents of certain topics from blogs and mass media web sites and so on.

Before starting the experiments, we have developed an ontology of the domain of feed readers from 17 documents, e.g. manuals of similar existing systems, using our ontology creation tool in 3 hours. It included 178 ontological concepts and 270 relationships. We call these 17 documents *original source documents*.

4.1 Hypotheses

At first, we have the following three hypotheses to show our method and tool can be useful to compensate lacks of domain knowledge for requirements analysts.

H1. With the help of our tool, almost any analysts without domain knowledge can get the same results of eliciting requirements.

H2. Our tool enables an analyst without domain knowledge gets the results same as an analyst with domain knowledge does.

H3. Even if an analyst can refer to original source documents of domain knowledge, which had been used for ontology creation, she or he cannot elicit requirements more effective than she/he does using our tool.

If H1 proves to be true, we can conclude that our tool and ontology used in the tool contribute to requirements elicitation by usual analysts in the similar way. If H2 is true, we can conclude that our tool plays a role of domain knowledge in domain experts. If H3 is true, we can conclude that our tool with the ontology is more useful than just a chunk of domain knowledge, and in other words, our technique to model ontologies can systemize domain knowledge well so that the requirements analysts can use easily it.

4.2 Experimental Design

To confirm the hypotheses mentioned above, we had 5 experiments below. For each experiment, one analyst elicits requirements for a feed reader based on an initial requirements list within 2 hours. The initial list contained 25 requirements. An analyst in each experiment and a customer were students in software engineering course, thus they knew fundamental knowledge about requirements engineering. Each analyst may communicate with the customer freely during elicitation. Analysts may also refer to public information on the Web sites. Note that the same student played a customer's role in all experiments, but a different student played an analyst's role in each experiment. The output of an elicitation task is the list of requirements, i.e., final requirements list. Table 1 shows the differences among the experiments. In experiments T1 and T2, these two subjects could use our tool, but they did not have domain knowledge of feed readers and they could not refer to the documents that we used for ontology creation (the original source documents). In the same way as T1, the subject of D1 had domain knowledge but she/he could not use our tool or could not refer to the original source documents during the experiment D1. In fact, she/he was a heavy user of the feed reader and had experiences in developing feed reader software, and could be considered as a domain

Table 1. Differences among Experiments

Experiment	having domain knowledge	using our tool	using original source documents
T1	-	Yes	-
T2	-	Yes	-
D1	Yes	-	-
R1	-	-	Yes
R2	-	-	Yes

expert. In experiments R1 and R2, the subjects without the domain knowledge could use and refer to the original sources instead of the ontology via our tool.

To investigate the hypotheses H1, H2 and H3, we focus on the differences of the obtained final requirements lists. To identify such differences, we categorize requirements in a list into the following six disjoint types by investigating and analyzing semantic contents of the elicited requirements and the ontology.

a1: Requirements that existed in both the initial and final requirements list.

a2: Requirements added or updated due to the help of our tool.

a3: Requirements added or updated due to the original sources documents.

a4: Requirements added or updated due to the proposal of a customer.

a5: Requirements added or updated due to the proposal of an analyst.

a6: Requirements added or updated due to public resources other than the original source documents.

Note that we abbreviate a1 + a2 + a3 + a4 + a5 + a6 to a0, that is the whole requirements in a requirements list.

If the number of the requirements of type a2 (simply, type a2 requirements) is almost the same in experiment T1 and T2, we can assume H1 is true. If the number of type a0 requirements in D1 is almost the same as the number in T1 and T2, we can assume H2 is true. If the number of type a0 requirements in T1 or T2 is more than the number of R1 and R2, we can assume H3 is true.

4.3   Results and Discussion

Table 2 shows the result of categorizing the final requirements into the categories mentioned above. For example, 23 requirements out of 56 were elicited with the help of our tool in experiment T1. Because no original source documents were given in T1, T2 and D1, we put the mark "-" in the columns of the number of type a3 requirements. Similarly, as for the column of the number of type a2 requirements, we use the mark "-" in the rows of the experiments D1, R1 and R2, because no tool supports were given in them. Because there are not so many data for each type of experiments, we did not achieve any statistical tests for our hypotheses.

Based on the results of T1 and T2 in Table 2, the hypothesis H1 "With the help of our tool, almost any analysts without domain knowledge can get the same results of eliciting requirements" seems to be true because the number of a2 type requirements were almost the same. In addition, the number of a2 type requirements was the largest among the numbers of all types in T1 and T2 respectively.

Table 2. Number of requirements in each type for each experiment

Experiment	a1	a2	a3	a4	a5	a6	a0 (total)
T1	20	23	-	1	0	12	56
T2	18	22	-	2	3	16	61
D1	18	-	-	4	9	34	65
R1	16	-	0	4	8	12	40
R2	19	-	8	9	3	0	39

Based on the results of T1, T2 and D1 in Table 2, the hypothesis H2 "Our tool enables an analyst without domain knowledge gets the results same as an analyst with domain knowledge does" seems to be true because there was few differences of a0 among T1, T2 and D1. In D1 (a domain expert), requirements were mostly elicited based on the public information as shown in a6 column in Table 2. It suggests that he might already hold the overall image of the feed reader as his knowledge so that he could retrieve its detailed level of information from public information. The analyst D1 proposed to the customers relatively larger number of requirements than in other experiments as shown in a5 column in Table 2. In the case of T1 and T2, the results of using our tool (shown in a2 column) show that the contribution of the ontology by which T1 and T2 could elicit the requirements in the same level as D1.

Based on the results of R1, R2, T1 and T2 in Table 2, the hypothesis H3 "Even if an analyst can refer to original source documents of domain knowledge, which had been used for ontology creation, she or he cannot elicit requirements more effective than she/he does using our tool" seems to be true because the number of total requirements in R1 and R2 were relatively smaller than the number in T1, T2 as shown in a0 column. In R1 and R2, we were afraid that the original source documents played the same role as our tool. However, fortunately, the original source documents did not play such a role because the number of a3 type requirements in R1 and R2 were smaller than the number of a2 type in T1 and T2.

Note that since we focused only on the numbers of the elicited requirements in this section, we should evaluate the quality of the elicited requirements. We showed the elicited requirements to a distinguished expert, who had experiences in developing feed reader software, and asked him/her to evaluate the requirements. As a result, all of the requirements except one requirement elicited by R1 were *correct* one. Thus, in these experiments, the meaningful differences of quality of the requirements did not appear. However, further analyses on quality of requirements from various views such as completeness, unambiguity etc. are necessary.

5   Related Work

We could find several studies on the application of ontologies to requirements engineering. LEL (Language Extended Lexicon) [9] is a kind of electronic version of dictionary that can be used as domain knowledge in requirements elicitation processes.Although it includes tags and anchors to help analysts fill up domain knowledge, it has neither methods as guidance procedures nor semantic inference mechanisms. In addition, although it presented a manual process to create LEL, it did not consider how to support this ontology creation process. A feature diagram in Feature Oriented Domain Analysis can be considered as a kind of representations of a domain thesaurus, and in [10], a technique to analyze semantic dependencies among requirements by using features and their dependency relationships was proposed. The idea of this technique is similar to our approach in the sense that requirements can be semantically analyzed. However, the aim of our approach is the support for requirements elicitation, while the authors of [10] just aimed at modeling semantic dependencies lying behind a requirements specification. In [11], the authors combined several formal methods by using an ontology as their common concepts. This is another application of ontology to requirements engineering, especially method integration, but its goal is different from ours. The meta schema in Domain Theory approach [12] can be considered as coarse grained version of our meta model of domain ontologies shown in Figure 3. The generic models of an application domain are modeled using the meta schema and a domain matcher retrieves an appropriate generic model so that it can be reused as a template of requirements descriptions. This approach is based on reuse, different from ours. In [13], the authors proposed the ontological framework for integrating semantically requirements engineering modeling techniques such as viewpoints modeling, goal oriented approach and scenario analysis. Although its aim, i.e. method integration [14], is different from ours, we can list up the application of ontological techniques to Requirements Engineering.

6   Conclusions and Future Work

In this paper, we presented a supporting tool for requirements elicitation where a domain ontology is utilized as domain knowledge. The contribution of this tool is that the inference mechanism implemented with Prolog helps a requirements analyst to evolve requirements systematically by taking account of the semantic aspect of requirements. We have assessed the user-friendliness and effectiveness of our tool through the experimental studies. Although our experiments mentioned in section 4 was too small in the sense of practical setting to argue the generality of the experimental findings, we could find the possibility of supporting requirements elicitation processes with our tool. According to the results of interviews to our subjects, the user interface of our tool should be improved. A list of the many advices displayed to the subjects was not so easy to understand at a glance and it took more time to select suitable ones out of them.

The quality of requirements elicitation using our tool greatly depends on the quality of domain ontologies. In fact, none of ontologies that we used for the experimental studies included contradiction relationships and we could not confirm the detection of inconsistency requirements. The reason was that the documents we used for ontology creation did not contain any specific words denoting contradiction. We should explore more elaborated mining techniques together with good samples of documents. Similarly, we could not check if our tool could detect ambiguity, because of the following two reasons. 1) All of the mappings from words to ontological concepts that the subjects and our tool made were precise one-to-one except for the case of mapping synonyms and thus the rules for ambiguity were not applied, and 2) the documents that we created ontologies from did not have frequent occurrences of ambiguous words. The first reason suggests that the technique to making mappings should also be elaborated.

As mentioned in section 1, this ontological approach is a lightweight semantic processing, and we can apply it to semantic analysis for requirements. In fact, we have analyzed requirements changes from semantic view, and tried to mine the patterns of requirements changes from repositories. The extracted patterns will be very helpful to predict and guide requirements changes [15], and this is one of the future works.

References

1. Kaiya, H., Shinbara, D., Kawano, J., Saeki, M.: Improving the detection of requirements discordances among stakeholders. Requirements Engineering 10(4), 289–303 (2005)
2. Kaiya, H., Saeki, M.: Using domain ontology as domain knowledge for requirements elicitation. In: Proc. of 14th IEEE International Requirements Engineering Conference (RE 2006), pp. 189–198 (2006)
3. Maedche, A.: Ontology Learning for the Semantic Web. Kluwer Academic Publishers, Dordrecht (2002)
4. DAML Ontology Library, http://www.daml.org/ontologies/
5. KAON Tool Suite, http://kaon.semanticweb.org/
6. Kitamura, M., Hasegawa, R., Kaiya, H., Saeki, M.: An Integrated Tool for Supporting Ontology Driven Requirements Elicitation. In: Proc. of 2nd International Conference on Software and Data Technologies (ICSOFT 2007), pp. 73–80 (2007)
7. Sen - Morphological Analyzer written in Java (2006), https://sen.dev.java.net/
8. IEEE. IEEE Recommended Practice for Software Requirements Specifications. IEEE Std. 830-1998 (1998)
9. Breitman, K., Leite, J.C.S.P.: Ontology as a Requirements Engineering Product. In: Proc. of 11th IEEE Requirements Engineering Conference (RE 2001), pp. 309–319 (2003)
10. Zhang, W., Mei, H., Zhao, H.: A Feature-Oriented Approach to Modeling Requirements Dependencies. In: Proc. of 13th IEEE International Conference on Requirements Engineering (RE 2005), pp. 273–284 (2005)
11. Petit, M., Dubois, E.: Defining an Ontology for the Formal Requirements Engineering of Manufacturing Systems. In: Proc. of ICEIMT 1997 (1997)
12. Sutcliffe, A.G., Maiden, N.A.M.: The domain theory for requirements engineering. IEEE Trans. Software Eng. 24(3), 174–196 (1998)
13. Lee, S.W., Gandhi, R.: Ontology-based active requirements engineering framework. In: Proc. of 12th Asia-Pacific Software Engineering Conference (APSEC 2005), pp. 481–488 (2005)
14. Brinkkemper, S., Saeki, M., Harmsen, F.: Meta-Modelling Based Assembly Techniques for Situational Method Engineering. Information Systems 24(3), 209–228 (1999)
15. Kaiya, H., Saeki, M.: Ontology Based Requirements Analysis: Lightweight Semantic Processing Approach. In: QSIC 2005, Proceedings of The 5th International Conference on Quality Software, pp. 223–230 (2005)

Pattern Detection in Object-Oriented Source Code

Andreas Wierda[1], Eric Dortmans[1], and Lou Somers[2]

[1] Océ-Technologies BV, P.O. Box 101, NL-5900 MA Venlo, The Netherlands
[2] Eindhoven University of Technology, Dept. Math. & Comp.Sc., P.O. Box 513
NL-5600 MB Eindhoven, The Netherlands
{andreas.wierda,eric.dortmans}@oce.com, wsinlou@win.tue.nl

Abstract. Pattern detection methods discover recurring solutions, like design patterns in object-oriented source code. Usually this is done with a pattern library. Hence, the precise implementation of the patterns must be known in advance. The method used in our case study does not have this disadvantage. It uses a mathematical technique, Formal Concept Analysis, and is applied to find structural patterns in two subsystems of a printer controller. The case study shows that it is possible to detect frequently used structural design constructs without upfront knowledge. However, even the detection of relatively simple patterns in relatively small pieces of software takes a lot of computing time. Since this is due to the complexity of the applied algorithms, applying the method to large software systems like the complete controller is not practical. It can be applied to its subsystems though, which are about 5-10% of its size.

Keywords: Pattern detection, formal concept analysis, object-oriented, reverse engineering.

1 Introduction

Architecture reconstruction and design recovery are a form of reverse engineering. Reverse engineering does not involve changing a system or producing new systems based on existing systems, but is concerned with understanding a system. The goal of design recovery is to "obtain meaningful higher-level abstractions beyond those obtained directly from the source code itself" [1].

Patterns provide proven solutions to recurring design problems in a specific context. Design patterns are believed to be beneficial in several ways [2,3,4], where knowledge transfer is the unifying element. Empirical evidence shows that developers indeed use design patterns to ease communication [5]. Given the fact that program understanding is one of the most time consuming activities of software maintenance, knowledge about applied patterns can be useful for software maintenance. Controlled experiments with both inexperienced [6] and experienced [7] software developers support the hypothesis that awareness of applied design patterns reduces the time needed for maintenance and the number of errors introduced during maintenance.

For an overview of methods and tools for architecture reconstruction and design recovery, see e.g. [8,9,10,11,12]. Architectural clustering and pattern detection are the most prominent automatic methods [13]. Pattern-based reconstruction approaches

J. Filipe et al. (Eds.): ICSOFT/ENASE 2007, CCIS 22, pp. 141–158, 2008.

detect instances of common constructs, or patterns, in the implementation. Contrary to the approach where one uses a library of known patterns to detect these in source code, we concentrate in this paper on the detection without upfront knowledge about the implemented patterns [14,15]. For this we use Formal Concept Analysis.

1.1 Formal Concept Analysis

Formal Concept Analysis (FCA) is a mathematical technique to identify "sensible groupings of formal objects that have common formal attributes" [16,17]. FCA is also known as Galois lattices [18]. Note that formal objects and formal attributes are not the same as objects and attributes in object-oriented programming!

The analysis starts with a *formal context*, which is a triple $C=(O,A,R)$ in which O is the finite set of formal objects and A the finite set of formal attributes. R is a binary relation between elements in O and A, hence $R \subseteq O \times A$. If $(o,a) \in R$ it is said that object o *has* attribute a. For $X \subseteq O$ and $Y \subseteq A$, the *common attributes* $\sigma(X)$ of X and *common objects* $\tau(Y)$ of Y are defined as [19]:

$$\sigma(X) = \left\{ a \in A \middle| \forall o \in X : (o,a) \in R \right\} \tag{1}$$

$$\tau(Y) = \left\{ o \in O \middle| \forall a \in Y : (o,a) \in R \right\} \tag{2}$$

A *formal concept* of the context (O,A,R) is a pair of sets (X,Y), with $X \subseteq O$ and $Y \subseteq A$, such that:

$$Y = \sigma(X) \wedge X = \tau(Y) \tag{3}$$

Informally a formal concept is a *maximal* collection of objects sharing common attributes. X is called the *extent* and Y the *intent* of the concept. The extents and intents can be used to relate formal concepts hierarchically. For two formal concepts (X_0, Y_0) and (X_1, Y_1) the subconcept relation $\leq$ is defined [19] as:

$$(X_0, Y_0) \leq (X_1, Y_1) \Leftrightarrow X_0 \subseteq X_1 \Leftrightarrow Y_1 \subseteq Y_0 \tag{4}$$

If p and q are formal concepts and $p \leq q$ then p is said to be a *subconcept* of q and q is a *superconcept* of p. The subconcept relation enforces an ordering over the set of concepts that is captured by the supremum $\bigsqcup$ and infimum $\bigsqcap$ relationships. They define the *concept lattice L* of a formal concept C with a set of concepts I [19]:

$$\bigsqcup_{(X_i, Y_i) \in I} (X_i, Y_i) = \left(\tau \left(\sigma \left(\bigcup_{(X_i, Y_i) \in I} X_i \right) \right), \bigcap_{(X_i, Y_i) \in I} Y_i \right) \tag{5}$$

$$\bigsqcap_{(X_i, Y_i) \in I} (X_i, Y_i) = \left(\bigcap_{(X_i, Y_i) \in I} X_i, \sigma \left(\tau \left(\bigcup_{(X_i, Y_i) \in I} Y_i \right) \right) \right) \tag{6}$$

where I is the set of concepts to relate. To calculate the supremum (smallest common superconcept) of a set of concepts their intents must be intersected and their extents joined. The latter set must then be enlarged to fit to the attribute set of the supremum. The infimum (greatest common subconcept) is calculated similarly.

A simple bottom-up algorithm is described in [20] that constructs a concept lattice L from a formal context $C=(O,A,R)$ using the supremum relation. It starts with the concept with the smallest extent, and constructs the lattice from that concept onwards. The algorithm utilizes that for any concept (X,Y) [21]:

$$Y = \sigma(X) = \sigma\left(\bigcup_{o \in X}\{o\}\right) = \bigcap_{o \in X}\sigma(\{o\}) \tag{7}$$

This equation enables calculating the supremum of two concepts by intersecting their intents. (8) gives a formalized description of the lattice construction algorithm, based on the informal description in [20]. The algorithm starts with the calculation of the smallest concept c_b of the lattice. The set of atomic concepts, together with c_b, is used to initialize L. Next the algorithm initializes a working-set W with all pairs of concepts in L that are not subconcepts of each other. The algorithm subsequently iterates over W to build the lattice using the supremum relation for each relevant concept-pair. The supremum of two concepts is calculated using (7). Recall that in this calculation the intents of the concepts c_1 and c_2 are intersected, after which τ is applied obtain the extent. If the calculated concept is new, it is added to L and the working-set is extended with relevant new concept pairs.

$$
\begin{aligned}
&c_b := \left(\tau(\sigma(\varnothing)), \sigma(\varnothing)\right)\\
&L := \{c_b\} \cup \left\{\left(\tau(\sigma(o)), \sigma(o)\right) \mid o \in O\right\}\\
&W := \left\{(c_1,c_2) \in L^2 \mid \neg(c_1 \le c_2 \vee c_2 \le c_1)\right\}\\
&\text{for each } (c_1,c_2) \in W \text{ do}\\
&\qquad c' := c_1 \amalg c_2\\
&\qquad \text{if } c' \notin L \text{ do}\\
&\qquad\qquad L := L \cup \{c'\}\\
&\qquad\qquad W := W \cup \left\{(c,c') \mid c \in L \wedge \neg(c \le c' \vee c' \le c)\right\}\\
&\qquad \text{od}\\
&\text{od}
\end{aligned}
\tag{8}
$$

The time complexity of algorithm (8) depends on the number of lattice elements. If the context contains n formal objects and n formal attributes, the lattice contains 2^n concepts [21]. This means the worst case running time of the algorithm is exponential in n. In practice however, the size of the concept lattice typically is $O(n^2)$, or even $O(n)$ [21,22,23]. This results in a typical running time for the algorithm of $O(n^3)$.

Algorithm (8) is a very simple lattice construction algorithm that does not perform very well. A comparison of algorithms [24] shows that for large contexts the Bordat algorithm [25] gives the best performance. For a concept lattice L with $|L|$ formal concepts and $|O|$ and $|A|$ formal objects and attributes of the formal context, the Bordat algorithm has a worst-case computational complexity of $O(|O| \cdot |A|^2 \cdot |L|)$.

1.2 Design Pattern Detection

The use of FCA to find recurring design constructs in object-oriented code is described in [22]. The key idea is that a design pattern amounts to a set of classes and a set of relations between them. Two different instances of a pattern have the same set of relations, but different sets of classes.

Let D be the set of classes in the design and T be the set of relationship-types between classes. For example $T=\{e,a\}$ defines the relationship types "extends" and "association". Then the set of inter-class relations P is typed $P \subseteq D \times D \times T$. To find pattern instances of k classes, the formal context $C_k=(O_k,A_k,R_k)$ is used with:

- O_k: set of k-sized sequences of classes in the design. More precisely

$$O_k = \left\{ (x_1,...,x_k) \mid x_i \in D \wedge i \in [1..k] \right\}$$

  where k is called the *order* of the sequence.
- A_k: set of inter-class relations within the sequences in O_k. Each is a triple $(x_i,x_j)_t$, where x_i and x_j are classes and t is a relationship-type. A_k is defined by

$$A_k = \left\{ (i,j)_t \mid (x_i,x_j)_t \in P \wedge i,j \in [1..k] \right\}$$

- R_k: "possesses" relation between the elements in O_k and in A_k.

A formal concept (X,Y) consists of a set of class-sequences X and a set of inter-class relations Y. Thus the intent Y specifies the pattern and the extent X specifies the set of pattern-instances found in the code.

Before the lattice can be constructed from the context, this context must be generated from the class diagram. A simple inductive algorithm [22] is shown in (9). Recall that D is the set of classes and P the set of class-relations.

The initial step generates an order two context. This is done by collecting all pairs of classes that are related by a tuple in P; the set O_2 of formal objects of the order two context consists of all pairs of classes related by a tuple in P. This means that for all formal objects in O_2 a relation of type t exists from the first to the second class. Therefore, the set A_2 of formal attributes of the order two context consists of the tuples $(1,2)_t$ for which a tuple in P exists that relates two arbitrary classes by a relation of type t.

In the inductive step, the order of the context is increased with one. The construction of O_k appends one component, x_k, to the tuples in O_{k-1}. This x_k is defined as any class for which a tuple in P exists that relates x_k to some other class x_j that is present in the tuple of O_{k-1}. Next, A_k is constructed by extending A_{k-1} with two sets of tuples. The first set consists of the tuples $(k,j)_t$, for which j equals the index of the class x_j that allowed the addition of x_k during the construction of O_k, and a relation of type t exists in P from x_k to x_j. The second set is similar, with k and j exchanged.

Initial step:
$$O_2 = \left\{ (x,y) \mid (x,y)_t \in P \right\}$$
$$A_2 = \left\{ (1,2)_t \mid \exists x,y \in D : (x,y)_t \in P \right\}$$
Inductive step $(k>2)$:
$$O_k = \left\{ (x_1,...,x_k) \mid (x_1,...,x_{k-1}) \in O_{k-1} \wedge \right. \tag{9}$$
$$\left. \exists j, 1 \le j \le k-1 \wedge ((x_j,x_k)_t \in P \vee (x_k,x_j)_t \in P) \right\}$$
$$A_k = A_{k-1} \cup \left\{ (i,j)_t \mid \exists (x_1,...,x_k) \in O_k \wedge \right.$$
$$\left. ((i=k \wedge 1 \le j \le k-1) \vee (j=k \wedge 1 \le i \le k-1)) \wedge (x_i,x_j)_t \in P \right\}$$

Note that in (9) the order n context contains the order $n\text{-}1$ context in the sense that all lower-order sequences are initial subsequences of the objects in the order n context, and that all attributes are retained. The algorithm assumes that design patterns consist of connected graphs. This assumption holds for all patterns in [3], so provided that sufficient relationships between classes are extracted, it does not impose a significant restriction.

Algorithm (8) is used in [22] to construct the lattice. The concepts directly represent patterns, but redundancies can be present. For example, two concepts may represent the same pattern. The notions of equivalent patterns and equivalent instances to remove redundancies from the lattice, informally defined by [22], are given formally by equations (10) and (11).

Definition 1 (Equivalent Patterns). Let (X_1, Y_1) and (X_2, Y_2) be two concepts representing design patterns that are generated from the same order k context. (X_1, Y_1) and (X_2, Y_2) are equivalent patterns if an index permutation f on the index set $\{1..k\}$ exists such that:

$$X_2 = \left\{ \left(x_{f(1)},...,x_{f(k)} \right) \middle| \left(x_1,...,x_k \right) \in X_1 \right\} \wedge X_1 = \left\{ \left(x_{f^{-1}(1)},...,x_{f^{-1}(k)} \right) \middle| \left(x_1,...,x_k \right) \in X_2 \right\} \quad (10)$$

$(X_1, Y_1) \cong (X_2, Y_2)$ denotes that (X_1, Y_1) and (X_2, Y_2) are equivalent patterns. According to Definition 1 two patterns (X_1, Y_1) and (X_2, Y_2) are equivalent when X_2 can be obtained by reordering the classes in (some of) the elements of X_1 and vice versa. Consequently, each formal attribute in Y_1 can be transformed into one in Y_2 and vice versa.

Definition 2 (Equivalent Instances). Let $(x_{1,1},...,x_{1,k})$ and $(x_{2,1},...,x_{2,k})$ be two formal objects in the extent X of an order k concept (X, Y) that represents a design pattern. These formal objects represent equivalent instances within that concept if an index permutation g on the index set $\{1..k\}$ exists such that:

$$\left(x_{2,1},...,x_{2,k} \right) = \left(x_{1,g(1)},...,x_{1,g(k)} \right) \wedge \left(x_{1,1},...,x_{1,k} \right) = \left(x_{2,g^{-1}(1)},...,x_{2,g^{-1}(k)} \right)$$
$$\wedge Y = \left\{ \left(g(y_1), g(y_2) \right)_t \middle| (y_1, y_2)_t \in Y \wedge t \in T \right\} \quad (11)$$

Here, $(x_{1,1},...,x_{1,k}) \cong (x_{2,1},...,x_{2,k})$ denotes that $(x_{1,1},...,x_{1,k})$ and $(x_{2,1},...,x_{2,k})$ are equivalent instances. According to Definition 2, two formal objects in the extent X of a concept (X, Y) are equivalent within that concept if an index permutation exists that transforms them into each other, and when applied to the formal attributes in Y produces attributes that are also part of Y.

In [26] the method is applied to three public domain applications written in C++ (20-100 KLOC). Besides the static inter-class relations (inheritance and association), also dynamic inter-class relations (calls and delegates) and class attributes like member function definitions are taken into account. They report the detection of several recurring design constructs, like the Adapter pattern [3] in several variants. The order of the context was chosen between two and four, typically three. Higher-order patterns did not prove to be a good starting point because "they impose an increasing number of constraints on the involved classes and are therefore matched by few instances (typically just one)". For the order three context the number of formal objects was 1721 to 34147. The number of formal attributes was 10 in all cases.

2 Case Study

The subject for our case study is a printer controller. Such a controller consists of general-purpose hardware on which proprietary and third party software runs. Its main task is to control (physical) devices such as a print- and scan-engine, and act as an intermediate between them and the customer network.

The software running on the controller has been written in multiple programming languages, but mostly in C++. An as-designed architecture is available, but it is not complete and large parts of the architecture documentation are not consistent with the implementation.

Table 1 shows the characteristics of the controller and two of its subsystems, Grizzly and RIP Worker. Because of performance limitations it was not feasible to apply the design pattern detection to the complete controller. Instead, it has been applied to these two subsystems. The Grizzly subsystem provides a framework for prototyping on the controller. The RIP Worker subsystem transforms Postscript files into printable bitmaps, taking the print-settings the user specified into account ("ripping"). In [27] the architecture of this controller is reconstructed by detecting instances of architectural styles and design patterns in the source code by means of a pattern library.

Table 1. Software characteristics of the printer controller of the case study

	Controller	Grizzly	RIP Worker
Classes	2661	234	108
Header and source files	7549	268	334
Functions	40449	2037	1857
Lines of source code (*1000)	932	35	37
Executable statements (*1000)	366	18	16

2.1 Goals

Our case study investigates the detection of unknown structural design patterns in source code, without requiring upfront knowledge, using Formal Concept Analysis (FCA). We formulate the following hypothesis (H1): "*With Formal Concept Analysis frequently used structural design constructs in the source code of the controller can be detected without upfront knowledge of the expected structures.*"

The confirmation of H1 does not imply that the found design constructs represent a useful architectural view of the controller. We therefore formulate an additional hypothesis (H2): "*Knowledge of frequently used structural design constructs found with Formal Concept Analysis in the controller provides an architectural view that is useful to gain insight in the structure of the system.*"

The usefulness of knowledge on structural design constructs depends on the amount of information this knowledge gives. The number of classes in the pattern and the number of instances of the pattern are two important criteria for this. On average, the design patterns in [3] contain about four to five classes. Because we are reconstructing an architectural view and not a subsystem-design we want to find slightly larger patterns. Hence we decided the patterns must contain at least *six classes* to be useful for architecture reconstruction.

The other criterion, the minimal number of instances of a useful pattern, is difficult to quantify. To our knowledge no work is published on this subject, so we determine it heuristically. Because no pattern-library is used, maintainers need to invest time to understand the patterns before reaping the benefit of this knowledge. The benefit, easier program understanding, must outweigh this investment. Obviously this is not the case if the patterns have one instance. Because we search repeated structures and not named patterns (like library-based approaches do) the investment is relatively high. Hence, we decided that a pattern must have at least *four instances* to be useful to reconstruct an architectural view of the controller.

To confirm the two hypotheses H1 and H2, a prototype has been built that implements the approach Tonella and Antoniol proposed, described in section 1.2. Before applying the prototype to the complete controller it has been applied to two of its subsystems, namely Grizzly and the RIP Worker.

2.2 Pattern Detection Architecture

The architecture of the pattern detection prototype is based on the pipe and filter architectural style [28]. The processing modules have been implemented with two third party tools and XSLT transformations. XSLT has been chosen because it is a mature and platform independent language, the two third-party tools (Columbus and Galicia) both support XML export and import, and finally it allows functional programming. This is an advantage because one of the most important algorithms of the implemented approach is defined inductively by (9) and maps very well to a functional implementation.

Fig. 1 shows a view of the prototype's architecture. The blocks represent processing modules and the arrows directed communication channels between the modules. The latter are implemented with files. Below, these modules are discussed.

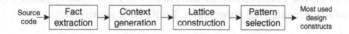

Fig. 1. Architectural view of the pattern detection prototype

Fact Extraction. The fact extraction module uses Columbus/CAN to extract structural information from the source code. Columbus uses the compiler that was originally used to compile the analyzed software, in this case Microsoft Visual C++. The extracted information is exported from Columbus with its UML exporter [29], which writes the information to an XMI file.

Because the XMI file has a relatively complex schema, the fact extraction module converts it to an XML file with a simpler schema. This file serves as input for the context generation module. It contains the classes and most important relationships between them. Three types of relations are extracted:

- *Inheritance*: The object-oriented mechanism via which more specific classes incorporate the structure and behavior of more general classes.
- *Association*: A structural relationship between two classes.
- *Composition*: An association where the connected classes have the same lifetime.

Context Generation. This module uses the inductive context construction algorithm (9) to generate the formal context that will be used to find frequently used design constructs. After algorithm (9) has been completed, the "context generation" module converts the formal context to the import format Galicia uses for "binary contexts".

Since XSLT does not support sets, the prototype uses bags. This, however, allows the existence of duplicates. The prototype removes these with an extra template that is applied after the templates that implement each of the initial and inductive steps. This produces the XSLT equivalent of a set.

Size of the output. The initial step of the context generation algorithm produces an order two context. The order of the context represents the number of classes in the patterns searched for. Each inductive step extends the order with one. This step is repeated until the desired order is reached. So in general the $(k-1)$-th step of the algorithm $(k \geq 2)$ produces a context $C_k = (O_k, A_k, R_k)$ of order k, where O_k is the set of formal objects, A_k the set of formal attributes, and R_k the set of relations between the formal objects in O_k and the formal attributes in A_k.

The number of formal attributes, $|A_k|$, is bounded by the number of different triples that can be made. Each formal attribute in A_k is a triple (p, q, t) where p and q are integers between 1 and k, and t is a relationship-type. The number of permutations of two values between 1 and k is bounded by k^2, so at most k^2 different combinations are possible for the first two components of the formal attributes. Therefore, if T is the set of relationship-types, and the size of this set is $|T|$, $|A_k| \leq |T| \cdot k^2$.

The number of formal objects, $|O_k|$, in the order k context is limited by the number of permutations of different classes of length k. If D is the set of classes, with size $|D|$, this means that $|O_k| \leq |D|^k$. So the number of formal objects is *polynomial* with the number of classes and *exponential* with the size of the patterns searched for. However, the fact that the connectivity of the classes in D is usually relatively low (and even can contain disconnected subgraphs), limits $|O_k|$ significantly.

Computational complexity. Let $P \subseteq D \times D \times T$ be the set of relations between classes, with D and T defined above. In the implementation the initial step is implemented with a template for the elements of P. Hence, if $|P|$ is the number of elements in P, the complexity of the initial step is $O(|P|)$.

The inductive step increases the order of the context with one. This is implemented with a template for the formal objects in the order $k-1$ context (the elements of O_{k-1}). This template extends each formal object $o \in O_{k-1}$ with a class that is not yet part of o and is related to one of the classes in o via a class-relation in P. Because every formal object in O_{k-1} consists of $k-1$ classes, the inductive step producing O_k has a computational complexity $O(|O_{k-1}| \cdot (k-1) \cdot |P|)$, which approximates $O(k \cdot |P| \cdot |O_{k-1}|)$.

Let $(x_1, ..., x_{k-1})$ be the sequence of classes represented by a formal object $o \in O_{k-1}$. Because in our implementation the previous inductive step appended classes to the *end* of this sequence, in the next inductive step only the last element x_{k-1} can lead to the addition of new classes to the sequence. Therefore, all but the first inductive steps do not have to iterate over all k-1 classes in the formal objects in O_{k-1}, but can only consider the most recently added class. This optimization reduces the computational complexity of the inductive step to about $O(|P| \cdot |O_{k-1}|)$. Because of limited implementation time this optimization has not been applied to the prototype, but is left as future work.

Because $|O_{k-1}|$ is polynomial with the number of classes in D, and in the worst case $|P|$ is quadratic with $|D|$, this optimization gives the inductive step a computational complexity that is polynomial with the number of classes in D. However, it is exponential with the size of the patterns searched for.

Lattice Construction. The prototype constructs the lattice with a third party tool, Galicia, an open platform for construction, visualization and exploration of concept lattices [30,31]. Galicia implements several algorithms to construct a lattice from a formal context. Because it is expected that the number of classes extracted from the source code, and hence the number of formal objects, will be relatively high, the Bordat algorithm [25] is best suited to generate the lattice, as explained in section 1.1.

Complexity of the lattice construction. Theoretically the size of the lattice, $|L|$, is exponential with the size of the context; if $|A|=|O|=n$ then $|L| \leq 2^n$. In practice the lattice size may be $O(n)$ [21], but this obviously depends on the properties of the formal context. Assuming that this is the case, and considering that in our case $|A|$ is much smaller than $|O|$, the computational complexity of the Bordat algorithm approximates $O(|O|^2)$. Thus, because the number of formal objects was polynomial with the number of classes and exponential with the size of the patterns searched for, this also holds for the computational complexity of the lattice construction.

Pattern Selection. The final module of the prototype filters the patterns in the lattice. Like the other data transformations in the prototype, this step is implemented with XSLT templates. Two filters are applied. First, sets of equivalent formal concepts, in the sense defined by (11), are replaced by one of their elements. Second, the concepts are filtered according to the size of their extent and intent (the number of formal objects and attributes respectively). In the remainder of this section these two filters are described more precisely.

The prototype does not filter for equivalent patterns in the sense defined by (10). It was planned to add this later if the output of the prototype proved to be useful. However, as is described in section 2.3, this was not the case.

Equivalent formal object filtering. Let X be the set of formal objects of some formal concept the lattice construction module produced, and let instance equivalence $\cong$ be defined by (11). Then, for every formal concept, the result of the first filter is the subset $X' \subseteq X$ that is the maximal subset of X that does not contain equivalent instances. If $|X'|$ and $|Z|$ refer to the number of elements in X' and another set Z respectively, this is defined as:

$$X' \subseteq X \wedge f(X') \wedge \neg \exists Z \subseteq X : f(Z) \wedge |Z| > |X'|$$
$$\text{with } f(X') \equiv \neg \exists x_1, x_2 \in X' : x_1 \neq x_2 \wedge x_1 \cong x_2 \tag{12}$$

This filter is implemented with two templates for the formal objects (the elements of X). The first template marks, for every formal concept, those formal objects for which an *unmarked* equivalent instance exists. Of every set of equivalent instances this leaves one element unmarked. The second template removes all marked formal objects. It is easy to see that this produces the maximal subset of X that does not contain equivalent instances.

Size-based filtering. The second filter removes all formal concepts with a small number of formal objects or attributes. Let p_x and p_y be two user-specified parameters that specify the minimum number of required formal objects and attributes respectively. Then the output of this filter only contains concepts with at least p_x formal objects and p_y formal attributes. This filter is implemented with a trivial template for the elements in the lattice.

Complexity of the pattern selection. Let $avg(|X|)$ and $avg(|Y|)$ represent the average number of formal objects and formal attributes respectively of the formal concepts. If $|L|$ represents the number of formal concepts in the lattice, the first filter has a time complexity of $O(|L|\cdot avg(|X|)\cdot avg(|Y|))$. If $avg(|X'|)$ represents the average size of the formal objects after equivalent instances have been removed by the first filter, the second filter has a computational complexity of $O(|L|\cdot(avg(|X'|)+avg(|Y|)))$. Because $avg(|X'|)$ is smaller than $avg(|X|)$, the pattern selection module has a total computational complexity of approximately $O(|L|\cdot avg(|X|)\cdot avg(|Y|))$.

We assume that the number of formal concepts $|L|$ is proportional to the number of formal objects (and the number of formal attributes, but that is much less). If every formal attribute is associated with every formal object, $avg(|Y|)$ equals the number of formal objects. Because we assume the number of formal attributes to be very small compared to the number of formal objects, $avg(|X|)$ is not relevant for the computational complexity. Therefore, the computational complexity of the filtering module is approximately quadratic with the number of formal objects. Because the number of formal objects was polynomial with the number of classes and exponential with the size of the patterns searched for, this again also holds for the complexity of the pattern-selection.

2.3 Results

The pattern detection prototype has been applied to the Grizzly and RIP Worker subsystems of the controller. The following sections give examples of patterns found.

Results for Grizzly. The application to the Grizzly source code (234 classes) produced a formal context and lattice with the characteristics shown in Table 2.

Table 2. Characteristics of the order four context for Grizzly and the corresponding lattice

Number of formal objects	40801
Number of formal attributes	37
Number of attribute-object relations	128065
Number of formal concepts	989

Recall from section 2.2 that the number of formal attributes $|A_k|$ of an order k context is bounded by $|A_k| \leq |T| \cdot k^2$, where $|T|$ is the number of relationship-types. In this case, $|T|=3$ and $k=4$ so the number of formal attributes is bounded by $3 \times 4^2 = 48$. Table 2 shows that the number of formal attributes (37) is indeed less than 48.

Recall from the same section that the upper bound of the number of formal objects of an order k context, $|O_k|$, is polynomial with the number of classes $|D|$. More specific $|O_k| \leq |D|^k$. Since the characteristics in Table 2 are of an order four context, $|O_k| = 234^4 \approx 3 \cdot 10^9$, which is clearly more than 40801. In fact, the number of formal objects is in the same order as $234^2 = 54756$. This large difference is due to the low connectivity of the classes.

The figures in Table 2 confirm the assumptions made in section 2.2. The number of formal attributes is indeed much lower than the number of formal objects. Furthermore, the number of formal concepts is not exponential with the size of the context. In fact, it is about one order smaller than the number of formal objects. This confirms our assumption that the size of the lattice is approximately linear with the number of formal objects.

With the user-specified filtering-parameters both set to four ($p_x = p_y = 4$), the prototype extracted 121 order four concepts from this context (with $p_x = p_y = 5$ only twelve remained). However, despite the filtering, many of the found patterns were very similar. The result even included several variants of the same pattern, for example with the associations organized slightly different.

The 121 concepts obtained with both filtering parameters set to four have been analyzed manually according to their number of formal objects and attributes. Fig. 2 shows two of the found patterns that were among the most interesting ones. Galicia generated the concept-IDs, which uniquely identify the concept within the lattice.

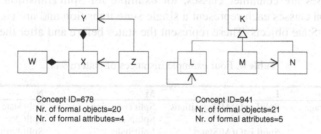

Concept ID=678
Nr. of formal objects=20
Nr. of formal attributes=4

Concept ID=941
Nr. of formal objects=21
Nr. of formal attributes=5

Fig. 2. Two patterns found in the Grizzly subsystem

Concept 678 represents a pattern with classes W, X, Y and Z, where Z has an association with X and Y. Both W and Y have a composition relationship with X. Analysis of the 20 instances of this pattern shows that for W fourteen different classes are present, for X and Y both two, and for Z three. This indicates that the instances of this pattern occur in a small number of source-code contexts.

Table 3 shows four example instances of this pattern. Examination of the Grizzly design documentation learned that the first instance in Table 3, with W=BitmapSyncContext, covers a part of an Interceptor pattern [28]. This pattern plays an important role in the architecture of Grizzly. The BitmapDocEventDispatcher class plays the role of event

Dispatcher, and the BitmapSyncContext the role of ConcreteFramework. The abstract and concrete Interceptor classes are not present in the detected pattern. (The designers of Grizzly omitted the abstract Interceptor class from the design.) The EventDispatcherT-est class is part of the Grizzly test code, and plays the role of the Application class in the Interceptor pattern. The Document class is not part of the Interceptor pattern. In the Grizzly design this class is the source of the events handled with the interceptor pattern.

Observe that the pattern in Fig. 2 does not contain the "create" relation between the BitmapDocEventDispatcher (Y) and the BitmapSyncContext (W) classes [28]. This does not mean that this relationship is not present; it is omitted from this pattern because the other pattern instances do not have this relationship.

Table 3. Four example instances of pattern 678

W	X	Y	Z
BitmapSyncContext SheetDocEventDispatcher	Document	BitmapDocEventDispatcher	BitmapDocEventDispatcherTest
FlipSynchronizer	BasicJob	BitmapDocSynchronizer	InversionWorkerJobInterceptor
StripeSynchronizer			BitmapDocSynchronizerTest

The other concept shown in Fig. 2 (with ID 941) represents a relatively simple pattern with four classes K, L, M and N. Class L, M and N inherit from K, L has a self-association, and M an association to N. Analysis of the 21 detected instances shows that in all cases K refers to the same class, L to three, and M and N both to six different classes. This indicates that all instances of this pattern are used in the same source-code context.

Table 4 shows four of the detected instances of pattern 941. SplitObjectStorage is an abstract class from which all workflow-related classes that store data inherit. The SplitList classes are container classes, for example for SplitTransition classes. The SplitTransition classes each represent a single state transition and are each associated with two SplitState objects. These represent the states before and after the transition.

Table 4. Four example instances of pattern 941

K	L	M	N
SplitObjectStorage	SplitListOfAllTransitions	SplitTransition	SplitState
		SplitNode	SplitDoc
	SplitListOfAllStates	SplitState	SplitAttribute
	SplitListOfAllDocuments	SplitDocPart	SplitImageSequence

Surprisingly, the Grizzly design documentation does not mention any of the classes listed in Table 4. Analysis of the code shows that these classes are concerned with workflow management in the controller, and represent points where Grizzly interfaces with the rest of the system. Strictly speaking these classes are not part of Grizzly but of the workflow-management subsystem of the controller. However, they are rede-fined in the Grizzly source-tree, and hence extracted by Columbus.

Results for RIP Worker. Applying the prototype to the RIP Worker (108 classes) produced a formal context and lattice with the characteristics shown in Table 5.

Table 5. Characteristics of the order four context for the RIP Worker and the corresponding lattice

Number of formal objects	52037
Number of formal attributes	41
Number of attribute-object relations	170104
Number of formal concepts	3097

The number of formal attributes, 41, is again less than the upper bound $|T|\cdot k^2$, which equals 48. The number of formal objects of the order k context, $|O_k|$, does not exceed the predicted upper bound: Table 5 represents an order four context, and $|O_k|=52037\leq|D|^4=108^4\approx1.4\cdot10^8$, so the number of formal objects is relatively low.

Like with Grizzly, the size of the lattice is approximately linear with the size of the context (one order smaller), and the number of formal objects is much higher than the number of formal attributes. With the user-specified size filtering parameters both set to five ($p_x=p_y=5$), the prototype produced 158 order four concepts (with $p_x=p_y=4$: 799). Like in Grizzly, the set of patterns found in the RIP Worker also contains a lot of similar patterns. Fig. 3 shows two of the patterns found.

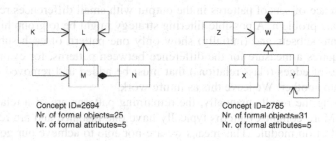

Concept ID=2694
Nr. of formal objects=25
Nr. of formal attributes=5

Concept ID=2785
Nr. of formal objects=31
Nr. of formal attributes=5

Fig. 3. Two patterns found in the RIP Worker

The output of the filtering module for concept 2694 shows that for class N 25 different classes are present, but for K, L and M all pattern instances have the same class. This indicates that all instances of this pattern are used in the same piece of the source code. Table 6 shows four examples of pattern 2694. All are concerned with job settings and the configuration of the system. The PJT_T_SystemParameters class stores information about the environment of the system, for example supported media formats and types. The PJT_T_JobSetting class represents the settings for a complete job, and is composed of the classes listed for N. The class for L, PJT_T_Product, is used to detect if the machine can handle a certain job specification.

Table 6. Four example instances of pattern 2694

K	L	M	N
PJT_T_SystemParameters	PJT_T_Product	PJT_T_JobSetting	PJT_T_MediaColor
			PJT_T_MediaWeight
			PJT_T_RunLength
			PJT_T_StapleDetails

Analysis of the 31 instances of the pattern for concept 2785 shows that in all cases W and Y refer to the same class. X refers to eight different classes and Z to four. This indicates that all instances of this pattern are used in the same source code context. Table 7 shows four example instances of pattern 2785. None of the listed classes are mentioned in the RIP Worker design documentation. Examination of the source code shows that all instances are part of a GUI library the RIP Worker's test tools use.

Table 7. Four example instances of pattern 2785

W	X	Y	Z
CWnd	CDialog	CFrameWnd	CCmdUI
	CButton		CDialog
	CListBox		CWinThread
	CEdit		CDataExchange

Quality of the Results. When examining the output for Grizzly and the RIP Worker, it is clear that better filtering is required. Recall that filtering for equivalent patterns, as defined by (10), has not been implemented in the prototype. The output contains many equivalent patterns, so in practice this filtering is desired too.

The occurrence of sets of patterns in the output with small differences represents a more significant problem. A possible filtering strategy might be to group highly similar patterns into subsets and (initially) show only one pattern of each subset of the user. This requires a measure for the difference between patterns, for example based on the number of edges (class relations) that must be added and removed to convert one pattern into another. We leave this as future work.

After filtering the results manually, the remaining patterns are of a relatively low complexity. More complex patterns typically have one instance and are removed by the pattern selection module. This means we are not able to achieve our goal of finding patterns that are useful to reconstruct architectural views (hypothesis H2).

Several publications report finding large numbers of design pattern instances in public domain code and few in industrial code [32,27]. We speculate that it could be the case that industrial practitioners structurally design software in a less precise way than public domain developers. Obviously, further experiments are needed to validate this statement, but it could explain why in our case study the number of instances of the found patterns remains fairly low.

Encountered Problems. During the fact extraction process several problems were encountered. First of all, Columbus crashed during the compilation of some source files. Because the compiler does not give errors for these files in normal use, the cause is an incompatibility between Columbus and Visual C++, or an error in Columbus itself. This problem occurred ten times for the full controller. In all cases, skipping the source file that triggered the error solved the problem. Because it only happened once for the RIP Worker, and not at all for Grizzly, it has little impact on the results.

The second problem occurred during the linking step of the fact extraction. No problems were encountered for the RIP Worker and Grizzly, but with the complete controller Columbus crashed. A few experiments revealed the size of the combined abstract syntax graphs, which is closely related to the size of the source files, as probable cause. Therefore it was not possible to extract facts from the full controller.

Execution Times. Both subsystems have been analyzed on the same test platform. Table 8 shows the characteristics of this platform and Table 9 shows the execution times for the RIP Worker and Grizzly subsystems for an order four context (wall-clock time). The lattice construction time includes the time to import the formal context into Galicia and to export the generated lattice to an XML file. For Grizzly the total execution time was 7:44:59 and for the RIP Worker 11:17:17 (hh:mm:ss).

Table 8. Test system characteristics

Processor	Pentium 4, 2 GHz
Memory	2 GB
Operating system	Windows 2000 SP4
Columbus	3.5
Galicia	1.2
Java	1.4.2_06

Table 9. Execution times (hh:mm:ss)

Process step	Grizzly	RIP Worker
1 Fact extraction	0:01:09	0:42:40
2 Context generation	0:26:00	0:36:00
3 Lattice construction	4:41:50	6:57:37
4 Pattern selection	2:36:00	3:01:00

The patterns detected in the Grizzly and RIP Worker source code are relatively simple. Possibilities to produce more interesting patterns are extending the size of the input to (for example) multiple subsystems of the controller, increasing the order of the context, or introducing partial matches. The last possibility requires fundamental changes to the method. If FCA would still be used, these changes would increase the size of the lattice significantly and hence also the execution time of the lattice construction step. The first two options have the disadvantage that they increase the size of the data processed. This affects the running time of all modules.

Because the computational complexity of the algorithms of each module is polynomial with the number of classes and exponential with the order of the context, we conclude that, given the executing times in Table 9, it is not practical to use the prototype to reconstruct architectural views of the complete controller.

3 Conclusions

Pattern detection methods based on a pattern library have been applied frequently and their properties are well known, but they require upfront knowledge of the patterns used and their precise implementation. Implementation variations make the latter difficult to specify. The pattern detection method we applied is based on Formal Concept Analysis and does not require a pattern library. The method proved to be able to detect frequently used design structures in source code without upfront knowledge of the expected constructs, thereby confirming our hypothesis H1 in section 2.1.

However, the detection of relatively simple structures in relatively small pieces of source code required a lot of calculations. For performance reasons no contexts of orders large than four could be analyzed, so the detected patterns consisted of four classes or less. Although large numbers of pattern instances were detected, these were typically confined to a few areas of the source code. Because it was not possible to detect patterns with six classes or more, we failed to confirm hypothesis H2.

Since this is inherent to the used algorithms, the application of this technique to re-construct architectural views of large object-oriented systems, more specific, systems with the size of the controller, is not considered practical. It is possible to detect de-sign patterns in subsystems with a size of 5-10% of the complete controller.

Besides performance issues, the reduction of the large number of similar patterns in the output is also important. Based on the complexity of the patterns we filtered the output, but the results show that more advanced filtering is necessary. It might also be possible to group similar patterns into groups and show a single pattern of each group to the user. The similarity of patterns could be based on the number of edges that must be added and removed to transform them into each other.

Finding frequently used design constructs in the source code essentially finds fre-quently occurring subgraphs in the class graph. An alternative to the pattern detection currently used might be to use graph compression algorithms that are based on the detection of recurring subgraphs. We have built a small prototype that uses the Sub-due algorithm [33]. This algorithm creates a list of recurring subgraphs and replaces all occurrences of these subgraphs with references to this list. However, when this algorithm is used for pattern detection, the fact that the algorithm looks for perfectly identical subgraphs causes problems. The intertwining of structures often encountered in practice caused this prototype to find no patterns at all in two subsystems (Grizzly and the RIP Worker) of the controller. Lossy graph compression algorithms might introduce the required fuzziness.

References

1. Chikovsky, E.J., Cross, J.H.: Reverse Engineering and Design Recovery: A taxonomy. IEEE Software 7(1), 13–17 (1990)
2. Beck, K., Coplien, J.O., Crocker, R., Dominick, L., Meszaros, G., Paulisch, F., Vlissides, J.: Industrial Experience with Design Patterns. In: 18th Int. Conf. on Software Engineering (ICSE-18), pp. 103–114 (1996)
3. Gamma, E., Helm, R., Johnson, R., Vlissides, J.: Design Patterns: elements of reusable ob-ject-oriented software, 5th edn. Addison-Wesley, Reading (1995)
4. Keller, R.K., Schauer, R., Robitaille, S., Pagé, P.: Pattern-Based Reverse-Engineering of Design Components. In: 21st Int. Conf. on Software Eng. (ICSE 1999), pp. 226–235 (1999)
5. Hahsler, M.: A Quantitative Study of the Application of Design Patterns in Java. Technical report 1/2003, University of Wien (2003)
6. Prechtelt, L., Unger-Lamprecht, B., Philippsen, M., Tichy, W.F.: Two Controlled Experi-ments Assessing the Usefulness of Design Pattern Documentation in Program Mainte-nance. IEEE Trans. on Software Engineering 28(6), 595–606 (2002)

7. Prechtelt, L., Unger, B., Tichy, F., Brössler, P., Votta, L.G.: A Controlled Experiment in Maintenance Comparing Design Patterns to Simpler Solutions. IEEE Trans. on Software Engineering 27(12), 1134–1144 (2001)
8. O'Brien, L., Stoermer, C., Verhoef, C.: Software Architecture Reconstruction: Practice Needs and Current Approaches. SEI Technical Report CMU/SEI-2002-TR-024, Software Engineering Institute, Carnegie Mellon University (2002)
9. van Deursen, A.: Software Architecture Recovery and Modelling, WCRE 2001 Discussion Forum Report. ACM SIGAPP Applied Computing Review 10(1) (2002)
10. Hassan, A.E., Holt, R.: The Small World of Software Reverse Engineering. In: 2004 Working Conf. on Reverse Engineering (WCRE 2004), pp. 278–283 (2004)
11. Sim, S.E., Koschke, R.: WoSEF: Workshop on Standard Exchange Format. ACM SIG-SOFT Software Engineering Notes 26, 44–49 (2001)
12. Bassil, S., Keller, R.K.: Software Visualization Tools: Survey and Analysis. In: 9th Int. Workshop on Program Comprehension (IWPC 2001), pp. 7–17 (2001)
13. Sartipi, K., Kontogiannis, K.: Pattern-based Software Architecture Recovery. In: Second ASERC Workshop on Software Architecture (2003)
14. Snelting, G.: Software Reengineering Based on Concept Lattices. In: European Conf. on Software Maintenance and Reengineering (CSMR 2000), pp. 1–8 (2000)
15. Tilley, T., Cole, R., Becker, P., Eklund, P.: A Survey of Formal Concept Analysis Support for Software Eng. Activities. In: 1st Int. Conf. on Formal Conc. Analysis (ICFCA 2003) (2003)
16. Siff, M., Reps, T.: Identifying Modules via Concept Analysis. Technical Report TR-1337, Computer Sciences Department, University of Wisconsin, Madison (1998)
17. Wille, R.: Restructuring lattice theory: An approach based on hierarchies of concepts. In: Rival, I. (ed.) Ordered Sets, pp. 445–470. NATO Advanced Study Institute (1981)
18. Arévalo, G., Ducasse, S., Nierstrasz, O.: Understanding classes using X-Ray views. In: 2nd Int. Workshop on MASPEGHI 2003 (ASE 2003), pp. 9–18 (2003)
19. Ganther, B., Wille, R.: Applied lattice theory: formal concept analysis. In: Grätzer, G. (ed.) General Lattice Theory. Birkhäuser Verlag, Basel (1998)
20. Siff, M., Reps, T.: Identifying Modules via Concept Analysis. In: Int. Conf. on Software Maintenance (ICSM 1997), pp. 170–179 (1997)
21. Snelting, G.: Reengineering of Configurations Based on Mathematical Concept Analysis. ACM Transactions on Software Engineering and Methodology 5(2), 146–189 (1996)
22. Tonella, P., Antoniol, G.: Object Oriented Design Pattern Inference. In: Int. Conf. on Software Maintenance (ICSM 1999), pp. 230–238 (1999)
23. Ball, T.: The concept of Dynamic Analysis. In: 7th European Software Engineering Conference, pp. 216–234 (1999)
24. Kuznetsov, S.O., Obëdkov, S.A.: Comparing performance of algorithms for generating concept lattices. In: 9th IEEE Int. Conf. on Conceptual Structures (ICCS 2001), pp. 35–47 (2001)
25. Bordat, J.P.: Calcul pratique du treillis de Galois d'une correspondance. Math. Sci. Hum. 96, 31–47 (1986)
26. Tonella, P., Antoniol, G.: Inference of Object Oriented Design Patterns. Journal of Software Maintenance and Evolution: Research and Practice 13(5), 309–330 (published, 2001)
27. Kersemakers, R., Dortmans, E., Somers, L.: Architectural Pattern Detection - A Case Study. In: 9th Int. Conf. on Software Engineering and Applications (SEA 2005), pp. 125–133 (2005)
28. Buschmann, F., Meunier, R., Rohnert, H., Sommerlad, P., Stal, M.: Pattern-Oriented Software Architecture: A System of Patterns. John Wiley and Sons, Chichester (1999)

29. Setup and User's Guide to Columbus/CAN, Academic Version 3.5. FrontEndART (2003)
30. Valtchev, P., Grosser, D., Roume, C., Hacene, M.R.: Galicia: an open platform for lattices. In: 11th Int. Conf. on Conceptual Structures (ICCS 2003), pp. 241–254. Shaker Verlag (2003)
31. Galicia Project, http://www.iro.umontreal.ca/~galicia/
32. Antoniol, G., Fiutem, R., Cristoforetti, L.: Design Pattern Recovery in Object-Oriented Software. In: 6th Int. Workshop on Program Comprehension, pp. 153–160 (1998)
33. Jonyer, I., Cook, D.J., Holder, L.B.: Graph-Based Hierarchical Conceptual Clustering. Journal of Machine Learning Research 2, 19–43 (2001)

Testing the Effectiveness of MBIUI Life-Cycle Framework for the Development of Affective Interfaces

Katerina Kabassi[1], Maria Virvou[2], and Efthymios Alepis[2]

[1] Department of Ecology and the Environment, TEI of the Ionian Islands
2 Kalvou Sq, Zakynthos, Greece
kkabassi@teiion.gr
[2] Department of Informatics, University of Piraeus
80 Karaoli & Dimitriou St., 18534 Piraeus, Greece
{mvirvou,talepis}@unipi.gr

Abstract. In this paper we test the effectiveness of a unifying life-cycle framework for the development of affective bi-modal interfaces. The life-cycle framework is called MBIUI and may be used for the application of many different multi-criteria decision making theories in intelligent user interfaces. As a case study we use an affective bi-modal educational system. More specifically, we describe the experimental studies that are proposed in MBIUI for designing, implementing and testing a multi-criteria decision making theory, called Simple Additive Weighting, in the educational application. The decision making theory has been adapted in the user interface for combining evidence from two different modes and providing affective interaction.

Keywords: Software life-cycle, multi-criteria decision making theories, affective interaction, user modeling.

1 Introduction

The actual process of incorporating multi-criteria analysis into an intelligent user interface is neither clearly defined nor adequately described in the literature. This is probably the main reason why decision making theories has not been used very often in knowledge-based software. Indeed, the actual process of incorporating multi-criteria analysis into an intelligent user interface involves several development steps that are not trivial. Furthermore, Hull et al. [1] point out that as the systems become more complex, their development and maintenance is becoming a major challenge. This is particularly the case for software that incorporates intelligence. Indeed, intelligent systems are quite complex and they have to be developed based on software engineering approaches that are quite generic and do not specialise on the particular difficulties of the intelligent approach that is to be used.

In view of the above, a life-cycle framework have been developed for the incorporation of a multi-criteria theory in an Intelligent User Interface (IUI). This framework is called MBIUI (Multi-criteria Based Intelligent User Interface) life-cycle framework [2] and involves the description of a software life-cycle that gives detailed information and guidelines about the experiments that need to be conducted, the design of the

J. Filipe et al. (Eds.): ICSOFT/ENASE 2007, CCIS 22, pp. 159–171, 2008.
© Springer-Verlag Berlin Heidelberg 2008

software, the selection of the right decision making theory and the evaluation of the IUI that incorporates a decision making theory.

In this paper, we test the effectiveness of this framework by applying it for the development of an affective bi-modal user interface. The particular interface is called Edu-Affe-Mikey and is an affective educational user interface targeted to first-year medical students. Emphasis has been given on the application of the MBIUI life-cycle framework for the application of a multi-criteria decision making method for combining evidence from two different modes in order to identify the users' emotions. More specifically, the Simple Additive Weighting (SAW) [3], [4] has been applied in the educational user interface for evaluating different emotions, taking into account the input of the two different modes and selecting the one that seems more likely to have been felt by the user. In this respect, emotion recognition is based on several criteria that a human tutor would have used in order to perform emotion recognition of his/her students during the teaching course.

A main difference of the proposed approach with other systems that employ decision making theories ([5], [6], [7], and [8]) is that the values of the weights of the criteria are not static. More specifically, the values of the criteria used in the proposed approach are acquired by user stereotypes and differ for the different categories of users. Stereotypes constitute a common user modelling technique for drawing assumptions about users belonging to different groups [9], [10]. In our case, user stereotypes have been constructed with respect to the different emotional states of users that are likely to occur in typical situations during the educational process and their interaction with the educational software and represent the weight of the criteria.

The main body of this paper is organized as follows: In section 2 we present and discuss the MBIUI life-cycle framework. In sections 3 and 4 we present briefly the experimental studies for requirements capture and analysis. Section 5 describes the design of the affective bi-modal educational application and section 6 its main characteristics. In section 7 we present and discuss the results of the evaluation of the multi-criteria model. Finally, in section 8 we give the conclusions drawn from this work.

2   MBIUI Life-Cycle Framework

MBIUI life-cycle framework is based on the Rational Unified Process (RUP) [11], which gives a framework of a software life-cycle that is based on iterations and maintains its phases and procedural steps. However, RUP does neither specify what sort of requirements analysis has to take place nor what kind of prototype has to be produced during each phase or procedural step. Such specifications are provided by our MBIUI framework concerning IUIs that are based on multi-criteria theories.

According to MBIUI framework, during the inception phase, the requirements capture is conducted. During requirements capture, a prototype is developed and the main requirements of the user interface are specified. At this point the multi-criteria decision making theory that seems most promising for the particular application has to be selected. This decision may be revised in the procedural step of requirements capture in the phase of construction.

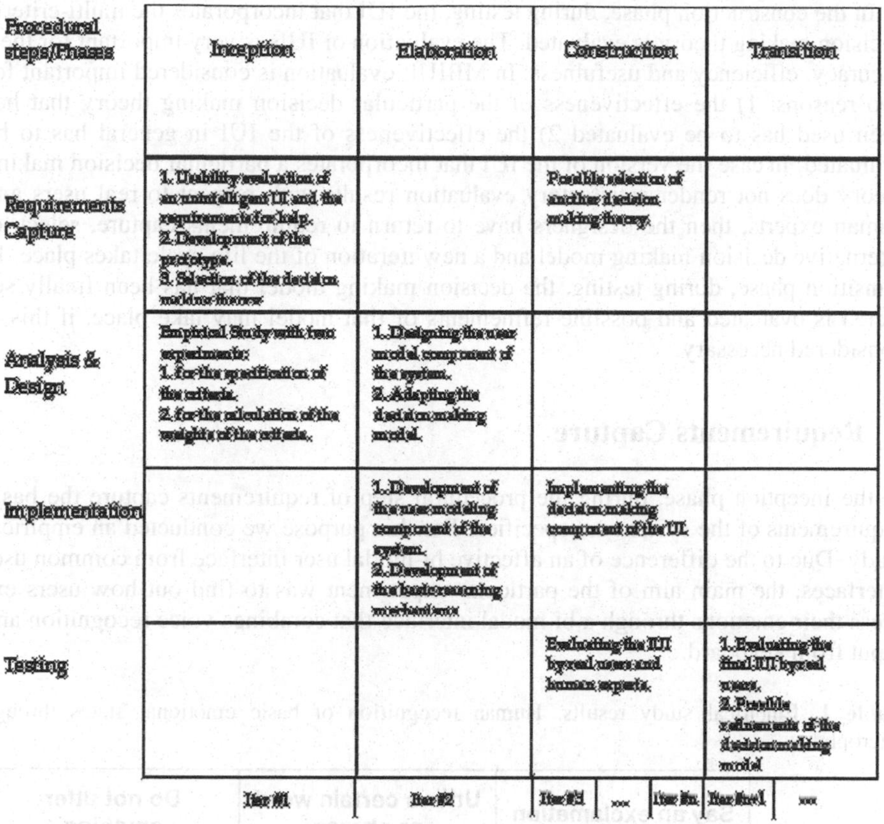

Procedural steps/Phases	Inception	Elaboration	Construction	Transition
Requirements Capture	1. Usability evaluation of an unintelligent UI and the requirements for help. 2. Development of the prototype. 3. Selection of the decision making theory		Possible selection of another decision making theory.	
Analysis & Design	Empirical Study with two experiments: 1. for the specification of the criteria. 2. for the calculation of the weights of the criteria.	1. Designing the user model component of the system. 2. Adapting the decision making model.		
Implementation		1. Development of the user modelling component of the system. 2. Development of the basic reasoning mechanism.	Implementing the decision making component of the IUI.	
Testing			Evaluating the IUI by real users and human experts.	1. Evaluating the final IUI by real users. 2. Possible refinements of the decision making model

Fig. 1. The MBIUI life-cycle framework

According to MBIUI, in the inception phase, during analysis, two different experiments are conducted in order to select the criteria that are used in the reasoning process of the human advisors as well as their weights of importance. The experiments should be carefully designed, since the kind of participants as well as the methods selected could eventually affect the whole design of the IUI. Both experiments involve human experts in the domain being reviewed.

The information collected during the two experiments of the empirical study is further used during the design phase of the system, where the decision making theory that has been selected is applied to the user interface. More specifically, in the elaboration phase, during design, the user modelling component of the system is designed and the decision making model is adapted for the purposes of the particular domain. Kass and Finin [12] define the user model as the knowledge source of a system that contains hypotheses concerning the user that may be important in terms of the interactive behaviour of the system.

In the elaboration phase, during implementation, the user modelling component of the system as well as the basic decision making mechanisms are developed. As a result a new version of the IUI is developed which incorporates fully the multi criteria decision making theory.

In the construction phase, during testing, the IUI that incorporates the multi-criteria decision making theory is evaluated. The evaluation of IUIs is very important for their accuracy, efficiency and usefulness. In MBIUI, evaluation is considered important for two reasons: 1) the effectiveness of the particular decision making theory that has been used has to be evaluated 2) the effectiveness of the IUI in general has to be evaluated. In case the version of the IUI that incorporates a particular decision making theory does not render satisfactory evaluation results with respect to real users and human experts, then the designers have to return to requirements capture, select an alternative decision making model and a new iteration of the life cycle takes place. In transition phase, during testing, the decision making model that has been finally selected is evaluated and possible refinements of that model may take place, if this is considered necessary.

3 Requirements Capture

In the inception phase, during the procedural step of requirements capture the basic requirements of the system are specified. For this purpose we conducted an empirical study. Due to the difference of an affective bi-modal user interface from common user interfaces, the main aim of the particular experiment was to find out how users express their emotions through a bi-modal interface that combines voice recognition and input from keyboard.

Table 1. Empirical study results. Human recognition of basic emotional states through microphone.

Emotions	Say an exclamation		Utter a certain word or phrase		Do not utter anything	
	Change in volume	Change in pitch	Change in volume	Change in pitch	Change in volume	Change in pitch
Neutral	6%		22%		72%	
	45%	18%	37%	12%		
Happiness	31%		45%		24%	
	40%	37%	55%	25%		
Sadness	8%		28%		64%	
	52%	34%	44%	14%		
Surprise	58%		24%		18%	
	66%	23%	60%	21%		
Anger	39%		41%		20%	
	62%	58%	70%	62%		
Disgust	50%		39%		11%	
	64%	43%	58%	38%		

50 users (male and female), of the age range 17-19 and at the novice level of computer experience participated the experiment. The particular users were selected because such a profile describes the majority of first year medical students in a Greek university that the educational application is targeted to. They are usually between the age of 17 and 19 and usually have only limited computing experience, since the background knowledge required for medical studies does not include advanced computer skills.

These users were given questionnaires concerning their emotional reactions to several situations of computer use in terms of their actions using the keyboard and what they say. Participants were asked to determine what their possible reactions may be when they are at certain emotional states during their interaction. Our aim was to recognise the possible changes in the users' behaviour and then to associate these changes with emotional states like anger, happiness, boredom, etc.

After collecting and processing the information of the empirical study we came up with results that led to the design of the affective module of the educational application. For this purpose, some common positive and negative feelings were identified. Table 1 illustrates the six basic emotions of our study and information we have collected through the audio mode of interaction.

The results of the empirical study were also used for designing the user stereotypes. In our study user stereotypes where built first by categorizing users by their age, their educational level and by their computer knowledge level. The underlying reasoning for this is that people's behaviour while doing something may be affected by several factors concerning their personality, age, experience, etc. For example, experienced computer users may be less frustrated than novice users or older people may have different approaches in interacting with computers, comparing with younger people. Younger computer users are usually more expressive than older users while interacting with an animated agent and we may expect to have more data from audio mode than by the use of a keyboard. The same case is when a user is less experienced in using a computer than a user with a high computer knowledge level. In all these cases stereotypes were used to indicate which specific characteristics in a user's behaviour should be taken to account in order make more accurate assumptions about the users' emotional state.

4 Analysis

According to MBIUI, during analysis, two different experiments are conducted. The first experiment aims at determining the criteria that are used in the reasoning process of the human advisors and the second aims at calculating their weights of importance.

4.1 Determining Multiple Criteria

Decision making theories provide precise mathematical methods for combining criteria in order to make decisions but do not define the criteria. Therefore, in order to locate the criteria that human experts take into account while providing individualised advice, we conducted an empirical study.

The empirical study should involve a satisfactory number of human experts, who will act as the human decision makers and are reviewed about the criteria that they take into account when providing individualised advice. Therefore, in the experiment conducted for the application of the multi-criteria theory in the e-learning system, 16 human experts were selected in order to participate in the empirical study. All the human experts possessed a first and/or higher degree in Computer Science.

The participants of the empirical study were asked which input action from the keyboard and the microphone would help them find out what the e-motions of the users were. From the input actions that appeared in the experiment, only those proposed by the majority of the human experts were selected. In particular considering the keyboard we have: a) user types normally b) user types quickly (speed higher than the usual speed of the particular user) c) user types slowly (speed lower than the usual speed of the particular user) d) user uses the backspace key often e) user hit unrelated keys on the keyboard f) user does not use the keyboard.

Considering the users' basic input actions through the microphone we have 7 cases: a) user speaks using strong language b) users uses exclamations c) user speaks with a high voice volume (higher than the average recorded level) d) user speaks with a low voice volume (low than the average recorded level) e) user speaks in a normal voice volume f) user speaks words from a specific list of words showing an emotion g) user does not say anything.1.3 Copyright Forms

4.2 Determining the Weights of Importance of the Criteria

During requirements capture the main categories of the users and emotions were identified. As a result the main stereotypes were designed. For the design of the body of the stereotype we have used the results of the empirical study described in section 4.1, in we categorized users' input actions in terms of the two modes of the bi-modal system. These actions would indicate possible changes in a user's emotional state while s/he interacted with the system. However, in order to identify the weights of the criteria (input action) another experimental study was conducted.

More specifically, 50 medical students were asked to use Edu-Affe-Mickey, which incorporated a user modelling component. The user modelling component recorded all users' actions as a filter between the user and the main educational application. Then these actions were classified by the six and seven basic input actions in regard to the keyboard and the microphone respectively.

The results of the empirical study were collected and analyzed. The analysis revealed how important each input action is in identifying each emotion. Therefore, the weight of each criterion (input action) for all emotions were identified and the default assumptions of the stereotypes were designed.

Tables 2 and 3 illustrate the values of the weights for two opposite (positive/ negative) emotions, namely the emotion of happiness and the emotion of anger. Variables k1 to k6 refer to the weights of the six basic input actions from the keyboard, while variables m1 to m7 refer to the weights of the seven possible input cases concerning interaction through the microphone. We may also note that for each emotion and for each mode the values of the weights have sum that equals to 1.

Table 2. Values for the stereotypic weights for the emotions of happiness and anger concerning input from the keyboard

Using the keyboard			
Emotion of happiness		Emotion of anger	
input action	weight	input action	weight
k1	0,4	k1	0,11
k2	0,4	k2	0,14
k3	0,1	k3	0,18
k4	0,05	k4	0,2
k5	0,05	k5	0,25
k6	0	k6	0,12

Table 3. Values for the stereotypic weights for the emotions of happiness and anger concerning input from the microphone

Using the microphone			
Emotion of happiness		Emotion of anger	
input action	weight	input action	weight
M1	0,06	m1	0,19
M2	0,18	m2	0,09
M3	0,15	m3	0,12
M4	0,02	m4	0,05
M5	0,14	m5	0,12
M6	0,3	m6	0,27
M7	0,15	m7	0,16

5   Design

In MBIUI the design of the running application is mainly concerned with the design of the user model and is divided into two major parts with respect to the application of a multi-criteria decision making theory: 1) design decisions about how the values of the criteria are estimated based on the information of the user model and 2) design of the embedment of the actual multi-criteria theory that has been selected into the system.

The input actions that were identified by the human experts during the first experimental study of analysis provided information for the actions that affect the emotional states that may occur while a user interacts with an educational system. These input actions are considered as criteria for evaluating all different emotions and selecting the one that seems more prevailing. More specifically, each emotion is evaluated first using only the criteria (input actions) from the keyboard and then only the criteria (input actions) from the microphone. For the evaluation of each alternative emotion the system uses SAW for a particular category of users. According to SAW, the multi-criteria utility function for each emotion in each mode is estimated as a linear combination of the values of the criteria that correspond to that mode.

In view of the above, for the evaluation of each emotion taking into account the information provided by the keyboard is done using formula 1.

$$em_{1e_11} = w_{1e_1k1}k_1 + w_{1e_1k2}k_2 + w_{1e_1k3}k_3 + w_{1e_1k4}k_4 + w_{1e_1k5}k_5 + w_{1e_1k6}k_6 \qquad (1)$$

Similarly, for the evaluation of each emotion taking into account the information provided by the other mode (microphone) is done using formula 2.

$$em_{1e_12} = w_{1e_1m1}m_1 + w_{1e_1m2}m_2 + w_{1e_1m3}m_3 + w_{1e_1m4}m + w_{1e_1m5}m_5 + w_{1e_1m6}m_6 + w_{1e_1m7}m_7 \qquad (2)$$

em_{1e_11} is the probability that an emotion has occurred based on the keyboard actions and em_{1e_12} is the probability that refers to an emotional state using the users' input from the microphone. em_{1e_11} and em_{1e_12} take their values in [0,1].

In formula 1 the k's from $k1$ to $k6$ refer to the six basic input actions that correspond to the keyboard. In formula 2 the m's from $m1$ to $m7$ refer to the seven basic input actions that correspond to the microphone. These variables are Boolean. In each moment the system takes data from the bi-modal interface and translates them in terms of keyboard and microphone actions. If an action has occurred the corresponding criterion takes the value 1, otherwise its value is set to 0. The w's represent the weights. These weights correspond to a specific emotion and to a specific input action and are acquired by the stereotype database. More specifically, the weights are acquired by the stereotypes about the emotions.

In cases where both modals (keyboard and microphone) indicate the same emotion then the probability that this emotion has occurred increases significantly. Otherwise, the mean of the values that have occurred by the evaluation of each emotion using formulae 1 and 2 is calculated.

$$\frac{em_{1e_11} + em_{1e_12}}{2}$$

The system compares the values from all the different emotions and determines whether an emotion is taking effect during the interaction.

As an example we give the two formulae with their weights for the two modes of interaction that correspond to the emotion of happiness when a user (under the age of 19 and novice with respect to his/her computer skills) gives the correct answer in a test of our educational application. In case of em_{1e_11} considering the keyboard we have:

$$em_{1e_11} = 0.4k_1 + 0.4k_2 + 0.1k_3 + 0.05k_4 + 0.05k_5 + 0k_6$$

In this formula, which corresponds to the emotion of happiness, we can observe that the highest weight value correspond to the normal and quickly way of typing. Slow typing, 'often use of the backspace key' and 'use of unrelated keys' are actions with lower values of stereotypic weights. Absence of typing is unlikely to take place. Concerning the second mode (microphone) we have:

$$em_{1e_12} = 0.06m_1 + 0.18m_2 + 0.15m_3 + 0.02m_4 + 0.14m_5 + 0.3m_6 + 0.15m_7$$

In the second formula, which also corresponds to the emotion of happiness, we can see that the highest weight corresponds to $m6$, which refers to the 'speaking of a word from a specific list of words showing an emotion' action. The empirical study gave us strong evidence for a specific list of words. In the case of words that express happiness, these words are more likely to occur in a situation where a novice young user gives a correct answer to the system. Quite high are also the weights for variables $m2$ and $m3$ that correspond to the use of exclamations by the user and to the raising of the user's voice volume. In our example the user may do something orally or by using the keyboard or by a combination of the two modes. The absence or presence of an action in both modes will give the Boolean values to the variables $k1...k6$ and $m1...m7$.

A possible situation where a user would use both the keyboard and the microphone could be the following: The specific user knows the correct answer and types in a speed higher than the normal speed of writing. The system confirms that the answer is correct and the user says a word like 'bravo' that is included in the specific list of the system for the emotion of happiness. The user also speaks in a higher voice volume. In that case the variables k1, m3 and m6 take the value 1 and all the others are zeroed. The above formulas then give us $em_{1e_11} = 0.4*1 = 0.4$ and $em_{1e_12} = 0.15*1 + 0.3*1 = 0.45$.

In the same way the system then calculates the corresponding values for all the other emotions using other formulae. For each basic action in the educational application and for each emotion the corresponding formula have different weights deriving from the stereotypical analysis of the empirical study. In our example in the final comparison of the values for the six basic emotions the system will accept the emotion of happiness as the most probable to occur.

6 Implementation

During implementation, in the elaboration phase the overall functionality and emotion recognition features of Edu-Affe-Mikey are implemented. The architecture of Edu-Affe-Mikey consists of the main educational application with the presentation of theory and tests, a programmable human-like animated agent, a monitoring user modeling component and a database.

While using the educational application from a desktop computer, students are being taught a particular medical course. The information is given in text form while at the same time the animated agent reads it out loud using a speech engine. The student can choose a specific part of the human body and all the available information is retrieved from the systems' database. In particular, the main application is installed either on a public computer where all students have access, or alternatively each student may have a copy on his/her own personal computer. An example of using the main application is illustrated in figure 2. The animated agent is present in these modes to make the interaction more human-like.

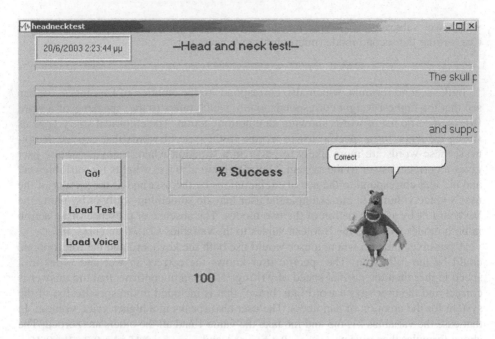

Fig. 2. A screen-shot of taking a test in Edu-Affe-Mikey educational application

While the users interact with the main educational application and for the needs of emotion recognition a monitoring component records the actions of users from the keyboard and the microphone. These actions are then processed in conjunction with the multi-criteria model and interpreted in terms of emotions. The basic function of the monitoring component is to capture all the data inserted by the user either orally or by using the keyboard and the mouse of the computer. The data is recorded to a database and the results are returned to the basic application the user interacts with.

Instructors have also the ability to manipulate the agents' behaviour with regard to the agents' on screen movements and gestures, as well as speech attributes such as speed, volume and pitch. Instructors may programmatically interfere to the agent's behaviour and the agent's reactions regarding the agents' approval or disapproval of a user's specific actions. This adaptation aims at enhancing the "affectiveness" of the whole interaction. Therefore, the system is enriched with an agent capable to express emotions and, as a result, enforces the user's temper to interact with more noticeable evidence in his/her behaviour.

7   Testing

In construction phase, during the procedural step of testing, the final version of the system is evaluated. When a user interface incorporates a decision making theory, the evaluation phase plays an important role for showing whether the particular theory is effective or not. In MBIUI life-cycle framework it is considered important to conduct the evaluation of a decision making model by comparing the IUI's reasoning with that of real users. Therefore, in this experiment it is important to evaluate how successful the application of the decision making model is in selecting the alternative action that the human experts would propose in the case of a user's error. For this reason, it has to be checked whether the alternative actions that are proposed by the human experts are also highly ranked by the application of the decision making model. In case this comparison reveals that the decision making model is not adequate, another iteration of the life-cycle has to take place and another decision model should be selected. This iteration continues until the evaluation phase gives satisfactory results.

In view of the above, an evaluation study was conducted. Therefore, the 50 medical students that were involved in the empirical study during requirements capture were also involved in the evaluation of the multi-criteria emotion recognition system. More specifically, they were asked to interact with the educational software and the whole interaction was video recorded. The protocols collected were presented to the same users in order to perform emotion recognition for themselves with regard to the six emotional states, namely happiness, sadness, surprise, anger, disgust and the neutral emotional state.

The students as observers were asked to justify the recognition of an emotion by indicating the criteria that s/he had used in terms of the audio mode and keyboard actions. Whenever a participant recognized an emotional state, the emotion was marked and stored as data in the system's database. Finally, after the completion of the empirical study, the data were compared with the systems' corresponding hypothesis in each case an emotion was detected. Table 4 illustrates the percentages of successful emotion recognition of each mode after the incorporation of stereotypic weights and the combination through the multi-criteria approach.

Table 4. Recognition of emotions using stereotypes and SAW theory

Using Stereotypes and SAW			
Emotions	Audio mode recognition	Recognition through key-board	Multi-criteria bi-modal recognition
Neutral	17%	32%	46%
Happiness	52%	39%	64%
Sadness	65%	34%	70%
Surprise	44%	8%	45%
Anger	68%	42%	70%
Disgust	61%	12%	58%

Provided the correct emotions for each situation identified by the user himself/herself we were able to come up with conclusions about the efficiency of our systems' emotion recognition ability. However, one may notice that there are a few cases where the proposed approach had a bit worse performance in recognizing an emotional state (e.g. neutral emotional state). Possible reason is the fact that some emotional states, e.g. neutral emotional state, give little evidence to certain modes (e.g. the keyboard mode). For the same emotions, other modalities (e.g. the visual mode) may give us significantly better evidence. However, the combination of the two modes in the multi-criteria model increases significantly the accuracy of the proposed approach.

Although, success rates may look at a first glance lower than expected, we should underline the fact that emotion recognition is a very difficult task for human and their success rates are also quite low. Therefore, the results of the evaluation study offer evidence for the adequacy of the multi-criteria multi-modal model for emotion recognition.

8 Conclusions

In this paper, we give evidence for the effectiveness of a general framework for applying a simple multi-criteria decision making theory in an adaptive bi-modal user interface. Indeed, we have described and used MBIUI life-cycle framework, which is a general framework that provides detailed guidelines for the application of a multi-criteria decision making theory in an IUI. More specifically, the MBIUI life-cycle framework facilitates the application of the multi-criteria decision making theory by providing detailed guidelines for all experimental studies during requirements capture and analysis as well as testing. This framework was initially designed for incorporating a decision making theory in a user interface that helps users during their interaction with a file-store system.

MBIUI life-cycle framework has been used for developing an affective bi-modal user interface that aims at providing medical education to first-year medical students and is called Edu-Affe-Mikey. In this system, the multi-criteria decision making theory, SAW, is used for combining evidence from two different modes in order to identify the users' emotions. SAW is used for evaluating different emotions, taking into account the input of the two different modes and selecting the one that seems more likely to have been felt by the user. The particular user interface offers affective bi-modal interaction and for this reason differentiates from the other user interfaces. The fact that the particular framework can be used in interfaces that differ in many ways strengthens MBIUI's generality.

The results of the evaluation of the affective bi-modal user interface prove the effectiveness of the multi-criteria decision making theory for combining evidence from two different modes and perform emotion recognition.

Acknowledgements. Support for this work was provided by the General Secretariat of Research and Technology, Greece, under the auspices of the PENED-2003 program.

References

1. Hull, M.E.C., Taylor, P.S., Hanna, J.R.P., Millar, R.J.: Software development processes-an assessment. Information and Software Technology 44, 1–12 (2002)
2. Kabassi, K., Virvou, M.: A Knowledge-based Software Life-Cycle Framework for the incorporation of Multi-Criteria Analysis in Intelligent User Interfaces. IEEE Transactions on Knowledge and Data Engineering 18(9), 1–13 (2006)

3. Fishburn, P.C.: Additive Utilities with Incomplete Product Set: Applications to Priorities and Assignments, Operations Research (1967)
4. Hwang, C.L., Yoon, K.: Multiple Attribute Decision Making: Methods and Applications. Lecture Notes in Economics and Mathematical Systems, vol. 186 (1981)
5. Naumann, F.: Data Fusion and Data Quality. In: Proceedings of the New Techniques and Technologies for Statistics (1998)
6. Schütz, W., Schäfer, R.: Bayesian networks for estimating the user's interests in the context of a configuration task. In: Schäfer, R., Müller, M.E., Macskassy, S.A. (eds.) Proceedings of the UM 2001 Workshop on Machine Learning for User Modeling, pp. 23–36 (2001)
7. Bohnenberger, T., Jacobs, O., Jameson, A., Aslan, I.: Decision-Theoretic Planning Meets User Requirements: Enhancements and Studies of an Intelligent Shopping Guide. In: Gellersen, H.-W., Want, R., Schmidt, A. (eds.) PERVASIVE 2005. LNCS, vol. 3468, pp. 279–296. Springer, Heidelberg (2005)
8. Kudenko, D., Bauer, M., Dengler, D.: Group Decision Making Through Mediated Discussions. In: Proceedings of the 9th International Conference on User Modelling (2003)
9. Rich, E.: Stereotypes and User Modeling. In: Kobsa, A., Wahlster, W. (eds.) User Models in Dialog Systems, pp. 199–214 (1989)
10. Rich, E.: Users are individuals: individualizing user models. International Journal of Human-Computer Studies 51, 323–338 (1999)
11. Jacobson, I., Booch, G., Rumbaugh, J.: The Unified Software Development Process. Addison-Wesley, Reading (1999)
12. Kass, R., Finin, T.: The role of User Models in Cooperative Interactive Systems. International Journal of Intelligent Systems 4, 81–112 (1989)

Developer Stories: Improving Architecture in Agile Practice

How Facilitating Knowledge Management and Putting the Customer in the Drivers Seat Enables Sound Architectural Design

Rolf Njor Jensen, Niels Platz, and Gitte Tjørnehøj

Department of Computer Science, Aalborg University
Selma Lagerlöffsvej 300a, 9220 Aalborg Ø, Denmark
{rolf,platz}@agil.dk, gtj@cs.aau.dk

Abstract. Within the field of Software Engineering emergence of agile methods has been a hot topic since the late 90s. eXtreme Programming (XP) ([1]) was one of the first agile methods and is one of the most well-known. However research has pointed to weaknesses in XP regarding supporting development of viable architectures. To strengthen XP in this regard a new practice: Developer Stories ([2]) was introduced in 2006 – mainly based on a theoretical argumentation.

This paper reports from extensive experimentation with, and elaboration of the new practice. Results from this experimentation shows that using Developer Stories increases the likelihood of developing a viable architecture through a series of deliberate choices, through creating disciplined and recurring activities that:
1) Facilitate sharing and embodying of knowledge about architectural issues, and
2) heighten visibility of refactorings for both customers and developers.

1 Introduction

Agile methods has been a hot topic since the late 90's ([3]), shifting system developments focus from processes and tools to individuals and interactions, from documentation to working software, from contract negotiation to customer collaboration, from following a plan to responding to change. eXtreme Programming ([1,4,5]) was one of the first methods within this paradigm and has become very popular within both research and industry. However XP has also been critiqued for not supporting the development of a viable architecture and hereby jeopardizing the quality of the developed systems.

What is design and what is architecture? No clear distinction can be made between the two in general, and in the case of XP, the two terms are frequently overlapping - i.e. both in terms of intention and scope. Architectural issues are dealt with through Enough Design Up Front combined with constantly designing and refactoring entailing that the architecture keeps improving but only on demand. The prevalent principle presented by XP to guide development of the architecture is "Simplicity", subject to the criteria: "Appropriate for the intended audience", "Communicative", "Factored" and "Minimal".

J. Filipe et al. (Eds.): ICSOFT/ENASE 2007, CCIS 22, pp. 172–184, 2008.

Does XP deliver a viable architecture? Embracing the values of XP, and following the practices in a disciplined manner in accordance with the principles, will lead to a sufficient and viable architecture ([5]).

However, discourses in literature and conference proceedings show that supporting development of a viable architecture still is a subject in XP. Some examples of the discourses are:

- The "Metaphor" practice that in the first book on XP ([1]) explicitly guided architecture, points to the need of focusing on a shared understanding of the system. The practice was found difficult to operationalize ([6]) and was therefore removed ([5]), but soon it was missed ([7,8]) and hints for operationalization was suggested.
- Several have proposed an introduction of some kind of requirements management into XP ([9,10]) when quality of the system under development is a concern.
- Reports of problems trying to integrate large-scale refactorings into the everyday work of XP projects ([11]). The quote "... aggressive refactoring probably will remain the most difficult, because it requires consistency, energy, and courage, and no mechanisms in the methodology reinforce it." ([12]), brings the deficiency of XP into focus, offering an explanation of the many attempts at amending XP.

Developer Stories – a new agile practice – was introduced to address the deficiency of XP concerning architecture and design ([2]). The new practice was inspired by the explicit focus on architecture in more traditional methods, but designed to fit perfectly in tune with the practices, principles and values of XP. Developer Stories manage non-functional and other architectural requirements for systems - in parallel with the feature-oriented user stories (effectively addressing functional requirements). The proposed practice was argued through a literature-based analysis ([2]).

Investigating how Developer Stories works in a concrete development project is the subject of this paper. Analyzing Developer Stories, we propose a hypothesis: Developer Stories essentially contributes to the development process by two means: 1) facilitating sharing and embodiment of knowledge concerning architectural issues, and 2) improving visibility of architectural changes – thereby giving the customer more leverage to steer the development process. This paper presents the results of an experiment with Developer Stories, which sets out to validate this hypothesis, providing evidence that these two means are in fact present. The experiment is designed as a combination of a field- and laboratory experiment, and evolves around an XP project of 3 iterations with 6 developers, a coach and an on-site customer. Concluding upon the experiment, we find that our hypothesis is true, and that Developer Stories improve knowledge sharing and heighten visibility by creating reoccurring, disciplined activities for exploring and processing possible architectural issues.

The remainder of this paper is organized as follows: Section 2 presents the new practice Developer Stories in some detail (more arguments can be seen in ([2])). Section 3 presents the experiment settings, data collection and analysis, while section 4 lists the findings from the experiment. Results are summarized in section 5, and then discussed in section 6. Finally we draw our conclusions in section 7, and hint possible future work in section 8.

2   Developer Stories

In the following we present in some detail the practice Developer Stories based on the original description of Developer Stories ([2]). For practical purposes some aspects were clarified before the experiment, but in all important aspects the practice has remained as the former proposition.

The overall goal of Developer Stories is to provide the development team with the opportunity and means to improve the architecture of systems in development, accomplishing greater business value, and a viable architecture.

The Developer Story practice is designed to fit into the synergetic mesh of XP's practices. It takes its place in the symbiotical relationships of the practices, relating to i.e. pair programming, incremental design, test-first programming and user stories ([2]), It is designed to both support and be supported by the values of XP, and in general follow the look and feel of XP.

2.1   The Artifact: A Developer Story

Developer Stories as a practice consists of a process – a number of activities, and artifacts. The artifacts are the developer stories, defined as such:

> The developer stories describe (changes to) units of developer-visible properties of the software. In contrast, user stories describe units of user-visible functionality of the software. The physical representation of a developer story is an index card, which may have another color than user stories, making it easy to distinguish them from each other. ([2])

In essence, a developer story is much like a user story - but where a user story describes features of the system and is written by the user, a developer story describes changes to the system that are often not visible to the user, but highly visible to the developers – and is therefore written by the developers.

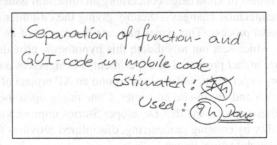

Fig. 1. Example of a developer story

Figure 1 depicts a developer story from the conducted experiment. The layout of a developer story is an index card, and the form of the content is not bound by any formalism. The extent of the content is at the leisure of the developers. Whether it should contain a description of the problem, solution, tests, tasks and estimate, or whether it (as is the case of our example from the experiment, Figure 1) only needs two lines of description.

2.2   The Process

The activities that constitute the rest of the practice is interwoven into the other activities of development work, especially the discovery of potential developer stories and implementation of the chosen stories. However the exploration of the architectural issues, the writing of the stories and the costumer choosing which to implement is organized as an autonomous activity; The Architectural Game.

Acknowledging Architectural Issues.   Integrated in the day-to-day activities of an iteration the developers acknowledge architectural issues that they experience. They either make a mental note, or preserve some artifact – a drawing on a flip-chart, a sticky-note or otherwise. Acknowledgement happens during pair-programming, shared discussions, lunch-breaks, etc., and serve as inspiration or reminders at the following Architectural Game.

The Architectural Game.   The Architectural Game (see Figure 2) is a two-hour meeting that occurs once every iteration just before or after the weekly planning meeting ([5]) (formerly known as the planning game ([4])) and may well be performed as a stand-up activity. The purpose of the game is to explicate and visualize refactoring needs as stories.

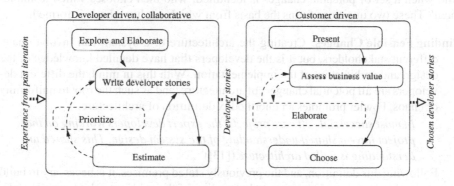

Fig. 2. The Architectural Game

As shown in Figure 2 the game is based on experiences (including the notes and artifacts mentioned above) from the past iteration(s). First the developers collaboratively explore and elaborate architectural issues and eventually write developer stories expressing refactoring needs. Then they estimate the stories and may prioritize them. The customer is optimally present at this part of the Architectural Game as bystander. The participants shift rapidly, fluent and imperceptibly between activities.

When the developers have expressed the refactoring needs that they have found in terms of developer stories, the stories are presented to the customer to choose from. He is responsible for gaining insight in the developer stories, so that he can assess the business value of the stories, based on his detailed business knowledge. This might require more elaboration or further explanations in non-technical terms by the developers.

Implementing Developer Stories. The chosen developer stories are handled similarly to the chosen user stories during the following iteration, e.g. developer stories are implemented and tested analogously to user stories. New unit tests are added and existing unit tests modified. Just like acceptance tests for user stories, acceptance tests for developer stories are written prior to implementation by the customer, though supported by the developers – to uphold the double checking principle ([5]).

3 Experiment

The overall aim of the performed experiment was to gain insight into the workings of Developer Stories, to understand how and why they possibly accommodate architectural improvement.

3.1 Hypothesis

We regard the architecture of a system, as something that at any given time can change in several different ways. Choosing which changes to implement, is a matter of assessing which changes (if any) that will provide the largest increase in the value of the system. But how do we determine which changes are possible and which are feasible? And when a set of potential changes is identified, who then chooses which to implement? These two questions forms the basis from which our hypothesis is formed.

Finding Feasible Changes. Creating the architecture of a system can involve many different stakeholders, but it is the developers that have detailed knowledge of the design and architecture of the implementation. With this in mind, the different developers are all potential changers of architecture, and as such the root to uniformity or chaos. Fowler promotes an unorthodox definition of architecture:

> *In most successful software projects, the expert developers working on that project have a shared understanding of the system design. This shared understanding is called architecture.*([13])

Following this definition, and the previously stated premises, it is necessary to build and maintain a shared, uniform understanding of the existing architecture, in order to identify which changes to the architecture may feasibly be implemented. Hereby it becomes imperative that any practice aiming to improve architecture facilitates knowledge sharing and embodiment.

Choosing Which Changes to Implement. While the developers (may) have an understanding of the architecture and possible changes, they do not posses all knowledge required to determine the change of value of the system entailed by effectuating a change. The missing knowledge is possessed by other stakeholders, which in XP are represented by the on-site customer. The required knowledge is i.a. the context in which the system is to be deployed, the context to which the system is developed, etc. If this knowledge is to be employed, the customer must be in a position to execute influence on which changes are implemented, and it follows that a practice aiming to improve the architecture must provide the customer with visibility of possible architectural changes.

In short, our hypothesis is that Developer Stories contribute to the development process with knowledge sharing and embodiment of architectural issues among the developers, and gives the customer visibility of the possible architectural changes.

3.2 Experiment Setting

The experiment was a full scale XP-development project integrated with the Developer Story practice. It was conducted in a laboratory but striving to share as many aspects as possible with a field experiment ([14]).

The field aspects were: 1) The laboratory was physically set up as a real XP room. 2) 11 out of 13 primary practices, and some secondary practices were employed in the software development process. 3) The development task was an industry request and the on-site customer was employed at the requesting company. 4) There were six developers, which is a normal size project. 5) We followed three one-week iterations, which is adequate to embody both ordinary as well as the new practice of XP and thus to study the effect. 6) The implemented software was real-world: a rather complex client- and server system applying new technology used to scan barcodes on parcels in a freight registration system for a truck fleet operator. 7) After the experiment the IT-system was actually implemented at the requesting firm.

The laboratory aspects of the experiment were: 1) The developers were fifth semester computer science students with some experience in developing systems, but unskilled in XP. 2) The researchers played the role of coach and one on-site customer and the project was conducted at the university – not in an industry setting.

3.3 Preparing the Experiment

Preparation for the experiment was done through three activities: Teaching the to-be XP developers XP, arranging for a laboratory in which to conduct development, and finally giving the developers insight into the product domain.

The developers were taught XP from scratch during a 20 hour course over a period of 5 days starting with and emphasizing the values of XP, later adding the principles, practices, and other features of XP.

The laboratory was set up as sketched in Figure 3 with one end occupied by three pair programming stations facing each other and the other end empty except from the information radiators – blackboards, bulletin boards (storywall, etc.), and a flip chart. This area also served as stand-up area. Finally a chair for the on-site customer was added. Compared to the recommended setup ([5]), denoted as "caves and common" ([12]) there is no private space for the developers to do solitary work. The need for caves was met by another room just down the corridor in which the developers also worked off-project hours.

3.4 Executing the Experiment

The experiment was conducted from mid-October to end-November. In total, three one-week iterations where completed. Development time was scattered throughout this period, concentrating more and more towards the end. The customer in cooperation with

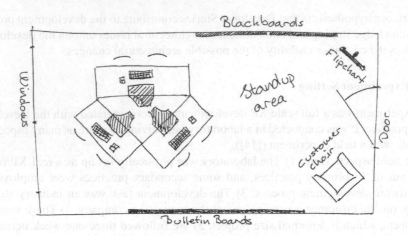

Fig. 3. Configuration of the development room

the researchers wrote the user stories for the project. The developers were enthusiastic about both XP and the product. The first iteration was a learning iteration with heavy involvement of the coach, but already in the second iteration the team, to a large degree, worked independently of the coach. The project was halted after the third iteration as planned only because the students had to refocus on other parts of their study. The product was however ready to be implemented and used by the requesting firm.

3.5 Collecting Data

We used four main sources for collecting data.

- A research log kept daily by the researchers focused on the themes from the hypothesis of the experiment, including observations and reflections of the ongoing development process.
- A questionnaire, that each developer answered three times per iteration (start, middle, end), 41 questionnaires in total. The questionnaire provided evidence on: A) How much Developer Stories relatively to the other practices contributed to each developers knowledge of the system, and B) Whether the systems architecture was perceived similarly by the developers, and how this perception evolved.
- The Architectural game was video taped along with the weekly meetings.
- As the development team employed a source revision control system, the source code at different revisions constituted the final data source.

The researchers were present during all work hours, taking actively part in the project, which constitutes a source of rich but informal data.

3.6 Analyzing Data

We analyzed data looking for support of the hypothesis (knowledge sharing and embodiment, and visibility of refactorings).

To provide a historical reference to the remainder of the analysis, we created a time-line using the logbook, and the artifacts produced throughout the development process (story index cards, flip charts, etc.).

We then reviewed the logbook and video recordings, mapping evidence to the main themes of the hypothesis; interactions between developers (sharing and embodiment of knowledge), interactions between developers and the customer (visibility of refactor-ings), and signs of architectural change (Developer Stories).

On the flip side of the questionnaire, the developers depicted their current perception of the architecture. We used these drawings to analyze how the perceptions evolved over time and how they converged or diverged between developers.

We analyzed the developers own subjective and relative rating of the contribution to their understanding of the system from Developer Stories, compared to other practices.

We even analyzed the different builds of the system with a code analysis tool (Struc-ture101 ([15])) to see if refactorings due to developer stories had effect on the actual architecture.

The data analysis was iterative going through several of the data sources more than once in the light of the findings from the other sources.

4   Findings

Presented below is the most significant findings grouped into four categories – three re-lating to different parts of the Developer Stories process (acknowledging architectural issues, the Architectural Game and implementing Developer Stories), and one with find-ings applicable to all parts.

4.1   Acknowledging Architectural Issues

We found that knowledge of implementation challenges, design problems and refactor-ings was shared through osmotic communication ([12]), and the use of the flip chart and blackboards. Typically a discussion about these architectural issues would evolve rapidly from involving one pair to engaging one or more persons from the other pairs, and at some point two or three people would move to the stand-up area and employ either the blackboard or the flip chart.

Considering the whole span of development, the discussions were spread evenly throughout the iterations, and all developers were found to engage actively in this shar-ing of knowledge.

The on-site customer, being present in the room, observed the discussions about architectural issues when they took place in the stand-up area, and occasionally was able to contribute some knowledge.

There were instances of the customer instigating a discussion of architectural issues among the developers, when some functionality was demoed to the customer, and the customer asked for a change in e.g. user interface design, functionality or otherwise.

4.2   The Architectural Game

The developers all engaged actively in discussions about architectural issues experi-enced in the previous iteration.

The flip charts and other artifacts that were created during the iteration leading up to a particular architectural game were used by the developers as reminders of the experienced architectural issues.

The first architectural game was guided by the coach, and we observed how the developers quickly and actively engaged in the process, effectively acting as a self-organizing team.

Video recordings from the architectural games showed a noticeably quick creation of a common and high level of understanding of the current system during the review of the flip chart. The developers obtained this understanding by actively seeking knowledge from each other about aspects of the system that were currently unclear to them.

Frequently we also observed developers feed the discussion with personal knowledge about i.a. a design pattern that had not been previously considered.

We observed the developers changing rapidly between different levels of abstraction – ranging from overall architectural patterns to implementation details.

We observed the customer choosing between the presented developer stories. When he was not immediately able to assess the business value, he challenged the developers, who where able to present the developer stories in terms that the customer could translate into business value.

4.3 Implementing Developer Stories

We found the knowledge of an architectural change happening due to a developer story spread very quickly among the developers. The black markings in Figure 4 shows how the knowledge of a particular architectural pattern described in a developer story spreads among the developers. On the 22nd of November two developers had acknowledged and depicted the architectural change. The 28th of November (one working day later), five developers had acknowledged the change. While the developers rotated in pairs frequently, not all the developers that acknowledged the architectural change had actually worked on that particular piece of code.

We found several situations where the customer was involved in the implementation of developer stories, just like user stories, either through discussions or direct questions concerning uncertain matters.

We analyzed a particular subset of the developed code in two different versions, and found a significant reduction of coupling and an enhancement of cohesion as an effect of a major refactoring in the third iteration. We see this as a sign of improvement of the architecture, according to the definition of architecture by the traditional quality criteria ([2]). The cause of the refactoring was clearly traceable to the implementation of a specific developer story.

4.4 Generally Applicable Findings

From the questionnaires, we found that the developers themselves subjectively rated the Developer Stories practice relatively higher than other XP-practices, as a mean to increase their personal knowledge of the system, and more so more and more over the iterations.

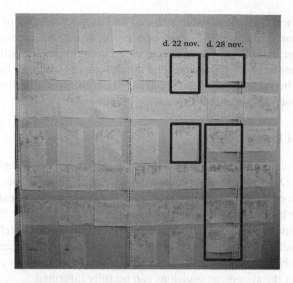

d. 22 nov.   d. 28 nov.

Fig. 4. Drawings of the system architecture hung on a wall. Each row corresponds to a developer, time increasing from left to right.

5   Results

Reflecting upon our findings, the following section summarizes the results of the experiment with regards to the hypothesis: "That Developer Stories contribute to the development process with knowledge sharing and embodiment of architectural issues among the developers, and gives the customer visibility of the possible architectural changes".

5.1   Knowledge Sharing and Embodiment

There is widespread evidence that the practice Developer Stories did heighten the knowledge sharing and embodiment throughout the whole project by providing occasion and tools for supporting and informally documenting the discussions. Even the developers themselves considered Developer Stories as a practice that contributed to their knowledge of the system, to a larger extent than other XP practices.

The knowledge sharing and embodiment processes was initiated during the iteration leading up to the architectural games. The process occurring during the iteration was characterized by being problem oriented, and not necessarily involving all developers. When the process continued in the architectural game, focus gradually shifted to become more solution oriented, and being communal, involving all developers. The rapid shifts in levels of abstraction throughout the discussion indicates that sharing and embodiment of knowledge was actually achieved ([16]).

When changing the architecture by implementing a developer story, the developers acknowledged the new architecture faster than otherwise expected. This indicates that the developers effectively embodied knowledge about architectural development that occurred due to developer stories.

The Architectural Game resulted in a shared and deep knowledge of the constituent parts of the system. Moreover, the developers perception of the architecture was uniform. Keeping the definition of architecture as a common understanding of the most important elements in mind, this means that the developers where in fact able to communicate about the same architecture, enabling them to make a uniform, disciplined and deliberate change of the architecture.

5.2   Visibility of Refactorings

The customer participated, through observation, instigation, control and sparring in all three parts of the Developer Stories process. This working pattern gave the customer an influence on both choice and implementation of developer stories equal to that of user stories, and some insight into the creation process.

The participation provided the customer with the insight needed for choosing what to include and to exclude (e.g. to avoid gold-plating) in the different iterations. This way the visibility combined with the customers knowledge of the systems context enhanced the architecture of the system, as decisions can be fully informed.

The act of choosing which developer stories to implement gave the customer a very tangible lever with which to control the direction of development in architectural matters.

As a positive side effect, the developers gained knowledge of the rationale on which the choice of developer stories was made, rendering them more empathetic towards the choices.

5.3   Effect of Developer Stories

When implementing developer stories, the architecture was evidently improved, but while we do conclude that the improvement in architectural quality we found, was due to implementing a developer story, we are not able to state for certain that the same refactoring had not happened in some other course of development.

However, based on this finding and on our general belief that communication raises the possibility of informed and thus wiser decisions, we do speculate that implementing the Developer Stories practice is likely to heighten quality of the architecture early in a development process by providing occasions for disciplined elaboration of the architecture.

6   Discussion

Our results from the experiment showed that the suggested practice Developer Stories had effect on how developers worked with architectural issues of the system and also on the architecture of the system itself.

We found that developer stories helped developer and customer to focus on needed non-functional or other architectural requirements in a consistent (reoccurring once every iteration), disciplined and structured (in terms of developer stories) manner balanced with the rest of the practices in XP. Based on this we argue that Developer Stories can strengthen XP, effectively accommodating the critique of the aggressive refactoring ([12]) and providing a more disciplined approach to design activities as suggested ([17]) while not jeopardizing the symbioses of the existing practices of XP.

From the suggested practice follows a high degree of visibility of needed refactoring tasks, of their value for the customer, of their cost (estimated) and of the customers eventual choice of stories. This visibility secures highly informed decisions on functionality and architecture and is thus likely to add greatly to the overall value of the system. Being continuous it could play some role of bounded requirements management, since all stories are known to the decision maker and he in turn is actively consulting and consulted by the developers.

We also found that Developer Stories affected knowledge sharing amongst the developer group and with the customer as well, due to his presence through the Architectural Game. Building and maintaining a uniform perception of the architecture is a prerequisite for making meaningful and consistent architectural changes. We also think this practice through supporting building of shared abstract common understanding of the system can add what was lost when the Metaphor practice was excluded from XP.

We can however not conclude anything regarding the actual quality of the architecture according to the classic criteria ([2]). Securing quality is not guaranteed by following the practice as such, but the practice makes it possible and maybe even more likely.

A consideration regarding the experiment setup is, that the on-site customer was technically knowledgeable. As such, we do acknowledge that communicating developer stories in terms that the customer may translate to business value is a major challenge for the involved parties. Considering the course of events during the experiment, we do however believe that this communication can be mastered by both developers and customer after a period of practice.

It should be recognized that a control group might have been employed to compare the performance of Developer Stories. Such a control group was not however within the scope of the research project. We do still consider the results of the experiment valid, due to the fact that we do only conclude upon the presence of knowledge sharing and embodiment, and visibility, not upon general positive or negative affects on the architecture when using Developer Stories.

7 Conclusions

We have experimented with integrating the practice of Developer Stories into XP to investigate the effect. The experiment was a full-scale XP development project of 3 iterations, 6 developers, one on-site customer and a coach, working on at real industry request. The product is now being implemented in the requesting firm. From the experiment we found that the practice Developer Stories contributes to the development process by: 1) Heightening visibility of refactorings for both customers and Developers, 2) facilitating sharing and embodying of knowledge about architectural issues and 3) creating occasions for a disciplined elaboration of the architecture. We speculate that by this contribution, Developer Stories give development teams better means for achieving viable architectures in developed systems.

8   Future Work

Investigating how Developer Stories may be used in agile development methodologies other than XP is an open question, and subject to ongoing work.

Gaining further insight into the effect of Developer Stories may be carried out by using Developer Stories in real-world projects. Experiences from any such activity would be very interesting, and may well be reported communally on www.developerstories.org

References

1. Beck, K.: Embracing change with extreme programming. Computer 32(10), 70–77 (1999)
2. Jensen, R.N., Møller, T., Sönder, P., Tjørnehøj, G.: Architecture and Design in eXtreme Programming; introducing Developer Stories. In: Abrahamsson, P., Marchesi, M., Succi, G. (eds.) XP 2006. LNCS, vol. 4044, pp. 133–142. Springer, Heidelberg (2006)
3. Beck, K., Beedle, M., van Bennekum, A., Cockburn, A., Cunningham, W., Fowler, M., Grenning, J., Highsmith, J., Hunt, A., Jeffries, R., Kern, J., Marick, B., Martin, R.C., Mellor, S., Schwaber, K., Sutherland, J., Thomas, D.: Manifesto for agile software development (2001), http://agilemanifesto.org
4. Beck, K.: Extreme Programming Explained: Embrace Change. Addison-Wesley, Reading (2000)
5. Beck, K.: Extreme Programming Explained: Embrace Change, 2nd edn. Addison-Wesley, Reading (2004)
6. Fowler, M.: Design: Reducing coupling. IEEE Software 18(4), 102–104 (2001)
7. Lippert, M., Schmolitzky, A., Züllighoven, H.: Metaphor design spaces. In: Marchesi, M., Succi, G. (eds.) XP 2003. LNCS, vol. 2675, pp. 33–40. Springer, Heidelberg (2003)
8. West, D.D., Solano, M.: Metaphors be with you. In: Agile2005 (July 2005)
9. Eberlein, A., do Prato Leite, J.C.S.: Agile requirements definition: A view from requirements engineering. In: TCRE 2002 (September 2002), http://www.enel.ucalgary.ca/tcre02/
10. Paetsch, F., Eberlein, A., Maurer, F.: Requirements engineering and agile software development. In: Twelfth International Workshop on Enabling Technologies: Infrastructure for Collaborative Enterprises, p. 308 (2003)
11. Lippert, M.: Towards a proper integration of large refactorings in agile software development. In: Eckstein, J., Baumeister, H. (eds.) XP 2004. LNCS, vol. 3092, pp. 113–122. Springer, Heidelberg (2004)
12. Cockburn, A.: Agile Software Development, The Cooperative Game. Pearson Education, Inc., London (2006)
13. Fowler, M.: Who needs an architect? IEEE Software 20(5), 11–13 (2003)
14. Galliers, R.D., Land, F.F.: Viewpoint - choosing appropriate information systems research methodologies. Communications of the ACM 30(11), 900–902 (1987)
15. Software, H.: Structure101 (accessed 29. December 2006), http://www.headwaysoftware.com/products/structure101/
16. Guindon, R.: Designing the design process: Exploiting opportunistic thoughts. Human-Computer Interaction 5, 305–344 (1990)
17. Fowler, M.: Is design dead (2004), http://martinfowler.com/articles/designDead.html

An Ontological SW Architecture Supporting Agile Development of Semantic Portals

Giacomo Bucci, Valeriano Sandrucci, and Enrico Vicario

Dipartimento di Sistemi ed Informatica
Università degli Studi di Firenze, Italy
{bucci,sandrucci,vicario}@dsi.unifi.it

Abstract. Ontological technologies comprise a rich framework of languages and components off the shelf, which devise a paradigm for the organization of SW architectures with high degree of interoperability, maintainability and adaptability. In particular, this fits the needs for the development of semantic web portals, where pages are organized as a generic graph, and navigation is driven by the inherent semantics of contents.

We report on a pattern-oriented executable SW architecture for the construction of portals enabling semantic access, querying, and contribution of conceptual models and concrete elements of information. By relying on the automated configuration of an Object Oriented domain layer, the architecture reduces the creation of a cooperative portal to the definition of an ontological domain model. Application of the proposed architecture is illustrated with reference to the specific case of a portal which aims at enabling cooperation among subjects from different localities and different domains of expertise in the development of a shared knowledge base in the domain of construction practices based on mudbrick.

Keywords: Pattern-Oriented SW Architectures, Semantic Web Technologies, Ontologies, Semantic Portals, SW Engineering.

1 Introduction

Technologies for ontological modelling and reasoning devise a new paradigm for the organization of software architectures with high degree of interoperability, maintainability and adaptability [1], [2].

In fact: ontologies natively allow representation of explicit semantic models combining non-ambiguity of technical specification with the understandability needed to fill the gap between technicians and stakeholders; ontologies enable the application of methods and tools off the shelf supporting reasoning for information search, model validation, and inference and deduction of new knowledge [3]; ontologies naturally fit into a distributed context, enabling creation of reusable models, composition and reconciliation of fragments developed in a concurrent and distributed manner [4], [5]; last, but not least, ontologies can model evolving domains, thus encouraging an incremental approach that can accompany the evolution towards shared models.

In particular, this potential appears well suited for the organization of web portals, where ontologies provide a paradigm to consistently design, manage, and maintain information architecture, site structure, and page layout [6]. This has crucial relevance

J. Filipe et al. (Eds.): ICSOFT/ENASE 2007, CCIS 22, pp. 185–200, 2008.

in the construction of portals with weblike organization [7], where pages are organized in the pattern of a generic graph, and navigation is driven by the inherent semantics of contents more than from a hierarchical upfront classification.

In [8], a semantic portal is presented which supports unified access to a variety of cultural heritage resources classified according to a public unifying ontology. The portal is developed using standard ontological technologies and SWI-prolog to support semantic search and presentation of retrieved data.

In [9], a portal based on ontology technologies is used as a basis to support the contribution of contents by a distributed community. The proposed architecture assumes that contribution is limited to individuals, and the investigation is focused on different techniques supporting the determination of a rank of relevance for the result-set of semantic queries.

In [10], a declarative approach to the construction of semantic portals is proposed, which relies on the definition of a suite of ontologies created by the portal programmer to define domain concepts and contents, navigation structure and presentation style.

In [11], the declarative approach of [10] is enlarged into a framework based on ontologies supporting the construction of a web application combining information and services. The framework implements the Model View Controller architectural pattern [12]. While the model is declared using ontologies, views are implemented with existing presentation technologies, and in particular JSP, which mainly rely on the OO paradigm. To fill the gap between the two paradigms [13], the developer is provided with a suite of predefined JSP components assuming the responsibility of the adaptation.

In this paper, we address the development of a portal enabling semantic access, querying, and contribution of both domain individuals and concepts. To this end, we propose an executable SW architecture based on standard languages and components off the shelf, which reduces the creation of a cooperative portal to the definition of an ontological domain model, and which is implemented with components that can be efficiently reused in a variety of data intensive and knowledge based applications. We report on the usage of the architecture in the case of a cooperative portal on mudbrick construction practices, that we call Muddy. In so doing, we also highlight the application programming model that permits the construction of a portal using the proposed architecture.

The rest of the paper is organized in 5 sections. After a brief overview of our perspective on the ontological paradigm in SW architecture, we introduce the Muddy case and we analyze its requirements, identifying abstract roles and use cases (Sect.2). We then expound the architectural design and some salient traits of its implementation, and we identify the roles involved in its application (Sect.3). Finally we describe the Muddy portal (Sect.4) and draw conclusions (Sect.5).

2   Requirements Analysis

2.1   The Ontological Paradigm

Ontological technologies mainly originate with the intent to contribute to the realization of the Semantic Web [14]. This denotes an evolution of the current web, in which information is sematically defined so as to enable automated processing.

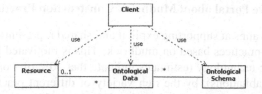

Fig. 1. Resources and meta-data relations

The Semantic Web paradigm thus emphasizes the relation between information and meta information, which distinguishes the concepts of Resource, Ontology Data, and Ontology Schema, as illustrated in Fig. 1. Resources are any information object on the web: a file, an HTML page, an image, a movie. In the semantic web perspective, each Resource will contain its Semantic Data. The Ontology Schema is a conceptualization shared among users, which captures the intensional part of the model, defining types of entities and their possible relations (concepts). Ontology Data (individuals), are the extensional part of the model, classifying Resources as realizations of the concepts in the Ontology Schema. Client is a Semantic Web reader and can be a human or an application. In both the cases, Client is interested to access Resources and their related semantic data.

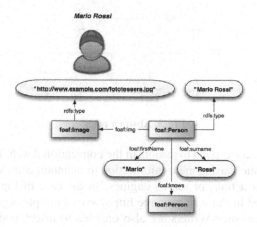

Fig. 2. Semantic annotation of a web resource

Fig. 2 exemplifies the concept: in this case the resource is a picture (fototessera.jpg) located at an http address (http://www.example.com). The public ontology named FOAF associates the resource with a set of concepts supporting interpretation by a SW agent: the picture (foaf:Image) is referred to (foaf:img) to a person (foaf:Person), with name (foaf:firstName) "Mario" and surname (foaf:surname) "Rossi", who is in relation with (foaf:knows) other persons (foaf:Person).

Ontological technologies comprise a rich framework of paradigms, languages, and components off the shelf, which can serve beyond the specific intent of the Semantic Web, and may become an effective pattern for the organization of complex SW architectures.

2.2   A Cooperative Portal about Mudbrick Construction Practices

The Muddy project aims at supporting explicit and shared representation of knowledge about construction practices based on mudbrick. This is motivated by the widespread use of this building technology (estimates indicate that 30% of world population still live in mudbrick habitations), by the rich variety of different practices developed in different localities, and by the high energetic efficiency and low environmental impact which make it a sustainable practice.

In particular, the Muddy project aims at developing a web portal based on ontological models, enabling cooperation among subjects from different localities and different domains of expertise, in the development of a shared model and in the contribution of concrete contents for it.

2.3   Abstract Roles

Analysis of functional requirements of the Muddy project in the light of the organization of information according to the ontological paradigm of Fig. 1, identifies roles, users' needs, and use cases generalized beyond the limits of the specific context of use [15]. These roles are outlined in Fig. 3

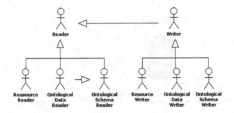

Fig. 3. Abstract roles

Resource Readers correspond to readers in the conventional web. They are interested in accessing information, using meta-information to maintain context and to access resources, through direct links or search engines. In the case of Fig. 2 they could for instance be interested in the web resource http://www.example.org/fototessera.jpg. In a similar manner, Resource Writers are also enabled to insert, update and delete resources.

Ontological Schema Readers take the role that [16] qualifies as second-level reader: they are interested in understanding the organization of concepts more then their concrete realizations. They need to navigate in ordered manner and search classes and properties of an ontological model. In the example of Fig. 2 they would be interested in the structure of the FOAF ontology more than in the concrete resource fototessera.jpg.

Ontological Schema Writers also modify models, taking part to the definition of the strategy of content organization. In particular, they may be interested in fixing errors, changing the model to follow some evolution, or extending the model by specialization and inheritance.

An Ontological Data Writer is a human or a SW indexing Resources with respect to the concepts of an Ontological Schema. Besides, an Ontological Data Reader is a human or, more frequently, a SW which exploits Ontological Data to access concrete resources in semantic querying [17]. Ontological data can also be formatted to be easily readable as well as a resource by human users [18].

In the example of Fig. 2 they will thus be interested in the instances of the FOAF ontology that refer to the resource "http://www.example.org/fototessera.jpg".

2.4   Use Cases in the Muddy Portal

In the specific context of the Muddy portal, the main goals of a Reader are browsing of pages derived from resources and semantic models, navigation of links due to relations in the ontological model, execution of semantic queries on the knowledge base (see Fig. 4).

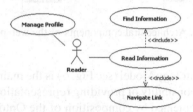

Fig. 4. Web portal Reader use cases

Besides, the Writer (see Fig. 5), extends the Reader capability to access information with the capability to contribute new knowledge in the form of a generic Ontological Schema or Data. Namely, the enabled user can contribute either the extensional or the intensional part of the ontological model for the web portal. Writers can also send/receive feedback and comments about the ontological model so as to encourage collaboration and cooperation among users. To contribute, writers will upload model files to ease the development of a portal prototype.

A further instrumental role is the Administrator, whose main responsibility is the assignment of readers and writers privileges.

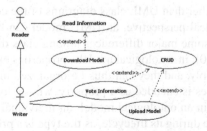

Fig. 5. Web portal Writer use cases

3 Architecture and Development Process

3.1 Architectural Components

The conceptual architecture of our semantic portal composes a variety of partecipants that can be effectively assembled using W3C supported specifications and components.

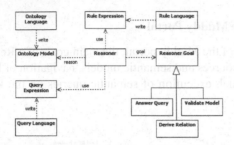

Fig. 6. Architectural components of the web portal

Ontology Model. The Ontology Model (see Fig. 6) is the main component of the architecture, with the main responsibility of providing representation of the domain model of the application. It is implemented by composition of the Ontology Schema and Ontology Data (see Fig. 1), both encoded using the W3C standard Ontology Web Language (OWL) [19]. In principle, the entire expressivity of OWL-full can be exploited, but successful termination of reasoning tasks included in portal services is guaranteed only under the assumption that the model is encompassed within OWL-DL reasoning.

In a logic perspective (see Fig. 7), the ontology model can be decomposed in three main parts: classes, properties and individuals. Classes in the Ontology Model are part of the Ontological Schema and play a role that can be compared to that of table definitions in a relational database. However, as opposed to relational tables, classes can be composed through delegation and inheritance. Properties are also part of the Ontological Schema, and they are used to define relations among classes. Individuals are realizations of concepts described by classes, and play a role that can be compared to that of records in a relational database.

It is worth noting that, in this architecture, the domain logic is modelled and specified using ontologies rather than UML class diagrams of the common practice of OO development. In a practical perspective, an UML model can be easily mapped to an ontology, provided that some major differences among class diagrams and ontologies are taken into account [20]: in ontological models, properties exist independently from classes to which they apply, and they can thus be organized in a hierarchy; a relation between two UML classes is encoded as a property, whose domain and range are the two classes themselves; in an ontological model, an individual can belong to multiple classes which can change during its lifecycle, as the type is a property itself of the individual; in ontological models, classes can be defined using set-theoretical constructs.

Fig. 8 reports the example of an ontological model: Book ed Author are classes associated through an oriented relationship capturing the property hasAuthor. Book, Author

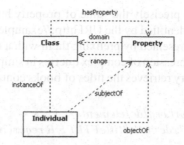

Fig. 7. Logical compontents of the ontology model

and hasAuthor comprise the intensional part of the ontological model. "Il Barone Rampante" is an instance of the class Book, while Italo Calvino is instance of the class Author. The two objects are related through an instance of the association hasAuthor. "Il Barone Rampante", Italo Calvino and their association comprise the extensional part of the ontological model.

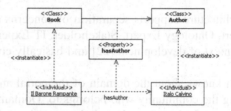

Fig. 8. An ontological model

Rules. To form the overall knowledge base, the ontological model is complemented with rules that extend the information explicitly represented with inference rules that let the system derive new knowledge. In general, a rule has the form of an implication $antecedent \Rightarrow consequent$ where both $antecedent$ and $consequent$ are conjunctions of atoms. Variables are indicated using the standard convention of prefixing them with a question mark (e.g., $?x$). For instance, the rule asserting that an uncle is the brother of a parent would be written as: $Person(?x)$ and $Person(?y)$ and $Person(?z)$ and $parent(?x, ?y)$ and $brother(?y, ?z) \Rightarrow uncle(?x, ?z)$. Note that $Person$ is a class in the ontological model while $parent$, $brother$ and $uncle$ are properties.

In our architecture, rules are represented using the W3C supported Rule Language SWRL. To circumvent limitations affecting open source reasoners, we internally represent the language using Jena rules [21].

Querying and Reasoning. Query on the knowlegde base are expressed using the W3C supported the SparQL. While the architecture supports the full expressivity of SPARQL, for the sake of usability, and in particular learnability, only a restricted fragment of the language is provided to the user. For instance, the query $SELECT?titleWHERE$

$\{< http : //example.org/book/book1 >< http : //purl.org/dc/elements/1.1/title >?title.\}$

retrieves the title (or, more precisely the value of property http://purl.org/dc/elements/
1.1/title) of the individual identified by the URI http://example.org/book/book1. Also in
SPARQL, variables are represented by prefixing them with a question mark.

SPARQL enables the expression of complex queries in simple and compact form. For
instance, the following query retrieves the titles of books containing the term Rampante
in their title:

$PREFIX\ dc:< http://purl.org/dc/elements/1.1/ >$
$SELECT?titleWHERE\{?xdc:title?titleFILTER\ regex(?title,"Rampante","i")\}$

The API Jena is used to drive reasoning and retrieve information by decidingSPARQL
queries on the model. The API is also used to validate the ontology model and derive
new knowledge using OWL axioms and inference rules. In general, any reasoner represents a trade-off between power and efficiency in computing. In particular, in the case
of our architecture termination of reasoning tasks is guaranteed only if the model and
the rules are encompassed within the boundaries of OWL-DL and SWRL, respectively.

3.2 Participants in the Development Process

The proposed SW architecture supports separation of concerns among four different
roles of Domain Expert, Ontology Expert, Stakeholder, IT Expert. These naturally fit
in a realistic social context of development [22] and basically correspond to the roles
identified in [4].

The Domain Expert knows about the domain of the portal and share partially formalized models among the community who belongs to. Domain Experts usually use
specific tools to do their analysis and produce their research result. It is often the case
that they don't know anything about ontologies and also they don't have opportunity
(no time available) to learn about them.

The Ontology Expert is able to use sematic modelling tools and can describe knowledge contributed by Domain Experts with an Ontology Model. In this way the information, that was heterogeneous and sometimes also tacit or embedded, becomes formalized, explicit, homogeneous and consistent [23].

The Stakeholder is interested in the domain logic but he/she is not necessarily expert.
For this role, it is useful to have an ontology model that can be read and studied and
that can be used to navigate through Domain Experts documents.

Finally, the IT Expert has to develop software tools needed by other roles so to let
them read and write resources and ontology models, see Fig. 1 and 3.

3.3 Salient Aspects of the Implementation

Layering. The source code of the web portal (see Fig. 9) is organized in three layers
[12], [24]. As usual, in SW architecture, layering separates presentation, domain logic
and persistence. In this case, layering also helps in filling the gap between ontological
and object oriented perspectives [13] [25], which somehow reproduces the well known
problem of Impedance Mismatch between Objects and relational databases [26].

Specifically: the presentation layer contains the logic to handle the interaction between the user and the software, relying on Java Server Faces (JSF) [27]; the domain

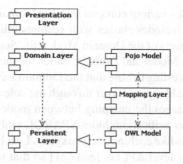

Fig. 9. Web portal layering

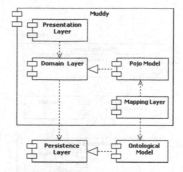

Fig. 10. Muddy Web portal layering

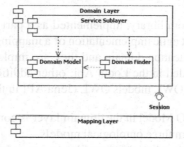

Fig. 11. Relation between domain logic and mapping layer

layer contains the application logic i.e. the domain model, implemented with Plain Old Java Objects (POJO); the persistent layer, is implemented as an ontology used to persist objects of the domain model.

Developers can refer to patterns such as Active Record or Mapper [24] and also use tools [28], [29] to let objects of the domain model correspond to individuals of the ontology model. For the web portal, a Mapping layer was used between domain and persistent layers because it allows better decoupling, easer testing and concurrent developing [30], [31].

The domain layer includes various components: the Domain Model (discussed later); the Domain Finder which includes classes with responsibility in querying the knowledge base to retrievethe objects of the Domain Model; the Service Sublayer which comprises a facade on Domain Model and DomainFinder for the purposes of higher layers of the application. It is interesting to note that the Domain Layer is able to access all the functions of the underlying Mapping Layer through the sole interface Session.

The Mapping layer manages the mapping between models and meta-models, elaborates complex relations i.e. reification, hides SPARQL embedded code and improves performances with methods like caching and proxying. Last but not least, only the mapping layer refers to the low level API i.e. Jena [21] so that is easer to change the used library [32] and that could be useful for a rapidly changing domain such as ontologies. To better understand its structure, notice that the Mapping Layer is comprised by three

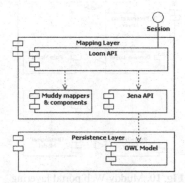

Fig. 12. Relation between mapping layer and persistence layer

major components: the Loom library, implemented as a part of our work, provides all basic functions that support agile implementation of a mapping layer; mappers and finders are specific to the application domain and they are implemented as sub-classes of abstract base classes included in the Loom Api; other additional libraries support low level access to ontological OWL models OWL (Jena API in the picture).

Domain Model. The POJO Model in the domain layer is composed by two Java packages which are derived from three ontological models.

The User package contains data about profiles, users and related information, and it is automatically derived from an ontology User model, so as to map ontology classes and properties to OO classes and attributes.

The Domain package has responsibility to manage information contributed by users and is derived from an ontological Domain model according to the architectural pattern of reflection [12]: all types in the ontological model are mapped into a single generic OO class (that would be a meta-level-class in the reflection pattern); this generic class has responsibility to manage relations among types defined by the users in the ontological model (that would comprise base-level-classes in the reflection pattern). This decouples the structure of the OO domain layer from the specific types defined in the Domain Ontology, thus enabling reuse of the OO layer for a variety of different ontological Domain models, defining different evolutions of a portal or different portals insisting on different application domains. Also, this is the feature that permits the cooperative portal to accommodate contributions not only in the individuals of the extensional part, but also in the concepts of the intensional part of the knowledge-base.

Derivation of the Domain package is also affected by an additional ancillary ontology defining directives for the presentation of data in the page layout. The individuals of this ontology are used by the mapping layer to determine the presentation and filtering of concepts defined in the ontological Domain model. This accomplishes a responsibility which is much similar to that of "site view graphs" in the OntoWebber framework [10].

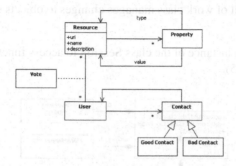

Fig. 13. The POJO Model of the web portal

Mapping Layer. Mapping between ontological and object-oriented models of the architecture was implemented following a pattern-oriented design [24], [12] aimed at building an extensible and reusable framework.

Mappers. The mapping layer includes a mapper class for each element of the OO domain model [24] (see Fig. 14). Each mapper class can read information form the ontological model to assign it to the object-oriented model and vice versa, and it is implemented as extension of the abstract DataMapper base class.

Mappers decouple the object-oriented Domain Layer from the ontological Persistence Layer, so that a client can ignore how objects are persisted in the ontology. For

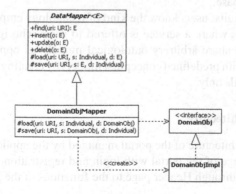

Fig. 14. Mappers

the sake of performance and and behavioral abstraction, DataMappers implement some specific patterns [24], [33]:

- **Identify Map (Cache):** mappers have a cache memory to speed up repeated access to the same object.
- **Proxy:** if possible, mappers substitute objects requested by the client with equivalent proxies. This delays the mapping operations until they are really needed.
- **LazyLoad:** mappers load objects of a list when they are used so the slow mapping operation is executed for useful objects only.
- **UnitOfWork:** unit of work class manages changes to objects that mappers have to persist.

Developers can use an instance of the class Session to access functions of the mapping framework (see Fig. 15).

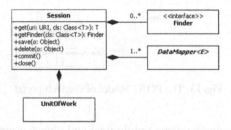

Fig. 15. The Session class

4   The Muddy Portal

Muddy is a web-based application implemented as an instance of requirements and architecture described so far. It allows reading and writing of concrete and conceptual information according to an ontological paradigm, providing the following user functions: navigate information following semantic links; execute semantic queries on the knowledge base; contribute new knowledge uploading model files; read/write feedback about the knowledge base.

As characterizing traits: users know the kind of technology employed to model data, as opposed to systems where a service is offered to users who ignore the underlying technology; users can share arbitrary ontological models, as opposed to applications where users interact with predefined conceptual models by creating, retrieving, updating and deleting individuals only.

4.1   The Portal Architecture

Fig. 16 depicts the architecture of the portal managed by the application.

Index is the first page of the portal with login and registration, giving access to the Home page and then, through Header page to the functions of the portal.

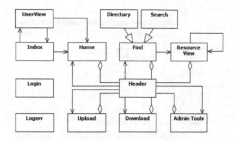

Fig. 16. Structure of the Muddy Portal

Users can be readers, writers and administrators and they are provided with different functions. A new user is always classified as reader and only administators can give users more privileges.

Find page is used to execute queryies on the knowledge base by users and it is specialized in Search and Directory pages. ResourceView page allows users to read information contained in the knowledge base. Upload and Download pages allow users to contribute new knowledge. Admintools page is for the administrator.

4.2 Find Pages

The Directory page (see Fig. 17) is used to execute pro-active search. The system shows to the user a list of categories that correspond to root classes of the ontological model managed by the portal. The user can select a category to get a page containing the list of instances of the category and the list of its direct subclasses. The user can navigate toward more specific classes or can inspect one of the instances found.

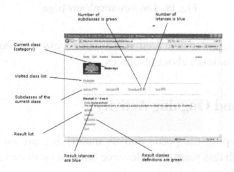

Fig. 17. The Directory search page

The Search page, see Fig. 18, implements an extension of full-text search methods. Users can specify one or more words that must be contained in desired resources. They can also specify the kind of relations that link desired resources to specified words. For instance, the expression "neededTools = sieve" lets a user require all resources that has a "sieve" among needed "tools".

This page tries to simplify the use of SPARQL to users.

Fig. 18. The Search page

4.3 Resource View Page

This page shows information about a resource (see Fig. 19), and allows users to speed up navigation towards semantic related resources.

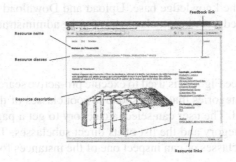

Fig. 19. The ResourceView page

The portal also allows users to give feedback about accessed resources which is used to calculate appreciation indexes about resources.

5 Conclusions and Ongoing Activity

We are further developing the portal and its underlying architecture, facing various interrelated issues, with less scientific relevance but crucial to tackle the transition phase towards the context of use:

- a usability cycle has been planned, to evaluate the capability of the portal to support the user in maintaining context in the navigation through the weblike structure of portal contents; in this perspective, the orientation towards change in the overall architecture, and in particular the concept of presentation ontology implemented in the mapper, provide major assets to face iterative refinement in design choices;
- preliminary performance profiling indicates that performance can be largely improved by the integration of a more elaborated RDF repository;

– functional extensions are being developed to implement the automated derivation of an editor of individuals based on the structure of ontology concepts and a tool for annotation of existing web resources with reference to the concepts of an ontological model.

References

1. Fayad, M., Cline, M.P.: Aspects of software adaptability. Communications of the ACM (1996)
2. ISO: Iso 9126 Software engineering – product quality (2004),
   http://www.iso.org/iso/en/ISOOnline.frontpage
3. Bozsak, E., Ehrig, M., Handschuh, S., Hotho, A., Maedche, A., Motik, B., Oberle, D., Schmitz, C., Staab, S., Stojanovic, L., Stojanovic, N., Studer, R., Stumme, G., Sure, Y., Tane, J., Volz, R., Zacharias, V.: Kaon tool suite (2007),
   http://kaon.semanticweb.org/frontpage
4. Tempich, C., Pinto, H.S., Sure, Y., Staab, S.: An argumentation ontology for distributed, loosely-controlled and evolving engineering processes of ontologies (diligent). In: Gómez-Pérez, A., Euzenat, J. (eds.) ESWC 2005. LNCS, vol. 3532, pp. 241–256. Springer, Heidelberg (2005)
5. Tempich, C., Pinto, H.S., Staab, S.: Ontology engineering revisited: an iterative case study with diligent. In: Sure, Y., Domingue, J. (eds.) ESWC 2006. LNCS, vol. 4011, pp. 110–124. Springer, Heidelberg (2006)
6. Garzotto, F., Mainetti, L., Paolini, P.: Hypermedia design analysis and evaluation issues. incomm. of the ACM. Communications of the ACM (1995)
7. Lynch, P.J., Horton, S.: Web style guide: Basic design principles for creating web sites. Yale University Press (2002)
8. Schreiber, G., Amin, A., van Assem, M., de Boer, V., Hardman, L., Hildebrand, M., Hollink, L., Huang, Z., van Kersen, J., de Niet, M., Omelayenko, B., van Ossenbruggen, J., Siebes, R., Taekema, J., Wielemaker, J., Wielinga, B.J.: Multimedian e-culture demonstrator. In: Cruz, I., Decker, S., Allemang, D., Preist, C., Schwabe, D., Mika, P., Uschold, M., Aroyo, L.M. (eds.) ISWC 2006. LNCS, vol. 4273, pp. 951–958. Springer, Heidelberg (2006)
9. Stojanovic, N., Maedche, A., Staab, S., Studer, R., Sure, Y.: Seal: a framework for developing semantic portals (2001)
10. Jin, Y., Decker, S., Wiederhold, G.: Ontowebber: Model-driven ontology-based web site management. In: 1st International Semantic Web Working Symposium. Stanford University, Stanford (2001)
11. Corcho, O., Lòpez-Cima, A., Gomez-Pérez, A.: A platform for the development of semantic web portals. In: ICWE 2006: Proceedings of the 6th international conference on Web engineering, pp. 145–152. ACM Press, New York (2006)
12. Schmidt, D.C., Rohnert, H., Stal, M., Schultz, D.: Pattern-Oriented Software Architecture: Patterns for Concurrent and Networked Objects. John Wiley & Sons, Inc., New York (2000)
13. Woodfield, S.N.: The impedance mismatch between conceptual models and implementation environments. In: Embley, D.W. (ed.) ER 1997. LNCS, vol. 1331. Springer, Heidelberg (1997)
14. Berners-Lee, T.: Semantic web roadmap (1998), http://www.w3.org/2001/sw/
15. ISO: Iso 9241-11 guidance on usability (1998),
    http://www.iso.org/iso/en/ISOOnline.frontpage
16. Eco, U.: Six walks in the fictional woods. Harvard University Press (1994)
17. Bonino, D., Corno, F., Farinetti, L.: Dose: a distributed open semantic elaboration platform. In: The 15th IEEE International Conference on Tools with Artificial Intelligence (ICTAI 2003), Sacramento, California, November 3-5 (2003)

18. Dzbor, M., Motta, E., Domingue, J.B.: Opening up magpie via semantic services. In: McIl-raith, S.A., Plexousakis, D., van Harmelen, F. (eds.) ISWC 2004. LNCS, vol. 3298, pp. 635–649. Springer, Heidelberg (2004)
19. W3C: Owl web ontology language (2004),
    http://www.w3.org/TR/owl-features/
20. Brockmans, S., Volz, R., Eberhart, A., Löffler, P.: Visual modeling of owl dl ontologies using uml. In: International Semantic Web Conference, pp. 198–213 (2004)
21. Company, H.P.D.: Jena a semantic web framework for java (2002),
    http://jena.sourceforge.net/
22. Cockburn, A.: The interaction of social issues and software architecture. Commun. ACM 39, 40–46 (1996)
23. Kryssanov, V.V., Abramov, V.A., Fukuda, Y., Konishi, K.: The meaning of manufac-turing know -how. In: PROLAMAT 1998: Proceedings of the Tenth International IFIP WG5.2/WG5.3 Conference on Globalization of Manufacturing in the Digital Communica-tions Era of the 21st Century, Deventer, The Netherlands, pp. 375–388. Kluwer, Dordrecht (1998)
24. Fowler, M.: Patterns of Enterprise Application Architecture. Addison-Wesley Longman Pub-lishing Co., Inc., Boston (2002)
25. Guizzardi, G., Falbo, R., Filho, J.: Using objects and patterns to implement domain ontolo-gies. In: 15th Brazilian Symposium on Software Engineering, Rio de Janeiro, Brazil (2001)
26. Ambler, S.: Agile Database Techniques: Effective Strategies for the Agile Software Devel-oper. John Wiley & Sons, Inc., New York (2003)
27. Mann, K.D.: JavaServer Faces in Action (In Action series). Manning Publications Co., Greenwich (2004)
28. Protégé.: Stanford-University (2006), http://protege.stanford.edu
29. Kalyanpur, A., Pastor, D.J., Battle, S., Padget, J.A.: Automatic mapping of owl ontologies into java. In: SEKE, pp. 98–103 (2004)
30. Heumann, J.: Generating test cases from use cases. The Rational Edge (2001)
31. Beck, K.: Test Driven Development: By Example. Addison-Wesley Professional, Reading (2002)
32. Aduna: Sesame: Rdf schema querying and storage (2007),
    http://www.openrdf.org/
33. Gamma, E., Helm, R., Johnson, R., Vlissides, J.: Design Patterns. Addison-Wesley Profes-sional, Reading (1995)

The Vcodex Platform for Data Compression

Kiem-Phong Vo

AT&T Labs, Shannon Laboratory
180 Park Avenue, Florham Park, NJ 07932, USA
kpv@research.att.com

Abstract. Vcodex is a software platform for constructing data compressors. It introduces the notion of *data transforms* as software components to encapsulate data transformation and compression techniques. The platform provides a variety of compression transforms ranging from general purpose compressors such as Huffman or Lempel-Ziv to structure related ones such as reordering fields and columns in relational data tables. Tranform composition enables construction of compressors either general purpose or customized to data semantics. The software and data architecture of Vcodex will be presented. Examples and experimental results will be given showing how the approach helps to achieve compression performance far beyond traditional approaches.

1 Introduction

Modern business systems manage huge amounts of data. As an example, the daily billing and network management data generated by an international communication company amount to hundreds of gigabytes [7]. Certain of such data might even be required by law to be kept online for several years. Thus, data compression is a critical component in these systems.

Traditional compression research mostly focuses on developing general purpose techniques that treat data as unstructured streams of objects. Algorithms based on pattern matching, statistical analysis or some combination thereof [9,11,22,23,24] detect and remove certain forms of information redundancy in such streams of data. Well-known compression tools such as the ubiquitous Unix Gzip or Windows Winzip are based on these general purpose techniques, i.e., the Lempel-Ziv compression method [23] and Huffman coding [9]. A recently popular compressor, Bzip2 [17], first reorders data by the famous Burrows-Wheeler Transform [4] to induce better compressibility before passing it on to other compressors.

Structures in data are sources of information redundancy that could be exploited to improve compression. For example, columns or fields in relational data tend to be sparse and often share non-trivial relationships. The Pzip table compression technique by Buchsbaum et al [3] achieves good results by grouping related the columns of a table together before invoking some conventional compressor such as Gzip for actual compression. The current best known table compression technique is by Vo and Vo [19,20]. It automatically computes dependency relations among columns and uses that to reduce information redundancy. The Xmill compressor by Liefke and Suciu [15] compresses XML documents, an important class of hierarchical data. It works in a similar spirit to

J. Filipe et al. (Eds.): ICSOFT/ENASE 2007, CCIS 22, pp. 201–212, 2008.

Pzip by grouping parts of a document with the same tags to be compressed by some conventional compressor.

Practical data often consist of ad-hoc mixtures of structures beyond just simple tables or single XML documents. For example, a log file generated from a network management system might contain a variety of records pertaining to different network events. However, records grouped by events often form tables due to their uniform formats. Thus, a log file may be compressed better by grouping records of same events first before invoking a table compressor on each group. Automatic exploitation of such data specific features for effective compression is too difficult to do in general but can be simple to carry out for particular data given some familiarity with their structures. As such, a challenging problem in dealing with large scale real world data is to provide a convenient way for practitioners to make use of data specific knowledge along with other existing techniques to achieve optimum compression results. This is the problem that the Vcodex data transformation platform addresses.

Central to Vcodex is the notion of *data transforms* or software components to alter data in some invertible ways. Although such transforms may represent general data processing techniques including encryption, portability encoding and others, we shall focus solely on compression here. For maximum usability, two main issues related to data transforms must be addressed:

- *Providing a Standard Software Interface to all Transforms*: As algorithms, both general purpose and data specific, are continually invented and improved, a standard software interface for data transformation eases application implementation and ensures code stability over time. A standard software interface also simplifies and encourages reuse of existing data transforms in implementing new ones.
- *Defining a Self-describing and Portable Standard Data Format*: As transforms may be continually developed by independent parties, a common data format is needed to accommodate arbitrary composition of transforms. Such a format should enable users to decode data without knowing how they were encoded. In addition, encoded data should be independent of OS and hardware platforms so that they can be transported and shared.

The rest of the paper gives an overview of the Vcodex platform and how it addresses the above software and data issues. Experiments based on the well-known Canterbury Corpus [1] for testing data compressors will be presented to show that Vcodex compression transforms could far outperform conventional compressors such as Gzip or Bzip2.

2 Software Architecture

Vcodex is written in the C language in a style compatible with C++. The platform is divided into two main layers. The base layer is a software library defining and providing a standard software interface for data transforms. This layer assumes that data are processed in segments small enough to fit entirely in memory. For file compression, a command tool Vczip is written on top of the library layer to enable transform usage and composition without low-level programming. Vczip handles large files by breaking them into suitable chunks for in-memory processing.

Transformation handle and operations

Vcodex_t – maintaining contexts and states
vcopen(), vcclose(), vcapply(), ...

Discipline structure	Data transform
Vcdisc_t	Vcmethod_t
Parameters for data processing,	Burrows-Wheeler, Huffman,
Event handler	Delta compression, Table transform, etc.

Fig. 1. A discipline and method library architecture for Vcodex

2.1 Library Design

Figure 1 summarizes the design of the base Vcodex library which was built in the style of the Discipline and Method library architecture [21]. The top part shows that a *transformation handle* of type Vcodex_t provides a holding place for contexts used in transforming different data types as well as states retained between data transformation calls. A variety of functions can be performed on such handles. The major ones are vcopen() and vcclose() for handle opening and closing and vcapply() for data encoding or decoding.

The bottom part of Figure 1 shows that each transformation handle is parameterized by an optional discipline structure of type Vcdisc_t and a required data transform of type Vcmethod_t. A discipline structure is defined by the application creating the handle to provide additional information about the data to be compressed. On the other hand, a data transform is selected from a predefined set (Section 2.2) to specify the desired data transformation technique.

```
1. typedef struct _vcdisc_s
2. {       void*    data;
3.         ssize_t  size;
4.         Vcevent_f eventf;
5. } Vcdisc_t;
6. Vcdisc_t disc = { "Source data to compare against", 30, 0 };
7. Vcodex_t* huff = vcopen(0, Vchuffman, 0, 0, VC_ENCODE);
8. Vcodex_t* diff = vcopen(&disc, Vcdelta, 0, huff, VC_ENCODE);
9. ssize_t cmpsz = vcapply(diff, "Target data to compress", 23, &cmpdt);
```

Fig. 2. An example of delta compression

Figure 2 shows an example of construction and use a handle for delta compression by composing two data transforms: Vcdelta and Vchuffman (Section 2.2). Here, delta compression [10] is a general technique to compress a *target data* given a related *source data*. It is often used to construct software patches or to optimize storage in revision control systems [18]. The transform Vcdelta implements a generalized Lempel-Ziv parsing technique for delta compression that reduces to conventional compression when no source data is given.

- Lines 1–6 show the type of a discipline structure and disc, an instance of it. The fields data and size are used to pass additional information about the data to be compressed to the handle, for example, source data in this case. Such additional data may be changed dynamically. The field eventf of disc optionally specifies a function to process events such as handle opening and closing.

- Lines 7 and 8 show handle creation and composition. The first and second arguments to vcopen() specify the optional discipline structure, e.g., disc on line 8, and the selected data transform, e.g., Vchuffman on line 7. Variants of a transform can be specified in the third argument (see the examples in Figure 5). The fourth argument is used to compose a transform with another one as seen on line 8 where huff is used to process data produced by diff. The last argument specifies encoding or decoding, for example, VC_ENCODE for encoding here.
- Line 9 shows the call vcapply() to compress data. The argument cmpdt returns the compressed result while the call itself returns the compression size. The Vcdelta transforms creates multiple output data segments for different types of data (Section 3). Each such segment would be separately passed to the continuing handle, huff, to generate a Huffman coding optimized by the characteristics of the given data.

The above example shows the encoding side. The decoding side is exactly the same except that VC_DECODE replaces VC_ENCODE in the vcopen() calls and the data passed to vcapply() would be some previously compressed data.

2.2 Data Transforms

Figure 3 shows Vcmethod_t, the type of a data transform. Lines 2 and 3 show that each transform provides two functions for encoding and decoding. An optional event handling function eventf, if provided, is used to process events. For example, certain transforms maintain states throughout the lifetime of a handle. The structures to keep such states would be created or deleted at handle opening and closing via such event handling functions. Each transform also has a name that uniquely identifies it among the set of all available transforms. This name is encoded in the output data (Section 3) to make the encoded data self-describing.

```
1. typedef struct _vcmethod_s
2. {      ssize_t (*encodef)(Vcodex_t*, void*, ssize_t, void**);
3.        ssize_t (*decodef)(Vcodex_t*, void*, ssize_t, void**);
4.        int     (*eventf)(Vcodex_t*, int, void*);
5.        char*   name;
6.        ...
7. } Vcmethod_t;
```

Fig. 3. The type of a data transform

Vcodex provides a large collection of transforms for building efficient compressors of general and structured data including a number that are application-specific. Below are brief overviews of a few important data transforms:

- Vcdelta: This implements a delta compressor based on a generalized Lempel-Ziv parsing method. It outputs data in the Vcdiff encoding format as described in the IETF Proposed Standard RFC3284 [13].

- Vcsieve: This uses approximate matching for various forms of compression including delta compression and compression of genetic sequences, the latter being a class of data notoriously hard to compress [16].
- Vcbwt: This implements the Burrows-Wheeler transform [4].
- Vctranspose: This treats a dataset as a table, i.e., a two-dimension array of bytes, and transposes it.
- Vctable: This uses column dependency [20] to transform table data and enhance their compressibility.
- Vcmtf: This implements the move-to-front data transform [2]. A variant uses prede-cessor-successor statistics to moves characters to the front sooner than their actual appearance.
- Vcrle: This provides the run-length encoder. Various variants encode certain common runs more compactly, e.g., zero-runs only [5].
- Vchuffman: This implements static Huffman coding [9].
- Vchuffgroup: This divides data into short segments of equal length and groups segments compress well together with a single Huffman code table.
- Vcama: This is an example of a data-specific transform. It processes telephone data of a type called AMA, hence the name. Records with the same length are collected together to form tables for compression via Vctable.
- Vcmap: This is an example of a data transform other than compression. It maps data from one character set to another, for example, translating data between ASCII and various versions of EBCDIC.

2.3 The Vczip Command

The Vczip command enables usage of the available data transforms without low level programming. It provides a syntax to compose transforms together to process a data file. In this way, end users can experiment with different combinations of transforms on their data to optimize compressibility.

```
1. Transform -> {delta, table, bwt, huffman, ama, ...}
2. Transform -> {delta, table, bwt, huffman, ama, ...} . Arglist(Transform)
3. Arglist(Transform) -> Nil
4. Arglist(Transform) -> Arg(Transform) . Arglist(Transform)
5. TransformList -> Nil
6. TransformList -> "-m" TransformList , Transform
```

Fig. 4. Language syntax to compose transforms

Figure 4 shows the transform composition syntax. Lines 1 and 2 show that each transform is specified by a *name*, e.g., delta or table, followed by an optional sequence of arguments depending on the particular transform. Lines 3 and 4 show that such arguments are separated by periods. Lines 5 and 6 show that the list of transforms can be empty to signify usage of a default compression method defined by Vczip. But if not, it must start with -m and follow by a comma-separated sequence of transforms.

Figure 5 shows an experiment to explore different ways to compress data using Gzip, Bzip2 and various compositions of Vcodex transforms. The file *kennedy.xls* was an Excel spreadsheet taken from the Canterbury Corpus [1].

```
1.    ls -l kennedy.xls
2.    -rw-------  4 kpv 1029744 Nov 11  1996 kennedy.xls

3.    gzip < kennedy.xls > out; ls -l out
4.    -rw-r--r--  1 kpv 206767 Apr 11 12:30 out

5.    bzip2 < kennedy.xls > out; ls -l out
6.    -rw-r--r--  1 kpv 130280 Apr 11 12:30 out

7.    vczip -mbwt,mtf.0,rle.0,huffgroup < kennedy.xls > out; ls -l out
8.    -rw-r--r--  1 kpv 129946 Apr 11 12:31 out

9.    vczip -mbwt,mtf,rle.0,huffgroup < kennedy.xls > out; ls -l out
10.   -rw-r--r--  1 kpv 84281 Apr 11 12:31 out

11.   vczip -mtable,mtf,rle.0,huffgroup < kennedy.xls > out; ls -l out
12.   -rw-r--r--  1 kpv 53918 Apr 11 12:31 out

13.   vczip -mtable,bwt,mtf,rle.0,huffgroup < kennedy.xls > out; ls -l out
14.   -rw-r--r--  1 kpv 35130 Apr 11 12:31 out

15.   vczip -u < out >x; cmp x kennedy.xls
```

Fig. 5. Experimenting with different combinations of data transforms

```
1.    ls -l ningaui.ama
2.    -rw-r--r--  1 kpv      kpv      73452794 Sep 12  2005 ningaui.ama

3.    gzip < ningaui.ama >out; lsl out
4.    -rw-r--r--  1 kpv      kpv      30207067 Apr 11 12:55 out

5.    bzip2 < ningaui.ama >out; lsl out
6.    -rw-r--r--  1 kpv      kpv      22154475 Apr 11 12:57 out

7.    vczip -mama,table,mtf,rle.0,huffgroup < ningaui.ama >out; lsl out
8.    -rw-r--r--  1 kpv      kpv      20058182 Apr 11 13:00 out
```

Fig. 6. Compression based on data-specific structures

- Lines 1–6 show the raw size of *kennedy.xls* and compression results by Gzip and Bzip2. Gzip compressed the data by roughly a factor of 5 while Bzip2 achieved a factor of 8 to 1.
- Lines 7–10 show Vczip instantiated using different variations of a BWT compressor. The 0 argument to the move-to-front transform mtf restricted it to moving a data byte only after accessing it. As that was analogous to Bzip2, about the same performance was seen. On the other hand, line 9 used a variant that aggressively predicted and moved data (Section 2.2) resulting in a 12 to 1 compression factor, much better than Bzip2.
- Lines 11-14 show the use of the table transform Vctable [20] to reorder data by column dependency to enhance compressibility. That achieved a compression factor of about 19 to 1. Inserting the Burrows-Wheeler transform after the table transform further improved the compression factor to nearly 30 to 1.
- Line 15 shows compressed data being decoded into a file x and compared against the original data. As seen, decompression did not require any knowledge of transforms used on encoding.

Figure 6 shows another experiment to compress a large file of telephone data in a proprietary format called AMA. This type of data consists of records in which the first 4 bytes of a record tell the length and the type of the record. The Vcama transform collected records of the same length together before passing them to the column dependency table transform Vctable. Again, this ability to exploit structures in data resulted in the best compression performance.

3   Data Architecture

For maximal usability, transformed data should be portable and self-described. Portability means that data coded on one platform can be transparently decoded on another. Self-description means that a coded data set should be decodable without users having to know how the data was encoded.

3.1   Portable Data Encoding

To support portability, Vcodex provides a variety of functions to encode strings, bits, and integers. String data are assumed to be in the ASCII character set. For example, a function vcstrcode() is available to transcribe string data from a native character set such as EBCDIC into ASCII. Identification strings of data transforms (Section 2.2) are coded in this way.

It is assumed that the fundamental unit of storage is an *8-bit byte*. The bits in a byte are position from left to right, i.e., the highest bit is at position 0, the second highest bit at position 1, and so on. For bit encoding, a sequence of bits would be imposed onto a sequence of bytes in that order. For example, the 12-bit sequence 101010101111 would be coded in the 2-byte sequence 170 240. That is, the first eight bits, 10101010, are stored in a byte so that byte would have value 170. The last four bits, 1111, are padded with 0's to 11110000 before storage so the representing byte would have value 240.

Integers are unsigned and encoded in a variable-sized format originally introduced in the Sfio library [12]. Each integer is treated as a number in base 128 so that each digit can be stored in a single byte. Except for the least significant byte, other bytes turn on their most significant bit, MSB, to indicate that the next byte is a part of the encoding. For example, the integer 123456789 is represented by four digits 58, 111, 26, 21 in base 128. Below are the coding of these digits shown in bits for clarity.

10111010	11101111	10011010	00010101
MSB+58	MSB+111	MSB+26	0+21

3.2   Self-describing Data

A large file might need to be divided into segments small enough to process in memory. Each such segment is called a *window*. The internal representation of a transformed window would depend on what data transforms were used. But at the file level, that can be treated simply as a sequence of bytes.

Figure 7 shows the structure of encoded data in which each file consists of a set of header data followed by one or more Window sections.

```
 1.   [ID_file:] 0xd6 0xc3 0xc4 0xd8
 2.   [Reserved:] Size Data
 3.   [Transforms:] Size
 4.        [Transform1:] ID_transform
 5.            [Argument:] Size Data
 6.        [Transform2:] ID_transform
 7.            [Argument:] Size Data
 8.            ...
 9.   [Window1:] Indicator
10        [Transformed data]: Size Data
11.   [Window2:] Indicator
12        [Transformed data]: Size Data
13.   ...
```

Fig. 7. The self-describing format of transformed data

- Line 1 shows that each file is identified by four header bytes, the ASCII letters V, C, D and X with high bits on.
- Line 2 shows the Reserved section usable by an application to store additional data. As shown here and later, each data element always starts with a Size to tell the actual number of bytes given in the following Data sequence.
- Lines 3–8 show the Transforms section which encodes the original composition of transforms. This starts with Size, the length in bytes of the rest of the section. Each transform consists of an identification string which, by convention, must be in ASCII followed by any additional data.
- Lines 9–13 show the list of Window sections. Each window section starts with an Indicator byte which can be one of the values VC_RAW, VC_SOURCEFILE and VC_TARGETFILE. VC_RAW means that the encoded data were not transformed. The latter two values are as defined in the Vcdiff Proposed Standard RFC3284 [13] to indicate that the data was delta compressed [10] against data from either the source or target file respectively. Following the Indicator byte is the encoded data.

Figure 8 gives an example of a compressed file produced by a delta compressor based on the transforms Vcdelta and Vchuffman. The top part of the figure shows the header data. As shown, the transform composition sequence is encoded in 16 bytes. The bottom part of the figure shows pictorially the various components in the file and the composition of each window of data. The three sections of data produced by Vcdelta are processed and stored separately by the Huffman coder Vchuffman.

4   Performance

An experiment was done to test compression performance by tailoring transform compositions to particular data. The small and large files from Canterbury Corpus [1] were used to compare Vczip, Gzip and Bzip2. Sizes were measured in bytes and times in seconds. Times were obtained on a Pentium 4, 2.8GHZ, with 1Gb RAM, running Redhat Linux.

Figure 9 presents compression and timing data in two tables. The top and bottom parts of each table show results of the small and large files respectively. The *Algs* entries show the algorithm compositions used to compress the files. Below are the algorithms:

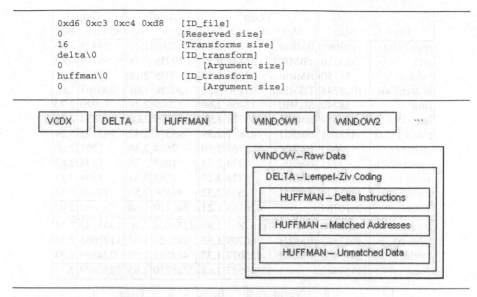

```
0xd6 0xc3 0xc4 0xd8     [ID_file]
0                       [Reserved size]
16                      [Transforms size]
delta\0                 [ID_transform]
0                          [Argument size]
huffman\0               [ID_transform]
0                          [Argument size]
```

Fig. 8. Header data and data schema of a delta compressor

- B: The Burrows-Wheeler transform Vcbwt.
- M: The move-to-front transform Vcmtf with prediction.
- R: The run-length encoder Vcrle with coding only runs of 0's.
- h: The Huffman coder Vchuffman.
- H: The Huffman coder with grouping Vchuffgroup.
- T: The table transform Vctable.
- S: The delta compressor Vcsieve using approximate matching.
- D: The transform Vcsieve with reverse matching and mapping of the letter pairs A and T, G and C, etc.

The *Bits* entries show the average number of bits needed to encode a byte with each row's overall winner in bold. Vczip only lost slightly to Bzip2 on *cp.html* and to Gzip on *grammar.lsp*. Its weighted compression rates of 1.21 bits per byte for the small files and 1.68 for the large files outdid the best results of 1.49 and 1.72 shown at the Corpus website. The pure Huffman coder Vchuffman outperformed its grouping counterpart Vchuffgroup on small files as the cost of coding groups became too expensive in such cases.

Vczip was slightly slower than Bzip2 and Gzip on most files. Although some speed loss was due to the general nature of transform composition, the major part was due to sophisticated algorithms such as move-to-front with prediction or Huffman with grouping. Faster variants to these algorithms with less compression performance are available so that applications can make appropriate choices to fit their needs.

For any collection of transforms, the number of algorithm combinations that make sense tend to be small and dictated by the nature of the transforms. For example, Huffman coding should never be used before other transforms as it destroys any structure in data. Thus, although not done here, it is possible to find an optimal composition sequence by trying all candidate composition sequences on small data samples.

File	Size	Vczip			Bzip2		Gzip	
		Algs	Cmpsz	Bits	Cmpsz	Bits	Cmpsz	Bits
alice29.txt	152089	BMRH	42654	**2.24**	43202	2.27	54423	2.86
ptt5	513216	TBMRH	33149	**0.52**	49759	0.78	56438	0.88
fields.c	11150	BMRh	3036	**2.18**	3039	2.18	3134	2.25
kennedy.xls	1029744	TBMRH	35130	**0.27**	130280	1.01	206767	1.61
sum	38240	SBMRH	12406	**2.60**	12909	2.70	12920	2.70
lcet10.txt	426754	BMRH	105778	**1.98**	107706	2.02	144874	2.72
plrabn12.txt	481861	BMRH	143817	**2.39**	145577	2.42	195195	3.24
cp.html	24603	BMRh	7703	2.50	7624	**2.48**	7991	2.60
grammar.lsp	3721	BMRh	1274	2.74	1283	2.76	1234	**2.65**
xargs.1	4227	BMRh	1728	**3.27**	1762	3.33	1748	3.31
asyoulik.txt	125179	BMRH	39508	**2.52**	39569	2.53	48938	3.13
Total	2810784		426183	**1.21**	542710	1.54	733662	2.09
E.coli	4638690	DH	1137801	**1.96**	1251004	2.16	1341243	2.31
bible.txt	4047392	BMRH	786709	**1.55**	845623	1.67	1191061	2.35
world192.txt	2473400	BMRH	425077	**1.37**	489583	1.58	724593	2.34
Total	11159482		2349587	**1.68**	2586210	1.85	3256897	2.33

File	Vczip		Bzip2		Gzip	
	Cmptm	Dectm	Cmptm	Dectm	Cmptm	Dectm
alice29.txt	0.05	0.01	0.03	0.01	0.01	0.01
ptt5	0.27	0.08	0.03	0.01	0.02	0.01
fields.c	0.01	0.01	0.01	0.01	0.01	0.01
kennedy.xls	0.27	0.18	0.23	0.07	0.11	0.01
sum	0.03	0.01	0.01	0.01	0.01	0.01
lcet10.txt	0.17	0.08	0.12	0.06	0.05	0.01
plrabn12.txt	0.21	0.11	0.15	0.09	0.08	0.01
cp.html	0.01	0.01	0.01	0.01	0.01	0.01
grammar.lsp	0.01	0.01	0.01	0.01	0.01	0.01
xargs.1	0.01	0.01	0.01	0.01	0.01	0.01
asyoulik.txt	0.04	0.01	0.02	0.01	0.01	0.01
E.coli	7.85	0.10	1.92	1.03	1.93	0.08
bible.txt	2.47	1.04	1.64	0.72	0.55	0.06
world192.txt	1.31	0.62	1.04	0.42	0.23	0.03

Fig. 9. Compression size (bytes), rate (bits/byte) and time (seconds) for the Canterbury Corpus

5   Related Works

Vcodex introduces the notion of *data transforms* to encapsulate compression algorithms for reusability. The same goal is behind the development of other library packages such as *zlib* [8] or *bzlib* [17]. However, except for their overall external compression interface, the constituent algorithms in these packages are well hidden in their implementation. For example, different and incompatible versions of Huffman coding were hard-coded in *zlib* and *bzlib*. By contrast, Vcodex data transforms are independent from one another and can be easily reused in writing new transforms. In particular, application systems can develop special transforms for proprietary data types and simply reuse other existing transforms as needed.

An important part of data compression often overlooked is the output data. Such data must be carefully constructed to be sharable across different hardware architectures. Unfortunately, few compression tools publish their coding formats. Two rare exceptions are the *Deflate* Format [6] used in *Gzip* and the *Vcdiff* Format [13] for delta encoding. Both have been standardized and sanctioned by the Internet Engineering Task Force to facilitate transporting of compressed data over the Internet. The specification of *Deflate* was relatively complicated since details of both Lempel-Ziv and Huffman coders must be defined together. By contrast, the self-describing data architecture of Vcodex enables the data formats of transforms to be defined independently from one another. This should ease any effort to specify and standarize such data formats.

6   Conclusions

Advances in compression are continually pulled in two opposite directions: *generalization* to devise techniques applicable to all data types and *specialization* to devise techniques exploiting specific structures in certain classes of data. General techniques such as Lempel-Ziv or Burrows-Wheeler Transform are easy to use and do perform well with most data. However, as seen in our experiments earlier, such techniques could seldom match the performance of algorithms specifically designed for classes of data such as tables, DNA sequences, etc. Unfortunately, the cost to develop data-specific compression tools can quickly become prohibitive considering the plethora of arcane data types used in large application systems. Vcodex provides a unique solution to this problem by applying a software and data engineering approach to compression. Its *data transform* interface frees algorithm designers to focus only on the data transformation task at hand without having to be concerned with other transforms that they may require. Further, users of compression also have the opportunity to mix and match data transforms to optimize the compressibility of their data without being locked into some compositional sequences picked by the tool designers. The best indication of success for Vcodex is that the platform has been in use for a few years in a number of large data warehouse applications handling terabytes of data daily. Tranform compositions properly tailored to data types help achieving compression ratios up to hundreds to one, resulting in significant cost savings. A variety of data transforms have been continually developed along with new data types without disrupting ongoing usage. Further, as the number of data types is limited per application system, simple learning algorithms based on training with small samples of data could often be developed to automatically find optimal combinations of transforms for effective compression. In this way, the platform has proven to be an effective tool to help advance compression not just in specialization but also in generalization.

References

1. Bell, T., Powell, M.: The Canterbury Corpus. Technical Report (2001),
   http://corpus.canterbury.ac.nz
2. Bentley, J., Sleator, D., Tarjan, R., Wei, V.: A Locally Adaptive Data Compression Scheme. Comm. of the ACM 29, 320–330 (1986)

3. Buchsbaum, A., Fowler, G.S., Giancarlo, R.: Improving Table Compression with Combinatorial Optimization. J. of the ACM 50(6), 825–851 (2003)
4. Burrows, M., Wheeler, D.J.: A Block-Sorting Lossless Data Compression Algorithm. Report 124, Digital Systems Research Center (1994)
5. Deorowicz, S.: Improvements to Burrows-Wheeler Compression Algorithm. Software—Practice and Experience 30(13), 1465–1483 (2000)
6. Deutsch, P.: DEFLATE Compressed Data Format Specification version 1.3. In: IETF RFC1951 (1996), http://www.ietf.org
7. Fowler, G.S., Hume, A., Korn, D.G., Vo, K.-P.: Migrating an MVS Mainframe Application to a PC. In: Proceedings of Usenix 2004. USENIX (2004)
8. Gailly, J., Adler, M.: Zlib. Technical report (2005), http://www.zlib.net
9. Huffman, D.A.: A Method for the Construction of Minimum-Redundancy Codes. Proc. of the IRE 40(9), 1098–1101 (1952)
10. Hunt, J.J., Vo, K.-P., Tichy, W.F.: Delta Algorithms: An Empirical Analysis. ACM Transactions on Software Engineering and Methodology 7, 192–214 (1998)
11. Jones, D.W.: Practical Evaluation of a Data Compression Algorithm. In: Data Compression Conference. IEEE Computer Society Press, Los Alamitos (1991)
12. Korn, D.G., Vo, K.-P.: SFIO: Safe/Fast String/File IO. In: Proc. of the Summer 1991 Usenix Conference, pp. 235–256. USENIX (1991)
13. Korn, D.G., MacDonals, J., Mogul, J., Vo, K.-P.: The VCDIFF Generic Differencing and Compression Data Format. Internet Engineering Task Force, RFC 3284 (2002), www.ietf.org
14. Korn, D.G., Vo, K.-P.: Engineering a Differencing and Compression Data Format. In: Proceedings of Usenix 2002. USENIX (2002)
15. Liefke, H., Suciu, D.: Xmill: an efficient compressor for xml data. In: Proc. of SIGMOD, pp. 153–164 (2000)
16. Manzini, G., Rastero, M.: A Simple and Fast DNA Compression Algorithm. Software—Practice and Experience 34, 1397–1411 (2004)
17. Seward, J.: Bzip2. Technical report (1994), http://www.bzip.org
18. Tichy, W.F.: RCS—a system for version control. Software—Practice and Experience 15(7), 637–654 (1985)
19. Vo, B.D., Vo, K.-P.: Using Column Dependency to Compress Tables. In: Data Compression Conference (2004)
20. Vo, B.D., Vo, K.-P.: Compressing Table Data with Column Dependency. Theoretical Computer Science 387, 273–283 (2007)
21. Vo, K.-P.: The Discipline and Method Architecture for Reusable Libraries. Software—Practice and Experience 30, 107–128 (2000)
22. Witten, I.H., Radford, M., Cleary, J.G.: Arithmetic Coding for Data Compression. Comm. of the ACM 30(6), 520–540 (1987)
23. Ziv, J., Lempel, A.: A Universal Algorithm for Sequential Data Compression. IEEE Trans. on Information Theory 23(3), 337–343 (1977)
24. Ziv, J., Lempel, A.: Compression of Individual Sequences via Variable-Rate Coding. IEEE Trans. on Information Theory 24(5), 530–536 (1978)

Part III

Distributed and Parallel Systems

Classification of Benchmarks for the Evaluation of Grid Resource Planning Algorithms

Wolfgang Süß, Alexander Quinte, Wilfried Jakob, and Karl-Uwe Stucky

Institute for Applied Computer Science, Forschungszentrum Karlsruhe GmbH
P.O. Box 3640, D-76021 Karlsruhe, Germany
{wolfgang.suess,alexander.quinte,wilfried.jakob,
uwe.stucky}@iai.fzk.de

Abstract. The present contribution will focus on the systematic construction of benchmarks used for the evaluation of resource planning systems. Two characteristics for assessing the complexity of the benchmarks were developed. These benchmarks were used to evaluate the resource management system GORBA and the optimization strategies for resource planning applied in this system. At first, major aspects of GORBA, in particular two-step resource planning, will be described briefly, before the different classes of benchmarks will be defined. With the help of these benchmarks, GORBA was evaluated. The evaluation results will be presented and conclusions drawn. The contribution shall be completed by an outlook on further activities.

Keywords: Grid Computing, Resource Management System, Resource Broker, Evolutionary Algorithm.

1 Introduction

It is the task of a resource management system to acquire all resources supplied by the grid [1] and to distribute the jobs of the users to these available resources in a reasonable manner. Ideally, planning and execution of these jobs take place with these resources at optimum costs and/or time in accordance with the wishes of the users, without the latter being burdened with unnecessary detailed knowledge about the resources. Other requirements on resource management are a good and cost-efficient load distribution and the capability of identifying, managing, and tolerating errors in order to ensure a error-free and stable operation.

The jobs are carried out in the form of workflows that contain all information on the working steps to be performed and the grid resources required for this purpose. To obtain a statement with respect to the performance of a resource management system, suitable benchmarks are required. Benchmarks are also needed for the development and selection of adequate optimization strategies for resource planning.

For this purpose, the resource management system GORBA (Global Optimizing Resource Broker and Allocator) [2] was developed. It uses various optimization algorithms for resource planning. To compare the performance of already implemented algorithms and later new developments, suitable benchmarks were constructed.

J. Filipe et al. (Eds.): ICSOFT/ENASE 2007, CCIS 22, pp. 215–227, 2008.

The resource management system GORBA shall be described briefly. The contribution will focus on the presentation of the systematic construction of benchmarks and on the evaluation of GORBA and the optimization strategies for resource planning using these benchmarks. The results of benchmark runs performed with various optimization strategies will be presented.

2 Resource Brokering for Complex Application

2.1 GORBA

As indicated by its name, GORBA (Global Optimizing Resource Broker and Allocator) represents a solution for the optimization of grid job planning and resource allocation in a grid environment. It was described in detail in a number of publications, e.g. in [2,3].

Resource management systems can be divided into *queuing systems* and *planning systems* [4]. The difference between both systems lies in the planned time window and the number of jobs considered. Queuing systems try to allocate the resources available at a certain time to the currently waiting job(s) being first in the queue(s). Resource planning for the future for all waiting requests is not done. In contrast to this, planning systems examine the present and future situation, which results in an assignment of start times to all requests. Today, almost all resource management systems belong to the class of queuing systems. Contrary to queuing systems, planning systems require more information, such as the duration of execution or costs, resource performance, long-term availability of resources, and others. Therefore, the implementation of queuing systems usually is much easier. However, a queuing system is only efficient in case of a low usage of the system. In the case of increased usage, the queuing system reveals considerable weaknesses with respect to the quality of services, resource usage, and execution time of the individual grid jobs. Additionally, no statements can be made about the presumable time of execution for waiting grid jobs. For these reasons, a user-friendly and future-oriented grid resource management system must be based on planning rather than on queuing only.

A special feature of GORBA is two-step job planning, where an evolutionary algorithms is combined with heuristic methods in order to provide the user with an optimum access to the available resources. In a first step, different heuristic methods are applied to provide rapid preliminary job plans under time-critical conditions. Based on the results of the first planning step, further improvements are made using an evolutionary algorithm, if necessary. Job planning in GORBA is dynamic. This means that in case of unforeseeable events, for example the failure or addition of resources, arrival of new jobs, change or premature deletion of jobs currently processed, a new planning cycle is initiated.

2.2 Workflow of the Grid Application

Usability and acceptance of a grid environment will largely depend on how the user has to formulate his grid application and to what an extent he is supported in doing so. The grid application shall be represented by a workflow that describes dependencies between elementary application tasks by predecessor relations. A workflow, called *application job*, consists of individual grid jobs that are basically described by the combination of various resources requirements. The resources are mainly hardware and software resources that execute the grid jobs. When specifying the resource requirement, the user is free to specify a certain resource he needs for his grid job or, less specifically, a certain resource type. In the latter case, the resources explicitly tailored to the grid job are allocated by the system. The less specific the resource is given by the user, the more planning alternatives result for the resource broker. According to the workflow concept, it is planned to support sequences, parallel splits, alternatives, concurrencies, and loops for the user to implement also dynamic workflows.

A workflow manager determines the relevant information from a user-specified workflow and supplies this information to GORBA for resource planning. It is concentrated on workflows that may be represented by DAGs (direct acyclic graphs). Fig. 1 presents examples of workflows of application jobs.

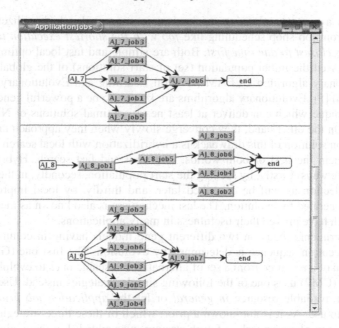

Fig. 1. Examples of application jobs

2.3 Resource Planning as an Optimization Problem

Resource planning in GORBA can only be accomplished when various information items are available. As use of the resources in the future is planned, the workflow and execution time normalized to a reference performance factor for each grid job have to be known. And, of course, it is essential to know which resources or resource types are needed by the grid job. GORBA also has to know the resources available in the grid and their performance and costs. Costs may vary according to day time, days of week or other time frames. The user can specify the earliest starting point, latest end, maximum costs, and weighing of time and costs for his application job. Planning problems like this belong to the class of NP-complete problems. This means that the optimum solution cannot be found within polynominal time. But this is not necessary, as long as a schedule is found, which fulfils all user requirements in terms of time and costs at least and the resources are used homogeneously in the sense of the resource supplier.

The quality of the schedule is determined by the fulfillment of different criteria, e.g. makespan, fulfillment of user requirements (time and costs) or resource utilization, which partly contradict each other. For all these criteria, a normalized quality function is defined and the resulting values are added up to a weighted sum. This weighted sum may be reduced by a penalty function which is applied in case of the violation of constraints. The weighted sum is used instead of pareto optimization, because alternative solutions make little sense in an automated scheduling process.

2.4 Optimization Strategies

In GORBA a two-step planning mechanism is suggested, which utilizes approved heuristics from job shop scheduling like *job with the shortest execution time first* or *job which is closest to due time first*. Both are simple and fast local optimizers. They are used to seed the initial population (set of start solutions) of the global optimizer, the evolutionary algorithm GLEAM (General Learning and Evolutionary Algorithm and Method) [5]. Evolutionary algorithms are known to be a powerful general optimization technique which can deliver at least nearly optimal solutions of NP-complete problems. On the other hand, they converge slowly when they approach an optimum. The common solution of this drawback is a hybridization with local search methods in order to obtain the best of both worlds: A global and fast search. Hybridization is done in three ways: Firstly, by seeding the start population, secondly, in the process of resource selection as will be described later, and thirdly, by local improvement of offspring generated by evolution. The last mechanism is also known as memetic algorithms which have proved their usefulness in many applications.

Our experiments focus on two different gene models having in common that the grid job execution sequence is determined by evolution. The first one (GM1) leaves the selection of a resource from a set of alternatively useable ones to evolution and the second one (GM2) uses one of the following simple strategies instead: Use the *fastest* or *cheapest* available resource *in general* or let the *application job priority* decide which one to use. As it is not known a priori which of these three strategies performs best for a given planning task, a fourth strategy was added: Let the evolution decide which of the three strategies to use for a generated solution. This means that the resource selection strategy is co-evolved together with the schedules.

3 Benchmarks

To evaluate scheduling algorithms, two types of benchmarks are used: Benchmarks modeled from real applications and synthetically produced benchmarks [6,7,8]. It is the advantage of application-oriented benchmarks that they are close to practice. Their drawbacks consist in a mostly small diversity and in the fact that their characteristic properties which will be described below cannot be influenced specifically.

Therefore, it was decided to use synthetically produced benchmarks to evaluate and improve the optimization strategies in GORBA.

Examples for other synthetically produced benchmarks can be found in [6,7]. These benchmarks are restricted to homogeneous resources and to single DAG scheduling. By contrast, the GORBA benchmarks include inhomogeneous resources with different performance factors, different costs, and availabilities. Another important aspect of GORBA is the possibility of planning and optimisation of multiple application jobs, each with its own individual optimisation goals (multiple DAG scheduling), which requires enhancements of the existing benchmarks. Multiple DAG scheduling is also treated in [9] and it is planned to examine these benchmarks and feed them to GORBA in the near future.

For the benchmarks, two parameters are defined, which describe their complexity. The parameter D denotes the degree of mutual dependency of the grid jobs, which results from their predecessor/successor relations. As the grid jobs usually have various resources requirements, which means that they cannot be executed on any resource, another parameter (R) describes the degree of freedom in the selection of resources. Both parameters are defined as follows:

$$\text{Dependence: } D = \frac{spj}{spj_{max}}$$

spj :	Sum of all predecessor jobs of all grid jobs.
$spj_{max} = \dfrac{n(n-1)}{2}$:	Maximum possible sum of spj
n :	Number of grid jobs

The dependence D yields the permutation of the orders of all grid jobs. The smaller D is, the larger is the number of permutations of all grid jobs. Fig. 2 shows various dependencies based on a simple example of three grid jobs. Depending on D, the set of execution sequences of the grid jobs and, hence, the number of planning alternatives varies. In the example, six possible permutations of grid jobs a, b, and c exist for $D = 0$ (Fig. 2a). For $D = 1/3$, the three permutations cab, acb, and abc are possible (Fig. 2b). For $D = 2/3$, there are only the two permutations of abc and bac (Fig. 2c). For $D = 1$, the execution sequence abc remains (Fig. 2d).

Consequently, a small D may result in a high parallelization capacity that depends on the degree of freedom in the selection of resources, however. The degree of freedom in resource selection is defined as:

$$R = \frac{tar}{n \cdot n_{res}}$$

tar :	Total of alternative resources of all grid jobs.
n :	Number of grid jobs.
n_{res} :	Number of grid resources.

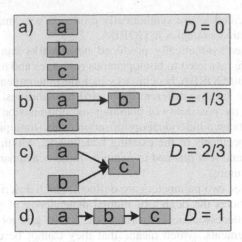

Fig. 2. Dependence D based on the example of $n = 3$

The degree of freedom in resources selection denotes the planning freedom with respect to the resources. A small value of R means a small mean possibility of selection of resources per grid job. A large value of R means a large mean selection per grid job. This means, for instance, that the class of benchmarks with small D and large R allows for the highest degree of parallelization and possesses the largest number of planning alternatives. In contrast to this, the class of benchmarks with high D and small R allows for the smallest degree of parallelization and resource alternatives.

To evaluate GORBA, four benchmark groups were set up, with a small and large dependence D, combined with a small and high degree of freedom R, respectively. For the comparison of different optimization strategies, it is sufficient in the first step to restrict resource usage to one resource per grid job, i.e. the coallocation capability of GORBA is not used.

Each of these four benchmark groups comprises three benchmarks with 50, 100, and 200 grid jobs, a total duration of 250, 500, and 1000 time units per performance factor (TU/PF), and 10, 20, and 40 application jobs, respectively. In Table 1 an overview of the different benchmarks is given.

All these benchmarks were based on the same grid environment simulated with a total of 10 hardware resources of varying performance factors (PF) and costs (C). In Table 2 an overview of the different hardware resources is given. The last column denotes the costs of the hardware resources related to the performance factor (C/PF).

As mentioned above, requirements by the user usually are made on the application jobs with respect to their maximum costs and their latest time of completion. The quality of resource planning depends on how well the requirements of the application jobs are met. As an optimum, all application jobs are within the given cost and time limits.

Apart from the characteristic parameters R and D defined above, the complexity of the planning problem also depends on the user requirements made on the application jobs, i.e. the influence of the user requirements on the benchmarks is not only expressed by R and D and has to be considered separately when constructing the benchmarks.

Table 1. Characteristics of benchmarks (sR means small value of R, lR means large R, sD means small D, and lD means large D)

Benchmark	No. Appl.jobs	No. Grid jobs	TU/ PF	R	D
sRsD-50	10	50	250	0.288	0.037
sRsD-100	20	100	500	0.304	0.019
sRsD-200	40	200	1000	0.303	0.009
sRlD-50	10	50	250	0.272	0.090
sRlD-100	20	100	500	0.278	0.044
sRlD-200	40	200	1000	0.28	0.022
lRsD-50	10	50	250	0.828	0.037
lRsD-100	20	100	500	0.842	0.019
lRsD-200	40	200	1000	0.843	0.009
lRlD-50	10	50	250	0.828	0.090
lRlD-100	20	100	500	0.828	0.044
lRlD-200	40	200	1000	0.832	0.022

Table 2. Characteristics of the resources

Hardware	PF	C	C / PF
HW_01	0.5	1	2
HW_02	0.5	1.1	2.2
HW_03	0.8	1	1.25
HW_04	0.8	1.4	1.75
HW_05	1	1.5	1.5
HW_06	1	1.5	1.5
HW_07	1.2	1.6	1.33
HW_08	1.5	1.8	1.2
HW_09	1.5	2.4	1.6
HW_10	1.5	2.5	1.67

As far as the user requirements are concerned, three classes of benchmarks can be distinguished. The first class comprises benchmarks that can be solved by the heuristic method already.

Hence, the more time consuming second planning step is not required. The second class of benchmarks includes benchmarks that can no longer be solved by the heuristic methods, but by the evolutionary algorithm of the second planning step. The third class includes benchmarks that cannot be solved at all because of too tight time requirements, for example. As this contribution mainly focuses on how the second planning step can be improved, benchmarks of the second class are of particular interest. Consequently, the time and cost requirements were defined, such that times or costs were exceeded in at least one up to four application jobs during the first planning step.

These benchmarks were used to determine the improvements achieved by the second planning step as compared to the first. As the evolutionary method GLEAM used in the second planning step is a non-deterministic method, 100 GLEAM runs were made for each benchmark in order to obtain a reasonable statistic statement. Each GLEAM run was limited to three minutes.

In the second planning step the two different gene models GM1 and GM2 were applied. The results of these benchmark studies shall be presented below.

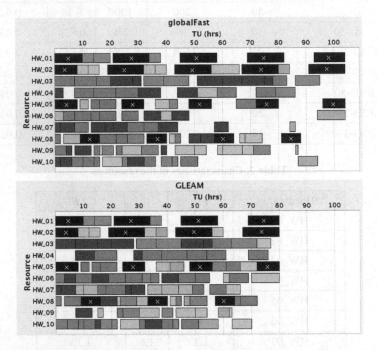

Fig. 3. Results of benchmarks sRID-100: On the top a schedule generated from the best heuristic planning. On the bottom a schedule generated from GLEAM.

4   Results

By way of example, Fig. 3 and Fig. 4 show the planning results of the benchmark sRID-100. The resource plans generated by both planning steps are shown in Fig. 3. The top plan shows the best result of the six heuristic algorithms integrated in the first planning step. The bottom plan represents the result of the second planning step, with the gene model GM1 being used by GLEAM. The plans show the allocation of the individual grid jobs to the resources. All grid jobs of an application job are marked by the same grey value. Black bars marked with an **x** indicate times, at which the resource must not be used (HW_01, HW_02, HW_05, and HW_08). Heuristic planning certainly has problems in allocation, which is reflected by large gaps in the plan. Compared to heuristic planning, GLEAM reaches a much more compact allocation of resources.

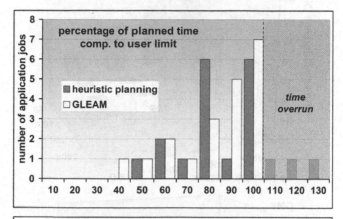

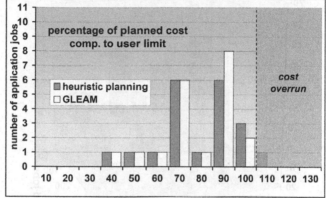

Fig. 4. Results of benchmark sRlD-100: Comparison of GLEAM and heuristic planning related to time and cost constraints

Fig. 4 shows the fulfillment of the time (top) and cost (bottom) requirements of the example from Fig. 3. The degree of fulfilling the requirement is given in percent on the X-axis. A value above 100% means that the requirements are exceeded. Values smaller or equaling 100% mean that the requirements are met or not even reached. The height of the bars represents the number of application jobs lying in the respective fulfillment range. It is aimed at all application jobs fulfilling the requirements.

The charts show that when using GLEAM, all application jobs meet the requirements. In heuristic planning three application jobs exceed the time limits and one application job the costs. The results of the four benchmark groups shall be presented below.

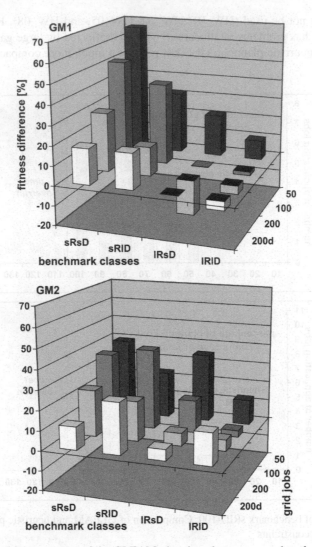

Fig. 5. Statistical improvement of the GLEAM planning phase compared to the heuristic planning phase. Use of gene model GM1 on the top and GM2 on the bottom.

They are based on 100 runs per benchmark due to the stochastic nature of GLEAM and the comparisons are based on averages. The time available to a GLEAM planning is limited to three minutes because planning must be done quickly for real applications. Fig. 5 compares the two gene models based on fitness differences between the best heuristic planning ignoring the penalty functions and the best GLEAM results. As this compares the aggregated quality values of the schedules while time and cost restrictions are ignored, fitness reductions can occur as it is the case with GM1 in Fig. 5.

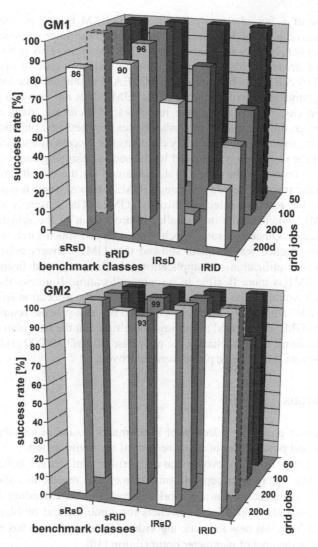

Fig. 6. Success rates of the GLEAM planning phase using gene model GM1 on the top and GM2 on the bottom. Dotted lines indicate runs, where only one population size yields 100% success. These runs are considered not to be robust.

The comparisons are based on unpunished fitness values, which are not seen by the evolutionary process. It is the first aim of the evolutionary optimization to overcome penalty functions and then improve the overall quality of the schedules. If this process is interrupted after the runtime of 3 minutes unpunished schedules with low quality can occur. It is remarkable that the impairment for GM1 of Fig. 5 are at largest for those benchmarks, which deliver poor results in Fig. 6.

In the case of a few alternative resources, GLEAM results in considerable improvements as compared to heuristic planning, with these improvements being better than in case of benchmarks with many alternative resources. This is because the heuristic planning already yields very good planning results in case of many alternative resources, which can hardly be improved by GLEAM within the time available.

Planning optimization with the gene model GM2 does not result in such high improvements in case of few alternative resources. If there are many alternative resources, however, optimization is somewhat better than heuristic planning.

Another topic of the benchmark study concerns the success rate which indicates the probability of the result being improved by the second planning step compared to the first one. Fig. 6 compares the success rates obtained for the two different gene models. Evolving the resource selection strategy (GM2) in most cases is equal to or better than evolving the resource selection directly (GM1). The reason is a larger search space for GM1, which results in a smaller improvement of the schedule within the given time frame. Other test runs which were stopped much later, when a certain degree of convergence was reached, showed that GM1 delivers better solutions in terms of resource utilization and application job cheapness and fastness. This was expected, as GM1 is more flexible in resource allocation. It allows the usage of resources, which would not be possible obeying one of the allocation strategies, as the decision is made individually for every grid job. But this process requires more time and, therefore, GM2 is preferred according to the rule that the best plan is useless, if it comes too late. In all cases, including the poor case of *lRsD* for 200 grid jobs of GM1, the schedules from the heuristic phase were improved.

5 Conclusions

It was shown that a suitable selection of benchmarks results in valuable information on the quality and possibilities of improvement of the optimization.

Global planning using an evolutionary algorithm can deliver better results than simple heuristics within an acceptable time frame. The results also show a need for improving the optimization. Current work concentrates on extending and enhancing GLEAM by newly developed local searchers for combinatorial problems. We expect a great benefit from this new memetic algorithm, as this approach has already proved its superiority in the area of parameter optimization [10].

So far, the benchmarks have been generated manually. At the moment, it is worked on a new concept for the construction and automatic generation of benchmarks. With this, the set of benchmarks will be extended considerably in order to improve the information quality. Moreover, it is planned to integrate other heuristic methods in the first planning step of GORBA. So actually we have integrated the Giffler-Thompson Algorithm [11], which is one of the most famous active schedule generation scheme for the classical job-shop problem. First results show that one of the already included heuristics of the first planning phase performs better than the Giffler-Thompson Algorithm. Detailed results are currently elaborated and will be published soon.

For GORBA, a modular setup is envisaged, with the optimization methods being tailored to the type of planning problem arising.

References

1. Foster, I., Kesselman, C., Tuecke, S.: The Anatomy of the Grid: Enabling Scalable Virtual Organisations. Intern. Journal of Supercomputer Applications 15(3), 200–222 (2001)
2. Süß, W., Jakob, W., Quinte, A., Stucky, K.-U.: GORBA: a global optimizing resource broker embedded in a grid resource management system. In: Proc. 17th IASTED Intern. Conference on Parallel and Distributed Computing Systems (PDCS), Phoenix, AZ, pp. 19–24 (2005)
3. Süß, W., Jakob, W., Quinte, A., Stucky, K.-U.: Resource Brokering in Grid Environments using Evolutionary Algorithm. In: Proceedings of the Int. Conf. on Parallel and Distributed Computing and Networks (IASTED), A, February 14-16 (2006)
4. Hovestadt, M., Kao, O., Keller, A., Streit, A.: Scheduling in HPC Resource Management Systems: Queuing vs. Planning. In: Feitelson, D.G., Rudolph, L., Schwiegelshohn, U. (eds.) JSSPP 2003. LNCS, vol. 2862, pp. 1–20. Springer, Heidelberg (2003)
5. Blume, C., Jakob, W.: GLEAM – An Evolutionary Algorithm for Planning and Control Based on Evolution Strategy. In: Conf. Proc. GECCO 2002, Vol. Late Breaking Papers, pp. 31–38 (2002)
6. Tobita, T., Kasahara, H.: A standard task graph set for fair evaluation of multiprocessor scheduling algorithms. Journal of Scheduling, 379–394 (2002)
7. Hönig, U., Schiffmann, W.: A comprehensive Test Bench for the Evaluation of Scheduling Heuristics. In: Proceedings of the 16th International on Conference Parallel and Distributed Computing and Systems (PDCS), Cambridge (USA) (2004)
8. Wieczorek, M., Prodan, R., Fahringer, T.: Comparison of Workflow Scheduling Strategies on the Grid. In: Wyrzykowski, R., Dongarra, J., Meyer, N., Waśniewski, J. (eds.) PPAM 2005. LNCS, vol. 3911, pp. 792–800. Springer, Heidelberg (2006)
9. Hönig, U., Schiffmann, W.: A Meta-algorithm for Scheduling Multiple DAGs in Homogeneous System Environments. In: Proceedings of the 18th International Conference on Parallel and Distributed Computing and Systems (PDCS), Dallas (USA) (2006)
10. Jakob, W., Blume, C., Bretthauer, G.: HyGLEAM -Towards a Generally Applicable Self-adapting Hybridization of Evolutionary Algorithms. In: Conf. Proc. of Genetic and Evolutionary Computation Conference (GECCO 2004), vol. Late Breaking Papers. CD, ISGEG, New York (published, 2004)
11. Giffler, B., Thompson, G.L.: Algorithms for solving production scheduling problems. Operations Research 8(4), 487–503 (1960)

A Disconnection-Aware Mechanism to Provide Anonymity in Two-Level P2P Systems

J.P. Muñoz-Gea, J. Malgosa-Sanahuja, P. Manzanares-Lopez,
J.C. Sanchez-Aarnoutse, and J. Garcia-Haro

Department of Information Technologies and Communications,
Polytechnic University of Cartagena,
Campus Muralla del Mar, 30202, Cartagena, Spain
{juanp.gea,josem.malgosa,pilar.manzanares,
juanc.sanchez,joang.haro}@upct.es

Abstract. In this paper, a disconnection-aware mechanism to provide anonymity in two-level P2P systems is proposed. As usual, anonymity is obtained by means of connecting the source and destination peers through a set of intermediate nodes, creating a multiple-hop path. The main contribution of the paper is a distributed algorithm able to guarantee the anonymity even when a node in a path fails (voluntarily or not). The algorithm takes into account the inherent costs associated with multiple-hop communications and tries to reach a well-balanced solution between the anonymity degree and its associated costs. Some parameters are obtained analytically but the main network performances are evaluated by simulation. We also quantify the costs associated with the control packets used by the distributed recovery algorithm.

1 Introduction

Peer-to-peer networks are one of the most popular architectures for file sharing. In some of these scenarios, users are also interested in keeping mutual anonymity; that is, any node in the network should not be able to know who is the exact origin or destination of a message. Traditionally, due to the connectionless nature of IP datagrams, the anonymity is obtained by means of connecting the source and destination peers through a set of intermediate nodes, creating a multiple-hop path between the pairs of peers.

There are various existing anonymous mechanisms with this operation, but the most important are the mechanisms based in Onion Routing [1] and the mechanisms based in Crowds [2]. The differences between them are the following: In Onion, the initiator can determine an anonymous path in advance to hide some identification information. When a user wants to establish an anonymous communication, he will forward its message to a *Onion proxy*. This *Onion proxy* will randomly select some nodes in the network and will establish the complete route (called *Onion*) between the initiator an the responder. This *Onion*, is a recursively layered data structure that contains the information about the route to be followed over a network. Every node can only decrypt its corresponding layer with its private key, therefore, the first node will decrypt the first layer to see the

J. Filipe et al. (Eds.): ICSOFT/ENASE 2007, CCIS 22, pp. 228–239, 2008.

information about next hop in the route, and this process will continue until the message reaches its destination. Tarzan [3], Tor [4], SSMP [5] and [6] are implemented based on Onion Routing to provide anonymous services.

On the other hand, in Crowds let middle nodes select the next hop on the path. A user who wants to initiate a communication will first send the message to its node. This node node upon receiving the request will flip a biased coin, to decide whether or not to forward this request to another node. The coin decides about forwarding the request based on probability p. If the probability is to forward, then it will forward to another node and the process continues. If the probability is not to forward, then it will directly forward the request to the final destination. Each node when forwarding to another node records the predecessors information and in this way a path is built, which is used for communication between the sender and the receiver. There are several works [7], [8], [9], based in the mechanism used by Crowds to provide anonymity.

The above procedure to reach anonymity has two main drawbacks. The first one is that the peer-to-peer nature of the network is partially eliminated since now, peers are not directly connected but there is a path between them. Therefore, the cost of using multiple-hop path to provide anonymity is an extra bandwidth consumption and a additional terminal (node) overhead.

In addition, as it is known, the peer-to-peer elements are prone to unpredictable disconnections. Although this fact always affects negatively the system performances, in an anonymous network it is a disaster since the connection between a pair of peers probably would fail although both peers are running. Therefore, a mechanism to restore a path when an unpredictable disconnection arises is needed but it also adds an extra network overhead in terms of control traffic.

[10] presented a comparative analysis about the anonymity and overhead of several anonymous communication systems. This work presents several results that show the inability of protocols to maintain high degrees of anonymity with low overhead in the face of persistent attackers. In [11], authors calculate the number of appearances that a host makes on all paths in a network that uses multiple-hop paths to provide anonymity. This study demonstrates that participant overhead is determined by number of paths, probability distribution of the length of path, and number of hosts in the anonymous communication system.

In this paper we propose an anonymity mechanism for a two-level hierarchical P2P network presented in a previous work [12] based on Crowds to create the multiple-hops path. However, our mechanism introduce a maximum length limit in the path creation algorithm used by Crowds, in order to retrict the partipant overhead.

The main paper contribution is a distributed algorithm to restore a path when a node fails (voluntarily or not). The algorithm takes into account the three costs outlined above in order to obtain an equilibrated solution between the anonymity degree and its associated costs. The parameters are obtained analytically and by simulation.

The remainder of the paper is as follows: Section 2 summarizes the main characteristics of our hierarchical P2P network. Section 3 and 4 deeply describes the mechanism to provide anonymity. Section 5 discusses the TTL methodology. Section 6 shows the simulation results and finally, Section 7 concludes the paper.

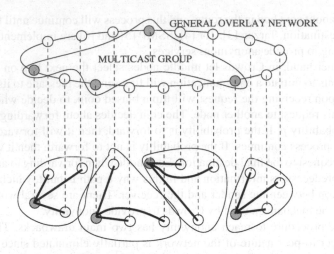

Fig. 1. General architecture of the system

2   A Two-Level Hierarchical P2P Network

In our proposal [12], peers are grouped into subgroups in an unstructured way. However, this topology is maintained by means of an structured lookup mechanism.

Every subgroup is managed by one of the subgroup members, that we call "subgroup leader node". The subgroup leader is the node that has the best features in terms of CPU, bandwidth and reliability. When searching for a content, the user will send the lookup parameters to its local subgroup leader. The local subgroup leader will resolve the complex query, obtaining the results of this query.

To allow the subgroup leaders to locate contents placed in remote subgroups, all the leaders of the network are going to be members of a multicast group. The structured P2P network can be used to implement an Application Level Multicast (ALM) service. CAN-multicast [13], Chord-multicast [14] and Scribe [15] are ALM solutions based on the structured P2P networks CAN [16], Chord [17] and Pastry [18], respectively. Anyone of these methods provides an efficient mechanism to send messages to all the members of a multicast group.

To implement the structured-based maintenance of the unstructured topology, all the nodes (peers) must be immersed into a structured network. Therefore, every node has an identifier (*NodeID*), and they have to contact with an existing node to join this network. Figure 1 describes the general architecture of the system.

The previous existing node also gives to the new node the identifier of the subgroup it belongs (*SubgroupID*), and using the structured lookup mechanism the new node will find the leader of that subgroup. This is possible because each time a node becomes a subgroup leader, it must contact with the node which *NodeID* fits with the *SubgroupID* and sends it its own IP address.

Initially, the new node will try to connect the same subgroup than the existing node. To do that, the new node must contact with the leader of this subgroup, that will accept its union if the subgroup is not full. However, if there is no room, the new node will be asked to create a new (randomly generated) subgroup or it will be asked to join the subgroup that the resquested leader urged to create previously. The use of different existing nodes allows the system to fill up incomplete subgroups. In addition, to avoid the existence of very small subgroups, the nodes will be requested by their leader to join another subgroup if the subgroup size is less than a threshold value.

When a new node finds its subgroup leader, it notifies its resources of bandwidth and CPU. Thus, the leader forms an ordered list of future leader candidates: the longer a node remains connected (and the better resources it has), the better candidate it becomes. This list is transmitted to all the members of the subgroup.

3   Providing Anonymity

In this section we propose a file sharing P2P application built over the previous network architecture which provides mutual anonymity. As it is defined in [19], a P2P system provides mutual anonymity when any node in the network, neither the requester node nor any participant node, should not be able to know with complete certainty who is the exact origin and destination of a message.

Our solution defines three different stages. On one hand, all peers publish their contents within their local subgroups (publishing). To maintain the anonymity during this phase, a random walk proceduce will be used. On the other hand, when a peer initiates a complex query, a searching phase will be carried out firstly (searching). Finally, once the requester peer knows all the matching results, the downloading phase will allow the download of the selected contents (downloading). In the following section these three stages are described in detail.

3.1   Publishing the Contents

To maintain the anonymity, our solution defines a routing table at each peer and makes use of a random walk (RW) technique to establish a random path from the content source to the subgroup leader. When a node wants to publish its contents, first of all it must choose randomly a connection identifier (great enough to avoid collisions). This value will be used each time this peer re-publishes a content or publishes a new one.

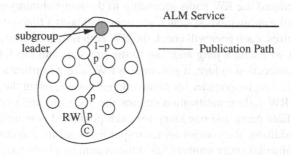

Fig. 2. Publishing phase

Figure 2 describes the random path creation proces. The content owner (C) randomly chooses another active peer (a peer knows all the subgroup members, which are listed in the future leader candidate list, see Section 2) and executes the RW algorithm to determine the next peer in the RW path as follows. The peer will send the *publish message* directly to the subgroup leader with probability $1 - p$, or to the randomly chosen peer with probability p. This message contains the connection identifier associated to the content owner and the published content metadata. The content owner table entry associates the connection identifier and the next peer in the RW path. Each peer receiving a *publish message* follows the same procedure. It stores an entry in its table, that associates the connection identifier, the last peer in the RW path and the next node in the RW path (which will be determined by the peer using the RW algorithm as it is described before).

To prevent a message forwarding loop within the subgroup, each initial *publish message* (that is generated by the content owner) is attached with a TTL (Time To Live) value. A peer receiving a *publish message* decrements the TTL by one. Then, if the resulting TTL value is greater than 1, the peer executes the RW algorithm. Otherwise, the message is sent directly to the subgroup leader. When the *publish message* is received by the subgroup leader, it just stores the adequate table entry that associates the connection identifier, the published content metadata and the last peer in the RW path.

The probability that the RW has n hops is

$$P(n) = \begin{cases} p^{n-1}(1 - p) & n < TTL, \\ p^{TTL-1} & n = TTL. \end{cases} \quad (1)$$

Therefore the mean length of a RW is

$$\overline{RW} = \frac{1 - p^{TTL}}{1 - p} \xrightarrow{p \to 1} TTL. \quad (2)$$

Thanks to the publishing procedure, each subgroup leader knows the metadata of the contents that have been published into its subgroup. However, this publishing procedure offers sender anonymity. Any peer receiving a *publish message* knows who is the previous node. However, it does not know if this previous peer is the content owner or just a simple message forwarder.

The RW path distributely generated in the publishing phase will be used during the searching and downloading phases, as it will be described later. Therefore, it is necessary to keep updated the RW paths according to the non-voluntary peer departures. From this joint to the anonymous system, each peer maintains a *timeout timer*. When its timeout timer expires, each peer will check the RW path information stored in its table. For each entry, it will send a *ping message* to the previous and the following peer. If a peer detects a connection failure, it generates a RW failure notification that is issued peer by peer to the subgroup leader (or to the content owner) using the RW path. Each peer receiving a RW failure notification deletes the associate table entries (only one entry in intermediate peers and one entry for each published content in the subgroup leader peer). In addition, the content owner re-publishes all its contents. On the other hand, if an intermmediate peer wants to leave the system in a voluntary way, it will get in touch with the previous and the following peer associated to each table entry in order to updated their entries.

An additional verification is carried out each time a new content is going to be published. Before sending the *publish messages*, the content owner peer makes a *totalPing* (a ping sent to the subgroup leader through the RW path). If the *totalPing* fails (any intermediate peer is disconnected), the content owner peer re-publishes all its contents initiating the creation of a new distributed RW path. The intermediate peers belonging to the failed RW path will update their table entries when their timeout expires.

3.2 Searching the Contents

The searching procedure to obtain requester anonymity is described in this section. Figure 3 describes this procedure. When a requester (R) wants to make a complex search, first of all it must select its connection identifier. If the peer already chose a connection identifier during a publish phase, this value is used. Otherwise, a connection identifier is chosen[1].

Then, the peer generates a new type of message called *lookup message*. A *lookup message*, that encapsulates the metadata of the searched content and the connection identifier associated to the requester peer (R), is forwarded towards the local subgroup leader using the RW path associated to this connection identifier (RW1). If the connection identifier is new, the RW1 path from the requester peer towards the subgroup leader is generated, distributely stored, and maintained as it was explained in the section before.

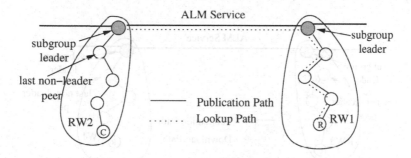

Fig. 3. Searching phase

Once the *lookup message* has been received, the local subgroup leader is able to locate the matching results within its subgroup. On the other hand, our system uses an ALM technique to distribute the lookup messages to all the subgroup leader nodes.

Each subgroup leader with a positive matching result must answer the requester's subgroup leader with a *lookup-response message*. This message contains the connection identifier established by the requester peer, the matching content metadata, the connection identifier established by the content owner peer and the identity of the last non-leader peer in the RW path (RW2) towards the content owner (C).

In fact, before sending the *lookup-response message*, the subgroup leader will check the complete RW2 path towards the content owner by means of a *totalPing*. If the RW2

---

[1] Each peer is associated to only a connection identifier.

path fails, the subgroup leader will send to all the peers in the subgroup a broadcast message containing the associate connection identifier. The peer that established this connection will re-publish all its contents (that is, it initiates the creation of a new RW for the connection identifier). Peers that maintain a table entry associated to this connection identifier will delete it. The rest of peers will take no notice of the message. Consequently, the subgroup leader only sends a *lookup-response message* to the requester subgroup leader after checking that the associate RW2 path is active.

Finally, the *lookup-response message* will be forwarded through the requester subgroup towards the requester peer using the adequate and distributed RW1 path stored in the tables.

3.3 Downloading the Contents

Once the requester peer receives the *lookup-response* messages, it is able to select which contents it wants to download. The anonymity is also important during this third phase: both the requester peer and the content owner peer should keep anonymous to the rest of peers in the system.

In the download phase (see Figure 4), the requester peer (R) generates a *download-request message* which initially encapsulates the required content metadata, the connection identifier established by itself, the connection identifier established by the content owner (C) and the identity of the last non-leader peer in the RW2 (in the content subgroup).

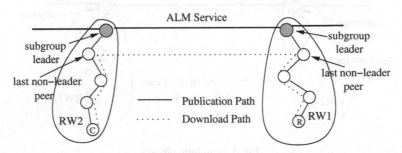

Fig. 4. Download phase

This message will be forwarded towards the subgroup leader using the distributely stored RW1 path. However, this message will not reach the subgroup leader. The last non-leader peer in the RW1 path (the previous to the leader) will be the responsible for sending the *download-request message* to the last non-leader peer in the RW2 (in the content subgroup). Before sending the message to the content owner subgroup, this peer (the last non-leader peer in RW1) must encapsulate an additional value: its own identity. The message will be forwarded through the content owner subgroup using the distributely stored RW2 path until it reaches the content owner.

Finally, the content owner peer will start the content delivery. The *download messages* encapsulate the requited content, the connection identifier established by the requester peer, the connection identifier established by itself and the identity of the last non-leader peer in the RW1 (in the requester subgroup). These messages will be forwarded towards the subgroup leader using the adequate RW2 path. However, they will not reach the subgroup leader. The last non-leader peer in the RW2 path will be the responsible for sending them to the node indicated in the message. Once in the requester subgroup, the messages will be forwarded using the RW1 path until they reach the requester peer. Therefore, the mean length of a download path (DP) is

$$\overline{DP} = 2(\overline{RW} - 1) + 1 = 2\frac{1 - p^{TTL}}{1 - p} - 1 \qquad (3)$$

4 Dynamic Aspects of the System

The peer dynamism or *churn* is an inherent feature of the P2P networks. Our solution exploits this property to improve the anonymity behavior of the system. As it has been described in the previous section, each peer establishes a RW path towards its subgroup leader during the publishing or the searching phases. If there was no churn, these RW paths (which are randomly established) would not change as the time passes. However, due to the peer departures and failures, the RW paths must be continuously updated. Consequently, the RW path information that could be obtained by attackers will be obsolete as the time passes.

The random election of each RW path is another interesting feature of our proposal. The main objective of this procedure is to reach the mutual anonymity. However, there is another important consequence. The first non-leader peer (closest to the leader) in each RW path changes from a RW to another. Consequently, the remote peers involved in the download process change depending on the content owner peer and the requester peer. As a consequence, the downloading load is distributed among all the peers. And something more important, the subgroup leaders don't participate during the download process making lighter their processing load.

As it has been described in the previous sections, the subgroup leaders have a key role during the publishing and the searching phases. Each subgroup leader maintains a list of the published contents in its subgroup that is used to solve the complex lookups. The subgroup leader management is defined by the network architecture. However, there is no doubt that the subgroup leader changes also affect the anonymous system. Each time a subgroup leader changes, all the RW paths would be obtained again. To optimize the system behaviour when a subgroup leader changes we propose some actions. If a subgroup leader changes voluntarily (another peer in the subgroup has become a most suitable leader or the leader decides to leave the system), it will send to the new leader a copy of its table. On the other hand, to solve the non-voluntary leader leaving, we propose to replicate the table of the subgroup leader to the next two peers in the future leader candidate list. Consequently, when the subgroup leader fails, the new subgroup leader will be able to update its table taking into account the published contents.

5 Comments about the TTL

A straightforward implementation to prevent a message forwarding loop within a subgroup is to use a time-to-live (TTL) field, initially set to S, and processing it like in

IPv4 networks [20]. However, there are multiple situations in which this implementation will immediately reveal to a "attacker" node whether the predecessor node is the initiator or not.

An approach to solve this problem is to use high and randomly chosen (not previously known) values for the TTL field. However, the objective is to limit the forwarding procedure, and high values for the TTL represent long multi-hop paths. Therefore, the TTL would have to be small. But, in this case, the range of possible random values for the TTL is too restricted, and it results in a similar situation to the one of a well-known TTL value among all the users. In this last case, corrupt nodes can easily derive whether the predecessor node is the origin of the intercepted message or not. We can conclude that the TTL methodology is not appropriate to limit the length of multi-hop paths.

To prevent corrupt nodes from being able to obtain information about the beginning of the multi-hop path, the first requirement of the used methodology is that it has to constrain the length of the paths in a non-deterministic way. In order to achieve that, an alternative is to ensure that the mean length of the paths is limited although it does not guarantee that the overlay paths have a maximum number of hops.

6 Simulations

We have developed a discrete event simulator in C language to evaluate our system. At the beginning, the available contents are distributed in a random way among all the nodes of the network. As it has been mentioned before, the subgroup leaders share information using an ALM procedure. The node life and death times follow a Pareto distribution using $\alpha = 0.83$ and $\beta = 1560$ sec. These paremeters have been extracted from [21]. The time between queries follows an exponential distribution with mean 3600 sec. (1 hour), and if a query fails the user makes the same request for three times, with an empty interval of 300 sec. among them. The *timeout timer* has a value of 300 sec. Finally, the simulation results corresponds to a network with 12,800 different contents and 6,400 nodes, with 128 subgroups.

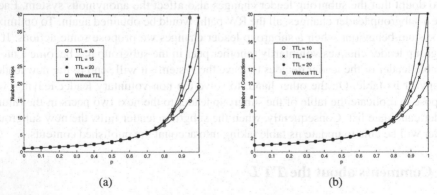

(a)   (b)

Fig. 5. (a) Average number of hops in a complete download path. (b) Average numer of connections that a node routes.

We have proposed a mechanism to provide anonymity to a two-level hierarchical P2P network. However, this service has a cost in terms of bandwith consumption and nodes overload. On the other hand, our anonymity mechanism also works fine under a join/leave or failure scenario. This service is required to maintain available paths along the time, but it involves an extra control packet interchange. It is necessary to quantify both costs.

Figure 5(a) represents the average number of hops in a complete download path, from the content owner to the requester node (download path length). This measurement helps us in order to estimate the bandwith consumption, since every hop in a path implies an extra bandwith consumption to carry out the download. This figure represents the results in function of p and TTL. We represent the results for 3 different TTL values (10, 15 and 20). When p is less than 0.8 the mean path length never exceeds 7, so the extra bandwith consumption is very limited. In addition, in this case the TTL guarantees that in any case the number of hops will be limited. If p is greater than 0.8 the extra bandwith consumption tends to infinity, but the TTL limits it to a reasonable value.

This simulation result corresponds with the analytical expresssion presented in Equation 3. Additionnaly, we represent the number of hops in a compelte download path without using a TTL limit. We observe than if p is greater than 0.9 the extra bandwith goes up quickly.

Figure 5(b) represents the average number of connections that a node routes in function of p and TTL. With $p = 0$ the number of paths is 0, because the connection is established directly between the owner and the requester node. If the value of p increases, the number of connections also increases. The influence of TTL doesn't appear until $p = 0.8$, as we observed in Figure 5(a). When $p > 0.8$ the number of connections a node routes does not tend to infinity thanks to the TTL.

This simulation result corresponds with the analytical expresssion presented in [11]. Additionnaly, we represent the the average number of connections that a node routes without using a TTL limit. We observe than if p is greater than 0.9 the number of connections goes up quickly. As a conclusion, a value of $p = 0.8$ is a trade off solution between anonymity efficience, extra bandwith consumption and nodes overload.

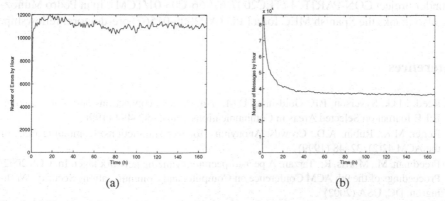

Fig. 6. (a) Number of request that cannot be carry out in an hour because at least one node in the download path is out. (b) Average number of control messages in function of time.

In the results represented in Figure 6(a) our system doesn't implement the reliability mechanism. The parameters used in the simulatons corresponds to a typical scenario: $TTL = 10$ and $p = 0.7$. It shows the number of requests that cannot be carry out in an hour because, at least one node in the download path is down when the download is performed. This result is represented in funcion of time (in hours). This number fluctuates around 11,000.

Therefore, it is clear than in a real anonymous P2P network (with unpredictable node failures) it is necessary to suply a mechanism to reconstruct efficiently the paths, although it entails an extra network overhead in terms of control packets.

In our network, a download process never fails since it is a reliable anonymous P2P network, but this feature involves an extra control traffic. Figure 6(b) represents the average number of control messages in function of time, in the previous scenario. Initially, due to simulation constraints, the *timeout timer* expires simultaneously in a lot of nodes, but in steady state the average number of control messages is under 40,000.

Usually, control messages have a small payload (about 10 bytes) and a IP/TCP header (40 bytes). If we suppose each control message has a length of 50 bytes, control traffic supposes only a 4.4 kbps traffic rate.

7 Conclusions

In this paper, we have proposed anonymity mechanisms for a two-level hierarchical P2P network presented in a previous work. Unfortunately, the mechanisms to provide anonymity entails extra bandwidth conssumption, extra nodes overload and extra network overhead in terms of control traffic. The simulations tries to evaluate the costs to provide anonymity in a real P2P scenario (with unpredictable node failures). As a conclusion, our proposal achieves mutual anonymity only with an extra bandwith consumption corresponding to 5 hops, 3 connections by node and a 4.4 kbps control traffic rate (in a typical scenario $TTL = 10$ and $p = 0.7$).

Acknowledgements. This work has been supported by the Spanish Researh Council under project CON-PARTE-1 (TEC2007-67966-C03-01/TCM). Juan Pedro Muñoz-Gea also thanks the Spanish MEC for a FPU (AP2006-01567) pre-doctoral fellowship.

References

1. Reed, M.G., Syverson, P.F., Goldshlag, D.M.: Anonymous connections and onion routing. IEEE Journal on Selected Areas in Communications 16(4), 482–494 (1998)
2. Reiter, M.K., Rubin, A.D.: Crowds: Anonymity for web transactions. Communications of the ACM 42(2), 32–48 (1999)
3. Freedman, M., Morris, R.: Tarzan: A peer-to-peer anonymizing network layer. In: CCS 2002. Proceedings of the 9th ACM Conference on Computer and Communications Security, Washington, DC, USA (2002)
4. Dingledine, R., Mathewson, N., Syverson, P.: Tor: The second-generation onion router. In: Proceedings of the 13th USENIX Security Symposium, San Diego, CA, USA (2004)

5. Han, J., Liu, Y., Xiao, L., Xiao, R., Ni, L.M.: A mutual anonymous peer-to-peer protocol design. In: Proceedings of the 19th International Parallel and Distributed Processing Symposium (IPDPS 2005), Denver, CO, USA (2005)
6. Xiao, L., Xu, Z., Shang, X.: Low-cost and reliable mutual anonymity protocols in peer-to-peer networks. IEEE Transactions on Parallel and Distributed Systems 14(9), 829–840 (2003)
7. Levine, B.N., Shields, C.: Hordes: A multicast-based protocol for anonymity. Journal of Computer Security 10(3), 213–240 (2002)
8. Mislove, A., Oberoi, G.A., Post, C.R., Druschel, P.: Ap3: Cooperative, decentralized anonymous communication. In: Proceedings of the 11th workshop on ACM SIGOPS European workshop: beyond the PC, New York, NY, USA (2004)
9. Lu, T., Fang, B., Sun, Y., Cheng, X.: Wongoo: A peer-to-peer protocol for anonymous communication. In: Proceedings of the 2004 International Conference on Parallel and Distributed Processing Techniques and Appliations (PDPTA 2004), Las Vegas, NE, USA (2004)
10. Wright, M., Adler, M., Levine, B.N., Shields, C.: An analysis of the degradation of anonymous protocols. In: Proceedings of the Network and Distributed Security Symposium (NDSS 2002), San Diego, CA, USA (2002)
11. Sui, H., Chen, J., Che, S., Wang, J.: Payload analysis of anonymous communication system with host-based rerouting mechanism. In: Proceedings of the Eighth IEEE International Symposium on Computers and Communications (ISCC 2003), Kemer-Antalya, Turkey (2003)
12. Muñoz-Gea, J.P., Malgosa-Sanahuja, J., Manzanares-Lopez, P., Sanchez-Aarnoutse, J.C., Guirado-Puerta, A.M.: A hybrid topology architecture for p2p file sharing systems. In: Shishkov, F.B., Helfert, M. (eds.) ICSOFT 2006. CCIS, 10, pp. 220–229. Springer, Heidelberg (2008)
13. Ratnasamy, S., Handley, M., Karp, R., Shenker, S.: Application-level multicast using content-addressable networks. In: Proceedings of the 3rd International Workshop of Networked Group Communication, London, UK (2001)
14. Ghodsi, A., Alima, L., El Ansary, S., Brand, P., Haridi, S.: Dks(n,k,f): A family of low communication, scalable and fault-tolerant infrastructures for p2p applications. In: Proceedings of the 3rd International Workshop on Global and P2P Computing on Large Scale Distributed Systems CCGRID 2003, Tokyo, Japan (2003)
15. Castro, M., Druschel, P., Kermarrec, A., Rowstron, A.: Scribe: A large-scale and decentralized application-level multicast infrastructure. IEEE Journal On Selected Areas in Communications 20(8), 100–110 (2002)
16. Ratnasamy, S., Francis, P., Handley, M., Karp, R., Shenker, S.: A scalable content-addressable network. In: Proceedings of ACM SIGCOMM, San Diego, CA, USA (2001)
17. Stoica, I., Morris, R., Liben-Nowell, D., Karger, D.: Chord: A scalable peer-to-peer lookup protocol for internet applications. IEEE/ACM Transactions on Networking 11(1), 17–32 (2003)
18. Rowstron, A., Druschel, P.: Pastry: Scalable, decentralized object location and routing for large-scale peer-to-peer systems. In: Guerraoui, R. (ed.) Middleware 2001. LNCS, vol. 2218, pp. 329–350. Springer, Heidelberg (2001)
19. Pfitzmann, A., Hansen, M.: Anonymity, unobservability and pseudomyity: a proposal for terminology. In: Proceedings of the Fourth International Information Hiding Workshop, Pittsburgh, PE, USA (2001)
20. Postel, J.: RFC 791: Internet Protocol (1981)
21. Li, J., Stribling, J., Morris, R., Kaashoek, M.F.: Bandwidth-efficient management of dht routing tables. In: Proceedings of the 2nd USENIX Symposium on Networked Systems Design and Implementation (NSDI 2005), Boston, MA, USA (2005)

Part IV

Information Systems and Data Management

Approximation and Scoring for XML Data Management

Giacomo Buratti[1] and Danilo Montesi[2]

[1] Department of Mathematics and Computer Science, University of Camerino
Via Madonna delle Carceri 9, Camerino, Italy
giacomo.buratti@unicam.it
[2] Department of Computer Science, University of Bologna
Mura Anteo Zamboni 7, Bologna, Italy
danilo.montesi@unibo.it

Abstract. XQuery Full-Text is the proposed standard language for querying XML documents using either standard or full-text conditions; while full-text conditions can have a boolean or a ranked semantics, standard conditions must be satisfied for an element to be returned. This paper proposes a more general formal model that considers structural, value-based and full-text conditions as *desiderata* rather than *mandatory* constraints. The goal is achieved defining a set of relaxation operators that, given a path expression or a selection condition, return a set of relaxed path expressions or selection conditions. Algebraic approximated operators are defined for representing typical queries; they return elements that perfectly respect the conditions, as well as elements that answer to a relaxed version of the original query. A score reflecting the level of satisfaction of the original query is assigned to each result.

1 Introduction

The widespread use of XML repositories led over the last few years to an increasing interest in searching and querying this data. Besides expressing database-style queries using languages like XPath and XQuery [1], users typically have the necessity to formulate IR-like queries [2], i.e. full-text and keyword search. These two paradigms converge to form the so-called *Structured Information Retrieval* (*SIR*), which is the area of investigation of many recent research works [3]. Many SIR languages have also been proposed, until W3C has published the working draft of XQuery Full-Text [4]. Our contribution to this convergence effort has been the definition of AFTX [5], an algebra for semi-structured and full-text queries over XML repositories. The algebra includes basic operators, which are used to restructure and filter the input trees, and full-text operators, which express boolean keyword searches (i.e. assigning a relevant / not relevant judgement to each input tree) or *ranked* retrieval (i.e. calculating a *score* value, reflecting how relevant the tree is with respect to the user query).

One of the main differences between structured and semi-structured paradigm is the *flexibility* of the schema. In fact the schema specifications for an XML document can define some elements as optional, or the cardinality of an element can differ from documents to documents. This flexibility poses interesting questions for what concerns answering to a query that imposes some constraints on the structure of XML fragments

J. Filipe et al. (Eds.): ICSOFT/ENASE 2007, CCIS 22, pp. 243–256, 2008.

to retrieve; it could be the case that such constraints are satisfied by a very small part of input documents. Nevertheless, there could be documents that are relevant to users, even if they do not closely respect some structure constraints.

XQuery Full-Text treats basic conditions (i.e. navigational expressions and constraints on value of elements/attributes) and full-text conditions in a non-uniform way. In fact a full-text condition can either have a boolean semantics or it can act as a ranked retrieval. On the contrary, basic conditions are always treated as *mandatory*: in order to be retrieved, an element *must* be reachable by exactly following the specified path expression, and all the conditions on values *must* be satisfied. In the effort of providing a uniform treatment of basic and full-text conditions, the key idea is to consider the searched path expression and the specified conditions on values as *desirable* properties to enjoy for an element to be returned, instead that considering them as mandatory constraints. Therefore, an element should be returned even if it does not perfectly respect basic conditions, and a *score* value should indicate how well such conditions are satisfied.

1.1   A Motivating Example

Consider the XML document shown graphically in Figure 1 and an XQuery expression containing the clause for $a in doc("bib.xml")/bib/book/author. The user need is probably to find all book authors, including those who just co-authored a book. The for clause, however, will find only those authors that are the single authors of at least one book. If the for clause has an *approximated* behavior, it could also return a subtree reachable by following a *relaxed* version of the original path expression, for example /bib/book//author. This relaxed query would find all book authors, but not paper authors, which could also be of interest. If this is the case, the query could be further relaxed, transforming the path expression into /bib//author.

Let us now consider the clause for $a in doc("bib.xml")/bib/book/title; it finds all book titles, but ignores paper titles. Taking into account the fact the some semantic relationship exists between the words *book* and *paper* (using some lexical database like

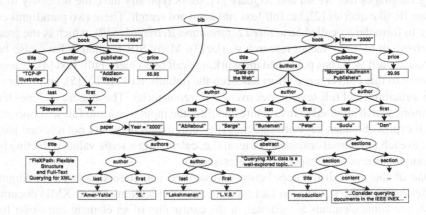

Fig. 1. Graphical representation of an XML document

[6] we can find that both these words are hyponyms of the word *publication*), a different kind of relaxation could treat the path /bib/paper/title as a approximated version of /bib/book/title and therefore include in the result also paper titles.

Let us now analyze a clause including the full-text operator ftcontains:

```
for $a in doc("bib.xml")//paper[//section/title ftcontains "INEX"]
```

We are looking for papers that include in a section title the word *INEX*. The paper shown in Figure 1 is not returned, because the titles of the various sections do not include the searched word. However, the word is included in the content of the first section of the paper; then a possible relaxation could transform the previous query by removing the last step in the path expression, thus obtaining:

```
for $a in doc("bib.xml")//paper[//section ftcontains "INEX"] .
```

Finally, let us consider the clause for $b in doc ("bib.xml") /bib/book [/price < 39], which finds the books with a price lower than 39. It could be the case that very few books satisfy such a constraint (in the document of Figure 1, no book satisfies the constraint); consequently, the user could also be interested in books having a slightly greter price. A relaxed version of the clause could be obtained by changing the comparison operator (/price $\leq$ 39) or even increasing the threshold price (/price < 45).

1.2 Our Contribution

The purpose of this paper is to formally define the notion of query relaxation. We begin in Sect. 2 by introducing the various *relaxation operators*, which apply to path expressions and predicates on element values, and return a relaxed version of them. Our proposal significantly extends previous works like [7], in the sense that we consider a wider spectrum of relaxations. We then incorporate the notion of approximation into an algebraic framework, introduced in Sect. 3, representing queries over XML. Approximated algebraic operators are based on the concept of *score*: a *relaxed answer* (i.e. an answer to a relaxed query) has a score which reflects how exact is the query that returns such an answer. In a certain way exact and relaxed queries play the same role of boolean and ranked retrieval in classical Information Retrieval (and in XQuery Full-Text): while exact queries classify each document fragment as either relevant (i.e. fulfilling constraints imposed on document structure and elements / attributes value) or not relevant, relaxed queries establish how relevant a fragment is.

The issue of score calculation is tackled in Section 4. In particular we concentrate on *structural score*, i.e. the score reflecting the level of satisfaction of conditions on structure and values; here we introduce the concepts of *Path Edit Distance* and *Comparison Satisfaction Ratio*. Finally in Sect. 5 we draw some conclusions.

2 Relaxation Operators

In this section we formally define a set of relaxation operators. We propose two kinds of relaxation: *path relaxations* and *comparison relaxations*.

2.1 Path Relaxations

We define four path relaxation functions: axis relaxation, internal step deletion, final step deletion, and name relaxation. Their goal is to obtain a set of relaxed path expressions, starting from an input path expression.

Axis Relaxation. P_A approximates a path expression λ_1/λ_2 by substituting a child axes with a descendant axes, thus obtaining $\lambda_1//\lambda_2$. Its goal has already been depicted in an example in Section 1, where the path expression /bib/book/author were transformed into /bib/book//author. Multiple relaxed expression can be obtained from the original expression, possibly recursively applying the relaxation function; in this example, other possible relaxed expressions are /bib//book/author, /bib//book//author etc.

Internal Step Deletion. P_S approximates a path expression $\lambda_1\alpha\beta//\lambda_2$ (where α is an axes and β is an element name) by eliminating the internal step $\alpha\beta$, thus obtaining $\lambda_1//\lambda_2$. As an example, consider the XML document in Figure 1 and the path expression /bib/book/authors/author. Using P_A we can transform it into /bib /book/authors//author, but this transformation is not yet sufficient to capture authors of a book having just one author; by further applying P_S we obtain /bib/ book //author, which captures single authors too.

Final Step Deletion. P_F removes from the original path expression $\lambda\alpha\beta$ the final step $\alpha\beta$, thus obtaining λ; for example, using P_F we can transform the expression /bib/book/content into the expression /bib/book. The behaviour of the final step deletion function is radically different to that of P_A and P_S; while an application of P_A or P_S results in a relaxed path expression that reaches all the elements that can be reached using the original path expression (plus some extra elements), an application of P_F results in a set of path expressions that reach a completely different set of elements.

Name Relaxation. P_N substitutes an element name β in a path expression $\lambda_1\alpha\beta\lambda_2$ with another name β', thus obtaining $\lambda_1\alpha\beta'\lambda_2$. For example, using P_N we could transform the expression//book/author into //paper/author. In the spirit of *data integration*, this operation should be intended as a way to manage heterogenous data sources disregarding possible name conflicts due to the usage, for example, of synonyms in the schema definition. Generally speaking, the Name Relaxation function should be thought of as a way to substitute a name with another one that has a certain degree of *similarity* with it; such a similarity could be calculated, as suggested in [8], using an ontology.

2.2 Comparison Relaxations

Typical operations on XML documents involve checking the satisfaction of a comparison predicate. For example, having selected a set of book elements, we could filter those element on the basis of the book price, using the predicate /price < 50. There are two possible relaxations we can perform on a predicate like this: operator relaxation and value relaxation.

Operator Relaxation. C_O substitutes an operator θ in a comparison expression $x\theta y$ with another operator θ', thus obtaining $x\theta' y$. For example, using C_O we could transform /price $<$ 50 into /price $\leq$ 50. Clearly, such a transformation makes sense only if the new operator guarantees a larger number of successful comparison than the original one. A partial ordering relation $\preceq$ can be defined between available operators in a certain domain, stating that $\theta \preceq_D \theta'$ if, $\forall x \in D$, $y \in D$, $x\theta y$ implies $x\theta' y$. For example, in the domain of reals, $< \preceq_R \leq$.

Value Relaxation. C_V substitutes an operand y in a comparison expression $x\theta y$ with another operand y', thus obtaining $x\theta y'$. For example, using C_V we could transform /price $<$ 50 into /price $<$ 55. This relaxation, along with C_O, permits to include in the list of satisfactory elements also those whose value slightly differs from the user request. As usual, its goal is therefore to expand the space of possible results. The choice of the new value y' should depend on the comparison operator. For example, given the comparison $x < y$, y' should be a value such that $y' > y$.

3 Approximated Algebraic Operators

In this section we define a set of approximated algebraic operators. For each of them, we briefly recall the semantics of the basic AFTX operator [5] and define the new operator on the basis of the relaxation operators that can be applied to the predicate of the algebraic operator.

3.1 Approximated Projection

AFTX projection π is a unary operator that takes is a forest and operates a *vertical* decomposition on it: every input tree contributes, with the subtrees of interest, to the projection output. The subtrees of interest are specified in the projection predicate through a *path expression* λ, a concept almost identical to that used in XPath, except for the fact that it can not contain selection conditions. For example, the AFTX expression $\pi_{/bib/book}($"bib.xml"$)$, where *bib.xml* is the XML document shown in Figure 1, returns a forest containing two trees, that correspond to the two subtrees rooted at book that can be found in the input tree.

The **approximated projection** $\pi^\star$ operator has the following behavior:

- calculate all possible relaxations of the path expression λ in the predicate, using Axis Relaxation P_A, Internal Step Deletion P_S, Name Relaxation P_N;
- for each relaxed path expression, execute the projection using that path expression;
- calculate the union of the projection results (including the result of the projection that uses the original path expression), eliminating duplicate trees (i.e. subtrees reachable by following two different relaxed path expressions).

As one could expect, the Final Step Deletion relaxation function P_F is not used here. In fact, such a relaxation would lead to results completely unrelated to those expected. For example, suppose we want to do a projection using the predicate /bib/book /author; by eliminating the final step we would obtain the predicate /bib/book; trees resulting from such a projection would represent books, a completely different concept to that of authors.

Example 1. Consider the XML document in Figure 1 and suppose to write the algebraic expression $\pi^*_{/\text{bib}/\text{book}/\text{author}}(\text{"bib.xml"})$. Using the basic AFTX projection this expression would return only the author *W. Stevens*. However, by applying P_A to the projection path expression we can obtain the relaxed path expression /bib/book //author, thus including in the result also *Serge Abiteboul*, *Peter Buneman*, and *Dan Suciu*. Moreover, by applying P_N we can substitute book with paper, thus obtaining the path expression /bib/paper//author and adding to the result also *S. Amer-Yahia* and *L.V.S. Lakshmanan*.

3.2 Approximated Selection

AFTX selection σ is a unary operator that takes is a forest and operates a *horizontal* decomposition on it: only the trees of interest contribute, with their entire content, to the selection output. The selection predicate is composed by a path expression λ and a *selection condition* γ. The path expression is used to operate a temporary projection on the input tree. Each subtree T' belonging to the temporary projection result is then checked: if at least one of them satisfies the selection condition, the original input tree is added to the selection output.

The selection condition is a conjunction of base selection conditions. The evaluation of each base condition γ_i depends on its form:

- if γ_i is the form λ', it is satisfied if exists at least one subtree T'_1 that can be reached from $root(T')$ by following λ';
- if γ_i is the form $\lambda' \equiv \lambda''$, it is satisfied if exists at least a pair of subtrees (T'_1, T'_2) that can be reached from $root(T')$ by following, respectively, λ' and λ'', and such that T'_1 is strictly equal to T'_2 (informally, strict equality between trees means that the two trees are two copies of *the same* tree);
- if γ_i is the form $\lambda'p\theta x$, where x is a constant and p is an element property (for example its value, the value of one of its attributes, etc.), it is satisfied if exists at least one subtree T'_1 that can be reached from $root(T')$ by following λ' such that $root(T'_1)p$ is in relation θ with x;
- if γ_i is the form $\lambda'p'\theta\lambda''p''$, it is satisfied if exists at least a pair of subtrees (T'_1, T'_2) that can be reached from $root(T')$ by following, respectively, λ' and λ'', and such that $root(T'_1)p'$ is in relation θ with $root(T'_2)p''$.

For example, the AFTX expression $\sigma_{/\text{book}[/\text{author AND}/\text{price.v}<40]}(\pi_{/\text{bib}/\text{book}}$ ("bib.xml")) returns a forest containing the subtrees rooted at book that have at least one author sub-element and such that the value of the price sub-element is less than 40.

The **approximated selection** σ^* operator does the following:

- calculate all possible relaxations of the path expression λ in the predicate, using Axis Relaxation P_A, Internal Step Deletion P_S, Name Relaxation P_N;
- calculate all possible relaxations of the selection condition γ, applying some relaxation function to each base selection condition depending on the kind of the base selection condition;

- for each relaxed path expression and selection condition, calculate the result of the selection;
- calculate the union of the selection results (including the result of the selection that uses the original selection predicate), eliminating duplicate trees.

If a base condition γ_i is the form λ', we can relax λ' as usual. For example, the selection predicate /book[/authors/author] can be relaxed into /book[//author]. Moreover, P_F can also be applied; for example /book[/authors/author] can be relaxed into /book[/authors], or even into /book[]; in practice, we relax (or even eliminate) the constraint on the presence of a subtree.

If γ_i is the form $\lambda' \equiv \lambda''$, we can apply to λ' and λ'' P_A, P_S and P_N. For example the selection predicate books[/csbook/authors/author $\equiv$ /mathbook /authors/author] can be relaxed into /books[/csbook//author $\equiv$ /mathbook//author].

If γ_i is the form $\lambda' p \theta x$, we can apply P_A, P_S and P_N on λ'; for example /book [/authors/author.count > 1] ("find all the books with more than one author") can be relaxed into /book [//author.count > 1]. Moreover, we can apply C_O and C_V on $p \theta x$; for example the predicate book[/price.v < 50] can be relaxed into book[/price.v $\leq$ 55].

Finally, if γ_i is the form $\lambda' p' \theta \lambda'' p''$, we can use all the relaxations seen for the previous case; moreover λ'' can also be relaxed using P_A, P_S and P_N. For example, the predicate /books[/csbook/totalprice < /mathbook/totalprice] could be relaxed into /books[/csbook/price $\leq$ /mathbook/price] using: 1) Name Relaxation on /csbook/totalprice, 2) Name Relaxation on /mathbook/totalprice, and 3) Operator Relaxation on <.

Example 2. Consider the XML document in Figure 1 and suppose to write the following algebraic expression:

$$\sigma^*_{/book[/author/last.v=\text{``Amer-Yahia''} \text{ OR } /price.v<60]}(\pi^*_{/bib/book}(\text{``bib.xml''})) \ .$$

Using non-approximated operators, projection returns a forest containing the two books, and the subsequent selection retains the book *Data on the Web*. The paper is not returned, even if, having *Amer-Yahia* among its authors, it is probably of interest for the user. However, using the approximated selection and projection operators:

- projection returns also the paper, because using P_N the path expression /bib /book can be transformed into /bib/paper;
- selection retains the paper in the result; using P_N and P_A the selection predicate is relaxed into /paper[//author/last.v = ``Amer-Yahia'' OR /price.v < 60];
- selection retains the book *TCP-IP Illustrated* also; in fact the selection base condition /price.v < 60 can be transformed into /price.v < 70.

3.3 Approximated Full-Text Selection

AFTX full-text selection ς behaves in a way similar to that of basic selection operator: it performs a horizontal decomposition of the input forest, retaining only those trees having at least one subtree satisfying the full-text selection predicate. The full-text selection predicate allows to search one or more words or phrases (specified by the parameter γ) into the full-text value of an element (i.e. the value of the element concatenated with the value of its sub-elements) or into the value of an attribute a. For example, the expression $\varsigma_{/\text{book/title}[\text{"XML" OR "Web"}]}(\pi_{/\text{bib/book}}(\text{"bib.xml"}))$ returns all the books that contain the word *XML* or the word *Web* in the title.

The **approximated full-text selection** operator ς^*, as usual, transforms the predicate using some relaxation operator and returns the union of the results of the relaxed full-text selections.

First of all, the path expression λ can be subject to P_A, P_S and P_N; for example, /book/chapter/section["XML"] can be relaxed into /paper//section["XML"], thus obtaining as result also those publications which are not divided into chapters.

Another relaxation function that is worthwhile applying is P_F, which *broadens* the search scope. For example, relaxing /book/title["XML" AND "Algebra"] into /book["XML" AND "algebra"] we obtain as result all the books that contain the searched words everywhere, instead that just in the title: we have broadened the search scope from the full-text value of /book/title to the full-text value of /book (that includes the full-text value of /book/title).

Example 3. Consider the XML document in Figure 1. Suppose we look for papers that include, in their title, the words *XML* and *INEX*. Then we write the following algebraic expression:

$$\varsigma_{/\text{paper/title}[\text{"XML" AND "INEX"}]}(\pi_{/\text{bib/paper}}(\text{"bib.xml"})) .$$

This expression would return an empty answer; in fact the paper title contains the word *XML*, while the word *INEX* is included only in the content of the first section. However, using P_F we can remove the title step in the path expression of the full-text selection predicate, thus obtaining the relaxed predicate /paper["XML" AND "INEX"]. Therefore the paper will be returned, because both searched words are found in its full-text value.

3.4 Approximated Full-Text Score Assignment

While full-text selection performs a full-text search using a *boolean* model (a tree either satisfies the selection condition or it does not satisfy the condition at all), full-text score assignment ξ does not perform a selection: each input tree is returned, without filtering. What it does is to assign to each tree a *full-text score*, that represents the level of satisfaction of the full-text condition. The full-text condition is specified in the score assignment predicate, in the same way as in the full-text selection predicate. However, a *weight* can be assigned to each searched word or phrase in order to specify which words (or phrases) should highly influence score calculation. The score is calculated by a full-text score function f_F specified in the score assignment predicate.

Consider the algebraic expression $\xi_{/\text{paper/title}[\text{"INEX"}],f_F}(\pi_{/\text{bib/paper}}(\text{"bib.xml"}))$. The expression would assign to the paper about *FleXPath* a score (calculated by the function f_F) of 0, because the word *INEX* is not found in the title; however this paper is probably of interest, because the searched word is included into the paper body. Using

approximated full-text score assignment, the path expression /paper/title can be relaxed using P_A, P_S, P_N, and P_F; this way it can be relaxed into /paper. Now a full-text score greater than zero is assigned to the paper, because *INEX* is present into the full-text value of paper.

3.5   Generalized Top-K and Threshold Selection

The four approximated operators have the goal to broaden the result space, by adding some trees that would have been discarded by applying the corresponding exact operator. Their usage is therefore valuable, because a strict interpretation of conditions imposed by the user query could discard trees which could be of interest for the user, even if they do not perfectly respect some conditions.

However, the usage of such relaxations could lead to the opposite problem: the user who poses the query could be overwhelmed by a huge amount of answers. What is needed is therefore a way to filter such results, retaining only those that best match the user needs.

A similar problem has already been tackled in basic non-approximated algebra, regarding the full-text score assignment operator ξ. After using this operator, a user typically wants to receive results in score order, disregarding those with a lower score. The solution has been found in the introduction of two derived operators: top-K full-text selection $\top$ and threshold full-text selection ω. They assign a score to each input tree and return, respectively, the k trees with highest scores and those trees whose score is higher than a defined threshold τ; in both cases trees are returned in descending score order.

Having introduced relaxed operators into our algebra, now such a process of filtering and ordering could be based on two kinds of score:

- the full-text score, which represents the level of satisfaction of full-text conditions; such a score is calculated by full-text score assignment;
- a new *structural* score, which represents the level of satisfaction of non-full-text conditions.

Informally, the structural score should be the answer to the question *"how much have you relaxed my query in order to include this tree in the result?"*. We propose a way to calculate this score in Section 4 ([9] and [10] also deal with this issue); similarly, the way to combine the structural score with the full-text score is a quite interesting problem, and it is a candidate target for future research. However, supposing to have a set of *structural score calculation functions* and a set of *combined score calculation functions*, we can define a **generalized** version of **top-K** $\top^*_{f_1,f_2,k}(F)$ and **threshold** $\omega^*_{f_1,f_2,\tau}(F)$ operators, which operates as follows:

- calculate the structural score using the function f_1;
- combine the structural score just calculated and the full-text score (previously calculated by some score operator) using the function f_2;
- retain in the output, respectively, the k trees with highest score or the trees with a score higher than τ.

Example 4. Consider the XML document shown in Figure 1 and suppose to write the following algebraic expression:

$$\top^*_{f_1,f_2,1}\big(\xi_{/*/\texttt{title}[0.2\ \text{``Web''}\ \text{OR}\ 0.8\ \text{``XML''}]\texttt{f}}\big(\pi^*_{/\texttt{bib}/\texttt{book}}(\text{``bib.xml''}\big)\big)\big).$$

The projection find all books, then the score assignment calculates a score, using the scoring function f and considering the word *XML* more important than *Web*; finally the generalized top-K returns the book with the highest combined score.

By applying P_N to the path expression, the relaxed projection returns a forest containing, among the others, three trees corresponding to the two books and to the paper. Suppose the full-text scoring function f calculates a simple sum of the weights of the found words; then the full-text score of the two books are, respectively, 0 and 0.2, while the paper has a full-text score of 0.8.

Suppose now the structural scoring function f_1, when P_N is applied, assigns a structural score corresponding to the degree of similarity between the original word and the substitute, and suppose that the similarity between *book* and *paper* is 0.7. Therefore, the two books has a structural score of 1 (because no relaxation has been done for them), while the paper has a structural score of 0.7.

Finally, suppose the combined score calculation function f_2 returns a weighted sum of the structural score (with weight 0.2) and the full-text score (with weight 0.8). Then: the book *TCP-IP Illustrated* has a combined score of $1 * 0.2 + 0 = 0.2$; the book *Data on the Web* has a combined score of $1 * 0.2 + 0.2 * 0.8 = 0.36$; the paper has a combined score of $0.7 * 0.2 + 0.8 * 0.8 = 0.78$. Therefore, the generalized top-k operator returns the paper, which is the publication with the highest combined score.

4   Structural Score Calculation

In Sect. 3 we have defined four approximated algebraic operators. Even though the operators are parametric with respect to the scoring functions, in this section we present a possible way to calculate structural scores in the presence of path relaxations or comparison relaxation.

4.1   Path Relaxations

Let us consider a subtree T' of an input tree T, reachable from $root(T)$ by following a path λ', that is returned by the approximated projection $\pi^*_\lambda(F)$. In order to assign a structural score to T', we need a way to measure the *similarity* between λ' and the searched path λ. For example, suppose to execute $\pi^*_{/\texttt{bib}/\texttt{book}/\texttt{author}}$ over the XML document in Fig. 1. Which structural score should be assigned to the subtree rooted at author representing *Serge Abiteboul*? In other words, which is the similarity between the searched path /bib/book/author and the path /bib/book/authors /author? In order to answer this question, we define a novel similarity measure, called *Path Edit Distance* (*PED*).

Let us first introduce the concept of *Path Transformation System*, which is the tool that allows to transform a path expression into another one and to calculate the transformation cost. It is defined as a set of *transformation types*, each one having a cost. The

Path Transformation System we propose includes three transformation types: *Insert*, *Delete*, and *Substitute*.

Insert inserts a step into λ. The cost of an *Insert* is: 1 when inserting a step just before a child axis, because such a transformation extends the result space (e.g. transforming /bib/book/author into /bib/book/authors/author); 0 when inserting a step just before a descendant axis, because such a transformation does not extend the result space (e.g. transforming /bib/book//author into /bib/book/authors //author).

Delete deletes a step from λ. For example /bib/book/authors/author can be transformed into /bib/book/author by deleting the step /authors. The cost of a *Delete* is 1.

Substitute substitutes a name β with β'. For example /bib/book can be transformed into /bib/paper by substituting book with paper. The cost of a *Substitute* is $1 - x$, where x is the similarity between the approximated and the original element name.

These basic transformations can be composed in order to operate more complex transformations. In general there are more than one strategy (i.e. sequence of transformations) that can be followed for transforming a path expression λ into λ', and each of those sequences has a cost, given by the sum of the costs of the included basic transformations. The **Path Edit Distance** between λ and λ' is defined as the cost of the *cheaper* transformations sequence from λ to λ'.

PED clearly resembles classical String Edit Distance [11], with a difference: while in String Edit Distance we have a boolean judgement about equality between two letters, in PED we associate to each pair of element names (the equivalent of a letter in String Edit Distance) a score value representing the similarity between the names. A similarity of 1 means that the two names are equal, while a similarity of 0 means that the two names are completely unrelated.

The algorithm that calculates PED is a straightforward adaptation of the classical bottom-up dynamic programming algorithm for computing string edit distance; there is however a point to notice. We must recall that queries must be evaluated using a SVCAS semantics. The PED between two path expressions $\alpha_1\beta_1\alpha_2\beta_2\cdots\alpha_n\beta_n$ and $/\beta'_1/\beta'_2\cdots/\beta'_m$ is consequently calculated as the sum of: 1) the minimum distance between $\alpha_1\beta_1\alpha_2\beta_2\cdots\alpha_{n-1}\beta_{n-1}$ and $/\beta'_1/\beta'_2\cdots/\beta'_{m-1}$; 2) 1 minus the similarity between β_n and β'_m.

```
function PED(query pattern a[1]b[1]a[2]b[2]..,a[n]b[n],
  element pattern /c[1]/c[2].../c[m]
    for i=0 to n-1
      m[i,0]:=i;
    for i=0 to m-1
      if a[1]='/' then m[0,i]:=i else m[0,i]:=0;
    for i=1 to n-1
      for j=1 to m-1
        if a[i+1]='/' then  InsCost:=1 else InsCost:=0;
        m[i,j] := min(m[i,j-1]+InsCost, m[i-1,j]+1,
          m[i-1,j-1]+(1-Similarity(b[i],c[j])));
    return m[n,m] + (1 - Similarity(b[n], c[m]));
```

For example, consider the path expressions $\lambda = $ /bib/book//author and $\lambda' = $ /bib/paper/authors/paperauthor. Suppose that Similarity (*book*, *paper*)= 0.7,

Similarity(*author*, *paperauthor*) = 0.9, while the similarity between any other element name is 0.05. The function *PED* returns 0.4. In fact λ can be transformed into λ' by: 1) substituting *book* with *paper* (cost 0.3); 2) inserting *authors* before *//author* (cost 0); 3) substituting *author* with *paperauthor* (cost 0.1).

The concept of Path Edit Distance is well suited for representing the structural score of a tree returned by an approximated projection. However we need a score value between 0 and 1, so we should divide *PED*'s output by the maximum possible *PED* value, which is $\max\{|\lambda|, |\lambda'|\}$, where $|\lambda|$ is the number of steps in λ. In fact it is always possible to transform λ into any λ' by: 1) substituting any β_i in λ with β_i' in λ' (in the worst case the cost is 1 for each substitution); 2) inserting the remaining steps of λ' in λ or deleting the remaining steps in λ.

4.2 Comparison Relaxations

We have seen that comparison predicates can be subject to relaxation. For example the predicate price <= 50 can be relaxed into price <= 60. In this case, which structural score should be assigned to a book whose price is 59?

First of all, the scoring function should enjoy a common-sense property: the more a value is close to the searched value, the more the score should be high; in the previous example, if a book costs 51 and another one costs 58, the first one should have an higher score. Moreover, scoring should take into account the magnitude of the values; for example, if books have prices varying from 10 to 1000, a difference of 8 between searched price and found price is more acceptable than a difference of 0.1 in a domain having values varying from 0 to 1.

Taking into account these consideration, let us introduce the concept of **Comparison Satisfaction Ratio** (*CSR*). If a condition is satisfied, CSR is obviously equal to 1. If a condition is not satisfied CSR is 1 minus the ratio between 1) the difference between the found value and the searched value; 2) the maximum difference that can be found in the input set of values. The searched value is the *borderline* value that satisfies the condition; for example if the condition is $x \leq 50$ the borderline is 50. Now suppose that the original selection predicate is /book[/price.v <= 50] and that we are considering an element named totalprice. In this case the input set is the set of elements named *price* or *totalprice*. If the highest value in the input set is 100 and the value of the *totalprice* element is 60, then *CSR* is $1 - (60 - 50)/(100 - 50) = 0.8$.

If the values to compare are two strings and the comparison operator is =, CSR is instead the similarity between the two strings, as usual calculated using an ontology. An evaluation function must be defined for each kind of operator and values; for the sake of simplicity we do not discuss each possible case.

4.3 Putting Things Together

We have already discussed the fact that Path Edit Distance can be used for calculating the structural score of trees returned by approximated projection. Let us now analyze the other approximated operators.

The simplest form of selection predicate is $\lambda[p\theta x]$, where λ is a path expression, p is an element property (typically its value), θ is a comparison operator, and x is a constant. The system must find, for each input tree T, a subtree reachable by following an approximation of λ that satisfies an approximation of $p\theta x$. If F is the set of subtrees reachable by following an approximation of λ, a natural choice is to set the structural score of T to the maximum value among those calculated by combining (e.g. multiplying) the normalized PED between λ and λ', where λ' is the path from the root of the original tree to the root of the subtree we are considering, and the CSR of the predicate $p\theta x$.

In the case of full-text selection, where the predicate is of the form $\lambda[\gamma]$, we must check, for each input tree T, if there is some subtree that satisfies the full-text condition γ. If this is the case, the full-text score of T is set to 1, and the structural score is set to the maximum normalized PED among those of the subtrees that satisfies γ. In no subtree satisfies γ, the full-text score of T is set to 0, and the structural score is set to the maximum normalized PED value.

Finally, in the case of full-text score assignment, the full-text score assigned to each tree must range from 0 to 1, according to the level of satisfaction of the full-text condition. As in the case of full-text selection any subtree of an input tree T is analyzed, and T's full-text score is set to the maximum value of its subtree's full-text score, while T's structural score is set to the maximum structural score value among those of the subtrees with maximum full-text score.

5   Conclusions

In this paper we have presented an approximation-aware theoretical framework for full-text search over XML repositories. Some relaxation operators have been presented and used for defining approximated algebraic operators. The algebra is intended as a formal basis for the definition of an approximated query language for XML, which could be an extension of W3C's XQuery Full-Text.

We have also presented a possible way to calculate the structural score of a tree, which is based on the concepts of Path Edit Distance and Comparison Satisfaction Ratio; the algebra is parametric with respect to the structural scoring function, so different choices can be done. In any case a combined score calculation functions should return a global score by combining the structural score and the full-text score. Research should be directed towards establishing a set of properties that good scoring functions should enjoy. For example, if a combined scoring function relies, as one could expect, on *weights* assigned to the constituting scores, that function should be continuous on the weights, as discussed in [12].

Considering structural constraints of a query as a *desiderata* instead that a requirement also poses interesting performance issues. In fact, dealing with relaxation means transforming a query into a set of similar queries, each of which must be executed in order to calculate the final result. It is therefore needed a way to efficiently compute such answers. This problem is closely related to that of score calculation; again, the usage of scoring functions enjoying some properties (like monotonicity) should allow the definition of impacting optimization strategies, for example allowing to *prune* some part of the tree of the possible relaxed queries, which means avoiding to execute part of the relaxed queries.

References

1. W3C: XQuery 1.0: An XML Query Language, W3C Recommendation (2007),
   http://www.w3.org/TR/xquery/
2. Baeza-Yates, R., Ribeiro-Neto, B.: Modern Information Retrieval. Addison-Wesley, Reading (1999)
3. INEX: INitiative for the Evaluation of XML Retrieval (2006),
   http://inex.is.informatik.uni-duisburg.de/2006/
4. W3C: XQuery 1.0 and XPath 2.0 Full-Text, W3C Working Draft (2007),
   http://www.w3.org/TR/xquery-full-text/
5. Buratti, G.: A Model and an Algebra for Semi-Structured and Full-Text Queries (Ph.D. Thesis). Technical Report UBLCS-2007-03, University of Bologna (2007)
6. Princeton University, C.S.L.: Wordnet (2007), http://wordnet.princeton.edu/
7. Amer-Yahia, S., Lakshmanan, L.V.S., Pandit, S.: FleXPath: Flexible Structure and Full-Text Querying for XML. In: SIGMOD, pp. 83–94 (2004)
8. Theobald, A., Weikum, G.: The Index-Based XXL Search Engine for Querying XML Data with Relevance Ranking. In: Chaudhri, A.B., Unland, R., Djeraba, C., Lindner, W. (eds.) EDBT 2002. LNCS, vol. 2490, pp. 477–495. Springer, Heidelberg (2002)
9. Amer-Yahia, S., Koudas, N., Marian, A., Srivastava, D., Toman, D.: Structure and Content Scoring for XML. In: VLDB, pp. 361–372 (2005)
10. Marian, A., Amer-Yahia, S., Koudas, N., Srivastava, D.: Adaptive Processing of Top-K Queries in XML. In: ICDE, pp. 162–173 (2005)
11. Levenshtein, V.I.: Binary codes capable of correcting deletions, insertions, and reversals. Soviet Physics Doklady 10, 707–710 (1966)
12. Fagin, R., Wimmers, E.L.: A Formula for Incorporating Weights into Scoring Rules. Theoretical Computer Science 239, 309–338 (2000)

Quantitative Analysis of the Top Ten Wikipedias

Felipe Ortega, Jesus M. Gonzalez-Barahona, and Gregorio Robles

Grupo de Sistemas y Comunicaciones, Universidad Rey Juan Carlos
Tulipan s/n 28933, Mostoles, Spain
{jfelipe,jgb,grex}@gsyc.es
http://libresoft.es

Abstract. In a few years, Wikipedia has become one of the information systems
with more public of the Internet. Based on a relatively simple architecture it has
proven to be capable of supporting the largest and more diverse community of
collaborative authorship worldwide. Using a quantitative methodology, (analyzing public Wikipedia databases), we describe the main characteristics of the 10
largest language editions, and the authors that work in them. The methodology
is generic enough to be used on the rest of the editions, providing a convenient
framework to develop a complete quantitative analysis of the Wikipedia. Among
other parameters, we study the evolution of the number of contributions and articles, their size, and the differences in contributions by different authors, inferring
some relationships between contribution patterns and content. These relationships
reflect (and in part, explain) the evolution of the different language editions so far,
as well as their future trends.

Keywords: Wikipedia, quantitative analysis, growth metrics, collaborative
development.

1 Introduction

Wikipedia is one of the most important projects producing collaborative intellectual
work in the last years, and has gained the attention of millions of users worldwide. It is
also one of the most popular sites on the Internet (for instance, being ranked by Alexa
as the 11th most visited website, with over 50 million requests per day)[1].

Three reasons are usually mentioned to explain the success of Wikipedia. The first
one is that its articles and contents are based on the contribution of anyone willing to
improve them, with little to no restrictions. Many people would argue that this model
could not produce good quality compared to peer-review model generally found in scientific publications. Nevertheless, an article published in Nature [5] showed that the
accuracy of Wikipedia is very close to other *traditional* printed encyclopedias such as
Britannica. Therefore, if it were possible to create accurate articles with this open contribution model, it could be probably considered as a new method for collecting human
knowledge, with an unparalleled breadth and detail.

---

[1] Information extracted from http://www.alexa.com/search?q=wikipedia.org on March 23rd,
2007.

J. Filipe et al. (Eds.): ICSOFT/ENASE 2007, CCIS 22, pp. 257–268, 2008.
© Springer-Verlag Berlin Heidelberg 2008

The second reason is the ease of use of Wikipedia. Its contents are collected, presented and managed mainly with MediaWiki[2], a libre software[3] developed and maintained by the own project. This software offers simple-to-use and intuitive tools for editing articles, adding figures and multimedia content, and also for article reviews and discussions.

The third advantage of Wikipedia is that all textual contents are licensed using the GNU Free Documentation License (GNU FDL). This makes the content freely available to all users, and allows reprints by any third parties as long as they make them available under the same terms. Other contents, such as photographs and multimedia are subject to specific copyright notices, most of them sharing the same philosophy and principles of the GNU FDL.

However, although its success in sharing knowledge, Wikipedia may face some serious challenges in the near future. The most disturbing is the rapid growth in system requirements that has to be dealt with, mainly due to Wikipedia's enormous size. The English version has already surpassed the 1.5 million articles mark (1,697,653 articles as of March 21st 2007). The main consequence of this growth is that Wikipedia is beginning to consider how to expand its system facilities in order to not become a victim of its own success.

To evaluate to what extent Wikipedia is growing, and what scenarios the project will likely face in the future, a detailed quantitative analysis has to be designed and performed. This analysis would make it possible to know the evolution of the most important parameters of the project, and the construction and validation of growth models which could be used to infer those scenarios. In this paper, we propose a methodology for performing such kind of quantitative analysis. The growth of the whole project is affected by four different factors:

- *System Infrastructure.* Currently, most of the traffic served by Wikipedia comes from a cluster in Florida, maintained by the Wikimedia Foundation. In the past months, several mirror projects have been set up in Europe (France) and Asia (Yahoo! cluster in Singapore). Many other supporters maintain minor mirror sites all over the world. However, the size of Wikipedia seems to grow much faster than the project facilities, with the risk of overloading the servers and consequently producing a slowdown in the service.
- *Software Evolution.* MediaWiki is the core software platform of the Wikipedia Project. This tool, essentially developed in the PHP programming language, is responsible for retrieving the contents (text, graphics, multimedia...) from the database for a certain language, and delivering them to the Apache web servers, to satisfy the requests made by users. A very active community of developers supports the evolution of this software package, adding the functionalities users ask for and boosting the performance.
- *Evolution of Articles and Contents.* Articles are the core of Wikipedia. There is one article for each different topic, and topics are selected through consensus among

---

[2] http://www.mediawiki.org

[3] Through this paper, we will use the term libre software to refer both to free software and open source software (according to the respective definitions by the Free Software Foundation and the Open Source Initiative).

the wishes of users. Authors can also discuss article contents through special talk pages, thus reaching consensus about what should and should not be included in them. So far, the content of the articles may include text, graphics, photographs, math formulas and multimedia. They reflect the authors' interests and level of contribution (some languages gather more articles than others), so this factor is in close relationship with the last one.

– *Changes and contributions by the Community.* the expansion of Wikipedia is also affected by the contributions from editors and developers. Due to its collaborative nature, Wikipedia strongly depends on the work of volunteers to maintain its current rate of growth. If, for some reason, editors change their current behavior, or developers begin to decrease their rate of software contributions, this will definitely affect Wikipedia's future possibilities.

In this paper, we analyze Wikipedia focusing on the last two factors. We concentrate our efforts in gaining knowledge about the Wikipedia community of authors, in the ten most important language versions of the encyclopedia, and the evolution of the articles we find in each of them. The selection of the ten most important languages has been done regarding the total number of articles.

2 Background: Previous Research on Collaborative Projects

Although the process of collaborative content creation is relatively new, collaborative patterns have already been analyzed thoroughly in other technical domains. Libre software is a very good example of those collaborative environments. Several useful conclusions can be learned from a careful examination of their functional features. Wikipedia is, in some sense, a *libre contents* project. Its articles are subject to the GNU FDL, reflecting much of the same philosophy that we find in libre software. It should be interesting to check how much these two worlds show similar behaviors.

For example, a very popular concept introduced by Raymond [13] for the libre software development is the *bazaar*. Libre software projects tend to a development model that is similar to oriental bazaars, with spontaneous exchanges and contributions not leaded by a central authority, and without a mandatory scheduling. These methods can be seen as opposite to typical software development processes, as these are more similar to how medieval cathedrals were built, with very tight and structured roles and duties, and centralized scheduling.

But it was not until almost the year 2000 when the research community realized that lots of publicly available data about libre software could be obtained and analyzed. Some research works, including [4] and [10] showed that a small group of developers were the authors of a large amount of the available code. Mockus et al. [12] performed a research work about the composition of the developers communities of large libre software projects. They verified that a small group of developers (labeled as the *core group*) was in charge of the majority of relevant tasks. A second group, composed by developers who contribute frequently, is around one order of magnitude larger than the core group, while a third one, this one of occasional contributors, is about another order of magnitude larger.

Other interesting research works include [7] about the growth in size over time of the Linux kernel. Godfrey et al. showed that Linux grew following a super-linear model, apparently in contradiction to one of the eight *laws* of software evolution [11]. Although not yet confirmed, this may be indicative of a superior growth for open collaborative development environments than with closed industrial settings commonly used. Other research works have focused their attention on the study of Linux distributions, where hundreds to thousands of libre software programs are integrated and shipped. Especially the case of Debian [8,9,1,2] is very interesting in this regard as it is a distribution built exclusively by volunteers.

A methodological approach of how to retrieve public data from software repositories and the various ways that these data can be analyzed, especially from the point of view of software maintenance and evolution, can be fond in Gregorio Robles' dissertation [14]. This work puts special attention to developer-related (or social) aspects as these give valuable information about the community that is developing a software.

Specifically on the Wikipedia, we can find also some previous studies. [3] quantifies the growth of Wikipedia as a graph. The authors find many similarities among several language versions of Wikipedia, as well as with the structure of the World Wide Web. This should be no surprise, because in some way, wikis are simply another flavor of websites where content may be linked from other contents (by using HTML hyperlinks). Jakob Voss [16] introduced some interesting preliminary results about the evolution of contents and authors, mainly focusing on the German version of the Wikipedia: the number of distinct authors per article follows a power-law while the number of distinct articles per author follows Lotka's Law. Buriol et al. [3] showed that growth in number of articles and users were consistent with Voss' results, but this time in the English version.

Finally, Viegas et al. [15] found an alternative approach for studying contribution patterns to Wikipedia articles. They have developed a software tool, History Flow, that can navigate through the complete history of an article. This way, it is possible to identify periods of intense growth in the content of articles, acts of vandalism and other interesting patterns in users contributions.

3 Methodology

In this section, we present the methodology for a quantitative analysis of different language versions in Wikipedia. Firstly, we introduce some of the most relevant features of the database Wikipedia uses to store all its contents and edit information. Then, we briefly present the automatic system that the Wikimedia Foundation uses to create database dumps for all of its projects, and specifically for Wikipedia. Finally, we describe WikiXRay[4], our own tool developed for generating quantitative analysis of different language versions in Wikipedia automatically.

3.1 Wikipedia Database Layout

The MediaWiki software is currently strongly tied to the MySQL database software. Many functions and data formats are not compatible with other database engines. The

---

[4] http://meta.wikimedia.org/wiki/WikiXRay

logical model of the database that stores Wikipedia contents has suffered a deep transformation since version 1.5 of the MediaWiki software. The most up-to-date database
schema always resides in the *tables.sql* file in the Subversion repository.

The tables of the database logical model that are relevant to our purposes are:

- *Page*. One of the core tables of the database. In this table, each page is identified
  by its title, and provides some additional metadata about it. The name of each page
  refers to the namespace to which it belongs to.
- *Revision*. Every time a user edits a page, a new row is created in this table. This
  row includes the title of the page, a brief textual summary of the change performed,
  the user name of the article editor (or its IP address the case of an unregistered
  user) and a timestamp. The current timestamp support is somewhat basic as it is
  implemented using plain strings.
- *Text*. The text for every article revision is stored in this table. The text may be stored
  in plain UTF-8 compressed with gzip or in a specialized PHP object.

Database dumps do not only contain the articles, but other relevant pages that are used
by the user community of that language on a daily basis. To classify these pages, MediaWiki groups pages in logical domains known as namespaces. A tag included in the
page title indicates the namespace a page belongs to. Articles are grouped in the Main
namespace. Other relevant namespaces are User for the homepage of every registered
user, Meta for pages with information about the project itself, and Talk, User_talk
and Meta_talk for discussion pages related to articles, users and the project respectively. Most of our research work is focused on articles in the Main namespace.

3.2 Database Dumps

There are database dumps available for all Wikipedia versions through the web[5]. Some
major improvements have recently been included in the database dumps administration, the most relevant the automation of the whole dump process, including real-time
information about the current state of each dump. Other new features include the automatic creation of HTML copies of every article stored, and the upcoming system
for creating DVD distributions for different language versions. The tool employed to
perform database dumps is the Java-based mwdumper, also available in the Wikipedia
SVN repository[6]. This tool creates and recovers database dumps using an XML format.
Compression is achieved with bzip2 and SevenZip.

In our study we have retrieved a simplified version of the dumps which provides data
only for the page and revision tables of each language. An additional dump with the page
table alone had also to be downloaded, because we needed information about the length
in bytes of every single page. The simplified dump does not include that information.

3.3 Quantitative Analysis Methodology: WikiXRay

The methodology we have conceived to analyze the Wikipedia is composed of following steps: First, we collect the database dumps for the top-ten Wikipedia languages (in
number of articles, according to the list publicly available from Wikipedia main page).

---

[5] http://download.wikimedia.org
[6] http://svn.wikimedia.org/

We have therefore developed a Python tool, called WikiXRay, to process the database dumps, automatically collecting relevant information, and processing this information and proceed to an in-depth statistical analysis.

Quantitative results for each language can be obtained from two different points of view:

- *Community of Authors.* This is the first important parameter that affects the growth of the database. Relevant aspects include studying the total number of editors and contributions, number of contributions for each author over a certain period of time (i.e. contributions per month or per week) and correlating results with Wikipedia's own statistics.
- *Evolution of Articles.* We analyze the growth of the database from a different perspective, focusing on the evolution of the size of the articles over time, the distribution of articles sizes in general and how the evolution of articles correlates to the contributions made by users.

4   Case Study: The Top-Ten Wikipedia Languages

As case study for our methodology, we have considered convenient to analyze the database dumps of the top-ten largest language versions of the Wikipedia. At the time of writing, the most popular language corresponds to English, followed in this order by German, French, Polish, Japanese, Dutch, Italian, Portuguese, Swedish, and Spanish. Due to spaces limitations, we will not be able to include in this paper all the results obtained, but will show the most relevant ones[7]. Figure 1 is a graphic that shows the evolution over time of the number of contributions to articles for the top-ten languages. A contribution is considered any edition made by an user to an article. A logarithmic scale in the vertical axis to plot the graphics has been used, resulting in a clear common behavior at least since December 2004 for all languages.

For some language versions we can establish a strong correlation between relevant past events and abrupt increases of their growth rates for total contributions. For example, the Japanese Wikipedia experimented a quite remarkable growth of two orders of magnitude in its total number of contributions from February to March, 2003. In January 31, 2003, the Japanese online magazine Wired News covered Wikipedia. This has been reported as the first time Wikipedia was covered in the Japanese media[8]. So, we can infer a direct relationship between Wikipedia popularity and the number of contributions it receives.

4.1   The Community of Authors

One of the parameters we are interested in is the level of inequality that can be found for contributions. As already mentioned, previous research on the libre software phenomenon has shown that a relative small number of developers concentrate a large part of the contributions. Analyzing inequality will allow us to see if both phenomenons present similar patterns.

---

[7] http://meta.wikimedia.org/wiki/WikiXRay offers additional graphic results.
[8] http://en.wikipedia.org/wiki/Japanese_wikipedia

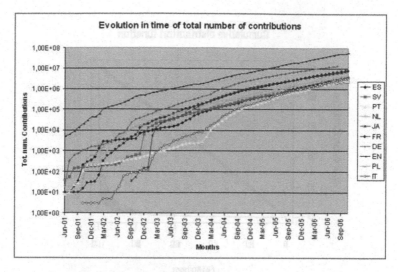

Fig. 1. Evolution over time of the total number of contributions

We will measure inequality by means of the Gini coefficient. This coefficient, introduced by Conrado Gini [6] to measure welcome inequality in economics, shows how unequal something is distributed among a group of people. To calculate the Gini coefficient we have first to obtain the Lorenz curve, a graphical representation of the cumulative distribution function of a probability distribution. Perfect distribution among authors is hence given by a 45 degree line. The Gini coefficient is given by the area between the two curves, providing how far the actual distribution is from the perfect equality. Figure 2 presents the Lorenz curve for all the languages under study. All of them present similar behaviors, with aproximately 90% of the users responsible all together for less than 10% of the contributions, (Gini coefficients ranging from 0.9246 in the Japanese version to 0.9665 in the Swedish version). Hence, we can state that as in the case of libre software, we also find a small amount of very active contributors.

4.2 Articles

An important parameter of articles is their size, as it gives the amount of content included in them. We have therefore plotted histograms for article sizes for all languages under study. This way we will be able of inferring different types of articles.

Figure 3 shows the histogram of the article size in the English and Polish Wikipedias. We take the decimal logarithm of the article size in bytes for this representation, which facilitates the identification of patterns. The solid black line plotted over the histograms represents the probability density function of the article size, giving about the same information but with better resolution.

After inspection, we can group articles attending to their size in two groups:

– *Tiny Articles.* The left side of the histograms shows a subpopulation conformed by those articles whose size varies from 10 bytes to 100 bytes. Some of those belong

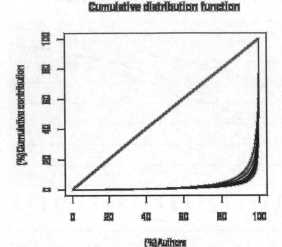

Fig. 2. Lorenz curves for contributions of authors to the top-ten languages

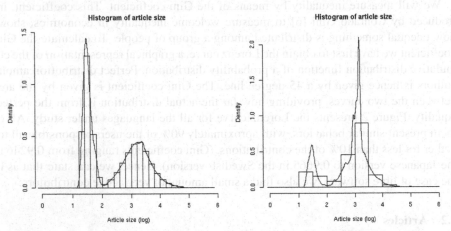

Fig. 3. Histogram for sizes of articles in the English (left) and Polish (right) languages

to a special category of articles known as 'stubs' in the Wikipedia jargon. Stubs are templates automatically created when a user requests a new article about some topic not previously covered. This way, the software makes it easier for any upcoming user interested in that topic to further contribute to the article. However, most of the articles in this subpopulation fall into another important category in Wikipedia: 'redirects'. One of the biggest problems for any encyclopedia is how to select an accurate entry name for each article, because many topics present alternative names users can also search for. Redirects are the perfect answer to deal with multiple names for the same article. They are special articles with no content at all, but a link that points to the main article for that topic. So, when users search for alternative

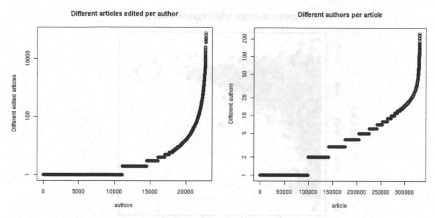

Fig. 4. Number of different articles edited per author (top) and number of different authors per article (bottom) for the Dutch Wikipedia

Table 1. Probability mass value for tiny articles and standard articles

Lang	Tiny Mass	Std Mass	Mass Ratio
English	0.51	0.49	0.96
German	0.38	0.62	1.63
French	0.3	0.7	2.33
Polish	0.18	0.82	4.55
Japanese	0.37	0.63	1.7
Dutch	0.29	0.71	2.44
Italian	0.23	0.77	3.34
Portuguese	0.24	0.76	3.16
Swedish	0.34	0.66	1.94
Spanish	0.33	0.67	2.03

names, the find the equivalent page that *redirect* them to the main article for that topic. Redirects also allow contents to be centralized in certain articles, thus saving storing capacity.

- *Standard Articles.* On the other hand, we can identify a second subpopulation on the right side of the histograms, corresponding to those articles whose size grows beyond 500 bytes, that is, articles that have a certain amount of content. Further research should be conducted to explain whether the community is more interested in those topics, or those articles have been on-line for a longer period of time, increasing the probability of receiving contributions.

We can extract interesting conclusions from the shape of the density function, as each subgroup of articles exhibits a Gaussian distribution. Its mean can be use to characterize the contributions of each user community to standard articles, and the average redirect size (for the tiny articles). We have calculated the ratio between the normalized mass of the density function for tiny and standard articles for all languages. Results, presented in Table 1, show some communities that are not very interested in creating redirects

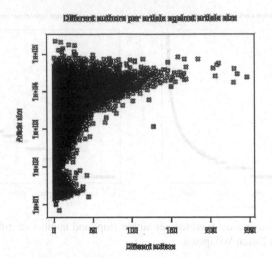

Fig. 5. Number of authors against article size (in bytes) for the Dutch Wikipedia

offering alternative entry names for their articles (for example Polish, Italian and Portuguese Wikipedias), while other ones (for example the English Wikipedia) generate more redirects balancing the probability mass of both subpopulations. Therefore, simply counting the number of different articles of a certain language version does not give us a clear picture about the size or quality of its articles. Some language versions may create a lot of redirects or stubs, while others may concentrate on adding real contents to existing articles. Further research about these results may lead to identify interesting content creation patterns in different communities of authors.

Other factors, such as robots that automatically create a bunch of new stubs from time to time, need to be taken under consideration. For example, the ratio for the English language is noticeable; a closer inspection has thrown as result that this is due to many automatically created articles e.g. with information from the U.S. census. Also noticeable is the case of the Polish Wikipedia. In July 2005 a new task was created for the tsca.bot, one of the bots of the Polish Wikipedia. It was programmed to automatically upload statistics from official government pages about French, Polish and Italian municipalities. This new feature introduced more than 40,000 new articles in the following months. That allowed the Polish Wikipedia to overtake the Swedish, Italian and Japanese language versions and become the 4th largest Wikipedia by total number of articles[9].

Figure 4 gives the number of articles edited by author (only registered authors are considered) and the number of authors per article for the Dutch version of Wikipedia. The article with the highest number of contributors accounts for over 10,000 registered users; around 50 have been the work of over 5000 authors. On the other hand, we can find authors that have contributed to more than 10,000 articles. Future research should focus on these results and explore if these type of authors concentrate in specific tasks, as for instance correcting errata or adapting the style to meet Wikipedia conventions.

---

[9] http://en.wikipedia.org/wiki/Polish_Wikipedia

On the other side, more than 250,000 articles of the Dutch version were contributed by less than 10 authors, corroborating that, in general, a majority of authors tend to focus their contributions only on a few articles.

Finally, in Figure 5 we represent the number of authors per article against the article size in the Dutch Wikipedia. We see that very few articles have been worked on by more than 150 authors. In general, article size correlates positively to the number of authors who have edited it, with smaller sizes for those articles with less contributors and larger sizes for those with wider number of authors. Despite these facts, it is also interesting to notice that the 10 largest articles reflect the work of less than 30 authors, and that the largest one has received contributions from less than 10 authors. So, the number of different authors should not be considered as the unique parameter affecting articles size.

5   Conclusions and Future Research

Some useful conclusions can be extracted from this research. We have shown that the top-ten language versions of Wikipedia present interesting similarities regarding the evolution of the contributions to articles over time, as well as the growth rate in the sum of article sizes. The Gini coefficients found for the studied languages present (as expected) big inequalities in the contributions by authors, with a small percentage being responsible for a large share of the contributions. However, the Gini values found for the languages could help to characterize the underlying author communities.

We have also identified certain patterns that could be used to characterize Wikipedia articles attending to the length (or size) of the articles. Two main subgroups (tiny articles and standard articles) represent the peculiarities of contributions behaviors in each language community. The ratio between them shows the interest of the corresponding communities in linking or opening new topics versus completing and improving existing ones.

Finally, we have found that there is no simple correlation between the number of authors that contribute to a certain article and the total size reached by that article. This leads us to think about additional factors that could affect the production process, including the nature of the topic and the level of popularity of that topic in the author community.

The methodology we have proposed provides an integral quantitative analysis framework for the whole Wikipedia project, a very ambitious goal that we confront for the near future.

References

1. Amor, J.J., Gonzalez-Barahona, J.M., Robles, G., Herraiz, I.: Measuring libre software using debian 3.1 (sarge) as a case study: preliminary results. Upgrade Magazine (2005a)
2. Amor, J.J., Robles, G., Gonzalez-Barahona, J.M.: Measuring woody: The size of debian 3.0. Technical Report. Grupo de Sistemas y Comunicaciones, Universidad Rey Juan Carlos. Madrid, Spain. Grupo de Sistemas y Comunicaciones, Universidad Rey Juan Carlos. Madrid, Spain (2005b)

3. Buriol, L.S., Castillo, C., Donato, D., Millozzi, S.: Temporal evolution of the wikigraph. In: Proceedings of the Web Intelligence Conference, Hong Kong. IEEE CS Press, Los Alamitos (2006)
4. Ghosh, R.A., Prakash, V.V.: The orbiten free software survey. First Monday (2000)
5. Gigles, J.: Internet encyclopedias go head to head. Nature Magazine (2005)
6. Gini, C.: On the measure of concentration with especial reference to income and wealth. Cowless Comission (1936)
7. Godfrey, M., Tu, Q.: Evolution in open source software: A case study. In: Proceedings of the International Conference on Software Maintenance, San Jos, California, pp. 131–142 (2000)
8. Gonzalez-Barahona, J.M., Ortuno-Perez, M., de-las Heras-Quiros, P., Gonzalez, J.C., Olivera, V.M.: Counting potatoes: the size of debian 2.2. Upgrade Magazine II(6), 60–66 (2001)
9. Gonzalez-Barahona, J.M., Robles, G., Ortuno-Perez, M., Rodero-Merino, L., Centeno-Gonzalez, J., Matellan-Olivera, V., Castro-Barbero, E., de-las Heras-Quiros, P.: Analyzing the anatomy of GNU/Linux distributions: methodology and case studies (Red Hat and Debian). In: Koch, S. (ed.) Free/Open Software Development, pp. 27–58. Idea Group Publishing, Hershey (2004)
10. Koch, S., Schneider, G.: Effort, cooperation and coordination in an open source software project: Gnome. Information Systems Journal 12(1), 27–42 (2002)
11. Lehman, M.M., Ramil, J.F., Sandler, U.: Metrics and laws of software evolution the nineties view. In: METRICS 1997: Proceedings of the 4th International Symposium on Software Metrics, p. 20 (1997)
12. Mockus, A., Fielding, R.T., Herbsleb, J.D.: Two case studies of open source software development: Apache and mozilla. ACM Transactions on Software Engineering and Methodology 11(3), 309–346 (2002)
13. Raymond, E.S.: The cathedral and the bazaar. First Monday 3(3) (1998)
14. Robles, G.: Empirical software engineering research on libre software: Data sources, methodologies and results. Doctoral Thesis. Universidad Rey Juan Carlos, Mostoles, Spain (2006)
15. Viegas, F.B., Wattengberg, M., Dave, K.: Studying cooperation and conflict between authors with history flow visualizations. In: Proceedings of the SIGCHI conference on Human factors in computing systems, Viena, Austria, pp. 575–582 (2004)
16. Voss, J.: Measuring wikipedia. In: Proceedings of the 10th International Conference of the International Society for Scientometrics and Infometrics 2005, Stockholm (2005)

A Semantic Web Approach for Ontological Instances Analysis

Roxana Danger[1] and Rafael Berlanga[2]

[1] Department of Information Systems and Computation
Technical University of Valencia, Spain
rdanger@dsic.upv.es
[2] Department of Computer Languages and Systems, Universitat Jaume I, Spain
berlanga@uji.es

Abstract. New data warehouse tools for Semantic Web are becoming more and more necessary. The present paper formalizes one such a tool considering, on the one hand, the semantics and theorical foundations of Description Logic and, on the other hand, the current developments of information data generalization. The presented model is constituted by dimensions and multidimensional schemata and spaces. An algorithm to retrieve interesting spaces according to the data distribution is also proposed. Some ideas from Data Mining techniques are included in order to allow users to discover knowledge from the Semantic Web.

1 Introduction

The *Semantic Web* is a new form of web conceived for allowing human and software tools to process and share the same sources of information. The Semantic Web relies on a set of standards which provide syntactic consistency and semantic value to all of its content. For example, Description Logic is used as the theoretic base for the description of web items, and the languages RDF and OWL for their syntactic representation. Description Logic defines a family of knowledge representation languages which can be used to represent, in a well-understood formal way, the knowledge of an application domain. This knowledge, known as ontology, ranges over the terminological cognition of the domain (the interesting object classes, or concepts, its *Tbox*) and its examples (the instances of the object classes, its *Abox*).

Data analysis in the Semantic Web will be the most important process when the population of ontologies[1] becomes a reality. Its final goal is the recognition of patterns amongst the values of the attributes of the ontological instances, which could turn into knowledge or allow its discovery. OWL tools allow users to create new concepts related to one or more existing ontologies, and to determine the instances associated to such concepts.

An additional tool is needed if we want to perform in a versatile way customized data analysis, either full or partial, so that each object can be studied from different points of view focusing on distinct particular features. Such a tool should allow users

---

[1] The population of ontologies is the process of adding instances to an ontology in order to enrich it with examples of its domain knowledge.

J. Filipe et al. (Eds.): ICSOFT/ENASE 2007, CCIS 22, pp. 269–282, 2008.
© Springer-Verlag Berlin Heidelberg 2008

to navigate through an instance set and its properties, being able to discriminate between relevant and superfluous information. Moreover, it should compute and display statistical indexes able to describe and report about the extracted patterns.

The formalization of such a tool, following the framework described below, is the purpose of the present work. Starting from an available ontology, this is enriched with information provided by the data analyst, who specifies the *atomic data combination functions*. These functions provide the way for combining atomic data to form a generalized instance representing a set of instances. Then, two main structures have to be built: the *conceptual dimensions* and the *multidimensional conceptual spaces*. The conceptual dimensions are partial order specifications between concepts, which allow browsing through their semantic relations. The multidimensional conceptual spaces can be seen as "intelligent object containers". They make use of a subset of conceptual dimensions, a specification of relevant abstraction levels and a set of atomic data combination functions in order to (re)construct appropriate generalized instances. Different statistical indexes (e.g. frequencies), associated to the conceptual dimensions, can be used to characterize patterns in the conceptual spaces. The most suitable data analysis technique for carrying out this proposal is *data warehousing*.

The present work is not the first attempt to formalize a data warehouse for the Semantic Web. Within the Data Warehouse Quality (DWQ) project [1] a formalization for the multidimensional modeling based on an extension of the constructors of Description Logic is proposed. In this way, new object classes could be described by specifying aggregability operations, and the traditional reasoning over ontological instances could be applied. However, the demonstration of the undecidability of minimal languages that operate with aggregate operators [2] makes the proposal of the DWQ project unfeasible in practice.

On the other hand, the ideas of the traditional data warehouse (and OLAP techniques) has been extended to object oriented modeling, [3,4,5,6,7]. Considering that description logic was designed as an extension to frames and semantic networks, the basis of object-oriented data warehouse could be applied in order to define a data warehouse for the Semantic Web. However, the flexibility of object-oriented formalization causes a more sparse structure in object-oriented databases that in traditional ones. Moreover, the restrictions of OLAP implementations drastically reduce the useful set of objects to be used in the analysis.

Unlike these previous works, this paper proposes a multidimensional model for the analysis of ontological instances that merge both approaches. The idea is the creation of meta-ontologies in order to enrich the knowledge of ontologies with data analysis information. This data analysis information focuses on the description of interesting object classes and on the aggregation process. The reasoning of description logic is used in a preliminary phase to 1) recover the satisfiable[2] concepts that can be used on analysis processes, 2) discover the hierarchical and aggregate orders between the concepts, and 3) assign each instance to the set of concepts to which it belongs.

This paper describe our proposal in detail. Firstly, the ideas underling our proposal are drawn in Section 2; and in Section 3 the data analysis information is introduced.

---

[2] A concept (or object class) is satisfiable if it is consistent and there exists an interpretation on which appears at least an instance of this concept.

Then, the proposed model is formalized, starting from the definition of dimensions and their operators (Section 4) and following with the specification of the multidimensional conceptual space (Section 5). The two following sections are focused on the extraction of interesting conceptual spaces and their use, respectively. The last section gives some conclusions and future work.

2   Proposal Framework

The framework of our proposal is depicted in Figure 1. A data analyst is in charge of describing the metadata of the analysis to perform. We assume that she is able to formalize concepts and enrich ontologies according to the analysis requirements. Two kinds of analysis metadata are described: new concepts representing the analysis goals, and atomic data combination functions (combination functions for short) specifying how to generalize instances (e.g. aggregation functions for measures). Further details on these two types of metadata are provided in Section 3.

As a second step, the *instance models* of the ontology concepts are extracted using reasoners. An instance model may be conceived as a directed graph starting from the described concept. Each node represents a concept which is linked to other concepts through arcs. Arcs are labeled with the properties which relate the linked concepts. Therefore, the *composition relations* between concepts in an ontology may be discovered through paths in an instance model. Reasoners can interpret an ontology at different depth levels depending on the kind of constructor properties they verify. For example, in [8] a deep interpretation is proposed which allows the recovering of all composition relations, included implicit ones. Simpler reasoners follow basic constructors to recover all explicit composition relations.

Dimensions are computed taking into account the previously defined instance models. A dimension is defined as a set of ontological concepts which maintains a partial order between them. This order may be defined by "is-a" relations (i.e. hierarchical semantic relations) and/or by properties tagging the arcs of the instance models. In this way, a dimension associated to a concept contains those concepts derived or reachable from it.

Composition relations (or features) and the dimensions associated to a given concept are the building blocks of a multidimensional conceptual schema. A *multidimensional conceptual schema* is defined by a set of orthogonal dimensions (associated to the arrival concepts through the features), along with the combination functions used to generalize their related concepts. In order to define the set of features to be used in the multidimensional schema, two methods can be used: a) specifying an instance selector (an instance model describing those features, with exact values or not, in which the analyst is interested) or b) using an algorithm to select useful sets of features, like the one proposed in Section 6.

Finally, a *multidimensional conceptual space* is generated instantiating the corresponding schema at a desired level. The multidimensional conceptual space can be used for knowledge discovery using two different methods: *Browsing the space* through traditional operations such as "roll-up" or "drill-drown", and using *Data mining* to discover implicit patterns in the space. Useful patterns that can be extracted are described in Section 7.

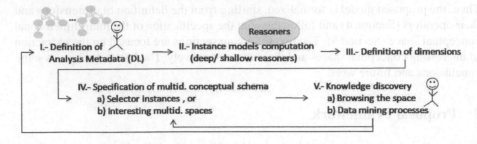

Fig. 1. Our proposal framework

All the described process is semi-automatic, since in some cases the data analyst has to specify or select the desired elements (such as the concepts in each dimension and the features to analyze). The analysis can be refined through a cyclic processing, using a simple feedback loop.

3   Analysis Metadata

Information descriptions useful for the analysis are those available in the ontologies in form of instances. However, they are not enough to both analyze data and discover patterns. New interesting concepts and particular issues related to the generalization process are essential in order to generate descriptions that represent relevant and realistic visions of the application domains of the analyzed ontologies. We call all this information *analysis metadata*, which comprises the following elements:

– *Description of New Concepts*, which introduces additional levels of abstraction in the concept hierarchies expressed in an ontology, and/or to link concepts from different ontologies. New concepts may be obtained by extending old ones via paths to previously unrelated concepts. They can also represent hierarchical clusters obtained using clustering algorithms.
– *Description of the Combination Functions* (see definition below); it is used to specify ways for generalizing sets of data of the same type during the instance generalization process. The data analyst is responsible for deciding the combination functions that are semantically suitable for a given data set. For example, the combination function which computes the average of a set of values is semantically suitable for a temporal sequence of temperatures of a town, but not for a set of temperatures of different towns.

Although it is perfectly plausible to define such descriptions for every new multidimensional conceptual space, a better solution is to keep this semantic information always available and to apply it according to the requirements of each case. This goal can be achieved building a meta-ontology containing the sort of information described above, again using Description Logic. In this way, analysts can proceed more efficiently as they can reuse the analysis metadata. Even more importantly, in this way the coherence of different studies is granted, providing an ontology with an intrinsic robustness toward analysis processes. Thus, further studies can be more easily performed by comparing different analysis on the same knowledge domain and/or the point of views of different analysts.

The description of the combination functions can be specified through instances associated to the notion of *Combinable Concept* of this meta-ontology:

$CombinableConcept \equiv$
$\equiv \exists hasConcept.URI \sqcap \forall hasRelation.URI \sqcap \exists hasCombinationFunction.CombinationFunction$
$CombinationFunction \sqsubseteq \exists hasName.String \sqcap \exists hasImplementation.URI$

Combinable concepts are those for which a combination function can be defined. A combinable concept can be a datatype, a named concept (defined via a URI), or a concept derived from a composition of relations beginning with a named concept (specified by the URI where the start concept is defined and the relations from it).

4 Dimensions and Their Operations

A dimension is described by both a set of concepts and the way to browse through them. Such browsing is performed using the operators of abstraction and generalization between ontological instances, and the selection operators defined below.

We consider that an abstract ontology is constituted only by the terminological knowledge. An ontology that contains a set of instance axioms (the *Abox of the ontology*, composed by axioms specifying the class C of an instance a -$C(a)$- and the relations between two instances a and b -$R(a, b)$-) is called concrete ontology. As it is usual in description logic, [9], the interpretation of the ontology is $(I = \Delta^I, .^I)$, where Δ^I denotes the set of instances belonging to an ontology $\mathcal{O}$, and $.^I$ the interpretation of the concepts defined on $\mathcal{O}$; $I \Vdash x$ represents that x is deduced from I; $\top$ represents the top concept: *thing*. We denote with $\mathcal{A}$ the Abox of $\mathcal{O}$, with $\mathcal{R}$ the set of axioms associated with the relations of an ontology, with N_C and N_R the set of named concepts and relations of the ontology, respectively. Besides, the interpretation of datatypes is defined by $I^D = (\Phi^D, .^D)$, where Φ^D denotes the set of all data belonging to datatypes, and the function $.^D$ associates each datatype Φ with a subset of data in Φ^D and specify the operations over the data on Φ^D. $\Delta^I \cap \Phi^D = \emptyset$. All representable data in the ontology belongs to the set $\mathcal{U} = \Delta^I \cup \Phi^D$.

The definition of path set between two concepts C and C' and that of dimensional partial order are given below. Intuitively, the former is the set of lists of relation-concept pairs that links C and C' by using consistent ontological definitions, and the latter is used to relate concepts using both the aggregate (as defined bellow) and the hierarchical order between concepts implicitly defined in a given ontology $\mathcal{O}$.

Definition 1. $Path(C, C') = \oplus_{1 \leq i \leq n} \langle RS_i, C_i \rangle$ *is an aggregation path from concept* C *to concept* C' *of the ontology* $\mathcal{O}$ *if* $C_1 = C$, $C_n = C'$, *and there exists an interpretation* $I = (\Delta^I, .^I)$ *of* $\mathcal{O}$ *such that* $\exists x_i \in \Delta^I, 0 \leq i \leq n$, *such that* $x_0 \in C^I$ *and* $x_i \in C_i^I$, $\langle x_{i-1}, x_i \rangle \in R_i^I$, *for* $1 \leq i \leq n$, $R_i \in RS_i$.

The process of path retrieval can be as exhaustive as the used DL reasoner allows. It is worth emphasizing that these paths not only describe the aggregation relations between two concepts, but also the aggregation order between all concepts of an ontology.

Definition 2. *Let $\mathcal{O}$ be an ontology. A dimensional partial order, denoted as $\overset{\sqsubseteq}{\to}$ is a partial order between all possible pairs of concepts $C, C' \in N_C$ defined according to the following constraints:*

- $C \overset{\sqsubseteq}{\to} C'$ *if $C' \sqsubseteq C$, or*
- $C \overset{\sqsubseteq}{\to} C'$ *if $\exists Path(C, C')$*

The symbol $\overset{\sqsubseteq^*}{\to}$ is the reflexive and transitive closure for relation $\overset{\sqsubseteq}{\to}$.

Definition 3. *Let $\mathcal{O}$ be an ontology. The pair $D = (C_d, \overset{\sqsubseteq}{\to})$ is a conceptual dimension, being C_d a set of satisfiable concepts in $\mathcal{O}$, $\top \in C_d$ and $\overset{\sqsubseteq}{\to}$ the relation of dimensional partial order for the elements in C_d.*

Example 1. In Figure 2 a *workplace* dimension which combines hierarchical and aggregation relations is shown. This dimension can be used to identify a specific place with different levels of granularity.

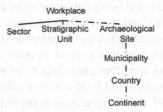

Fig. 2. *Workplace* dimension

Operations. The ontological instances of each dimension can be represented by using different point of views of (concepts associated with) the dimension. It is thus necessary to define two different kinds of operations over such instances. The first one is the selection operator, used to specify the interest portion of the instance that must be shown (for example, when the concept represented by a dimension is replaced by a concept related to the first one by an aggregate relation). The second important operation is the generalization, used to generalize a set of instances (for example, when the concept represented by a dimension is replaced by a concept related with the first one by a hierarchical relation). The following definitions formalize these operators.

Definition 4. *Let $\mathcal{O}$ be a concrete ontology with Abox $\mathcal{A}$; the description of an instance $a \in \mathcal{A}$ is the set $d(a) = \{R(a, b) \in \mathcal{A}\}$. This instance is said to be of type C, denoted by $a \in_* C$, if C is the most specific concept that can be deducted from I for a, i.e., $\forall C^*$ such that $I \Vdash C^*(a), C \sqsubseteq C^*$.*

Definition 5. *Instance a' is called the specialization of an instance a of class C towards class C', if its description $d(a \uparrow_{C'} a')$ is not undefined, and if $\mathcal{A} \backslash (C(a) \cup d(a)) \cup (\{C'(a') \cup d(a \uparrow_{C'} a'))$ is consistent. The description of $d(a \uparrow_{C'} a')$, is defined as the following set:*

$$\{R'(a',b')|R(a,b) \in d(a), fe(C,C')(R) = \{\langle R',C'\rangle\}, \quad \textit{if } \exists f_e(C,C'),$$
$$d(b \uparrow_{C''} b') \neq \textit{undefined}\}, \quad |d(a \uparrow_{C'} a')| = |d(a)|$$

$$\textit{undefined}, \qquad\qquad\qquad\qquad\qquad \textit{otherwise}$$

where f_e *is a specialization function of the concept* C *to the concept* C'. *This function defines how to transform each relation on the abstract concept to the appropriate relation on the specialized concept (its formalization is available in [8]).*

The operation of abstraction of an instance, denoted by $d(a \downarrow_{C'} a')$, can be defined in a similar way.

Definition 6. *Let* $\mathcal{O}$ *be an ontology. Let* ℓ *be an undefined data that represent any data in* $\mathcal{U}$. *A pseudo-instance* a^3 *of type* C *is a selector if its description,* $d(a)$, *satisfies that:*

$$\forall b \,|\, \exists \{R_1, ..., R_n\} \subseteq N_R, R_1(a,a_1), ..., R_{n-1}(a_{n-1}, a_n), R_n(a_n,b) \in .^I \Rightarrow$$
$$b \in \{\ell\} \bigcup_{\substack{\forall C', \exists Path(C,C') \text{ and } R_1, ..., R_n \\ \text{is the order of the relations on } Path(C,C')}} C'^I.$$

Definition 7. *Let* $\mathcal{O}$ *be a concrete ontology with Abox* $\mathcal{A}$, $a \in \mathcal{U}$. *Let* ℓ *be an undefined data that represent any data in* $\mathcal{U}$. *An instance* $a' \in \mathcal{U}$ *is selected by an instance selector* a *if:*

- $a' \in \Phi$, $a' \in \{a, \ell\}$ *or*
- $a \subset C^I$, $C \in N_C$ *and* a'' *computed for* $d(a' \downarrow_C a'')$ *is such that:*

$$\forall b \subset \Phi \cup \{\ell\} \text{ such that } \exists R_1(a,a_1), ..., R_{n-1}(a_{n-1}, a_n), R_n(a_n,b) \in .^I \Rightarrow$$
$$\exists R_1(a'',a''_1), ..., R_{n-1}(a''_{n-1}, a''_n), R_n(a''_n, b'') \in$$
$$\in d(a'') \wedge (b'' = b \vee b = \ell)$$

Example 2. In Figure 3 a fragment of an archeology ontology is represented. A selector instance a constituted by the set {*has_morphology(a, a'), has_group(a', "open"), has_order(a', ℓ), has_decoration(a, ℓ), has_color(a, "gray")* }
   allows the users to recover from the ontology the descriptive fragments of *ceramic artifacts* instances according to the properties *group, order, decoration* and *color*, but notice that the morphologic group of the ceramic must be *open*, and its color *gray*.

Definition 8. *Let* $c_{\Phi D}$ *be a function (called a combination of simple data) which allows each datatype* Φ *to be mapped to another function* rep_Φ, *which in turn maps subsets of* Φ *in a compact representation of the input subset[4]. Let* d, d' *be two data in* $\Delta^I \cup \Phi^D$. *A complete combination of data* d *and* d' *is the data* $d \cup_t d'$ *computed as follows:*

---

[3] We call a pseudo-instance because ℓ does not belong to the ontology, although in order to improve the clarity of the explanation we will call it instance selector.

[4] For example, if $\Phi^D = \mathcal{Z}$, $c_{\Phi D} = \{\langle \mathcal{Z}, rangeOfIntegerSets\rangle\}$, where the function $rangeOfIntegerSets$ has domain $\mathcal{Z}$ and as images the most compact representations of integer sets, $2^{\mathcal{Z}}$, using integer range sets, then $rangeOfIntegerSets(\{1,2,3,4,5,7\}) = [1,5] \cup [7,7]$.

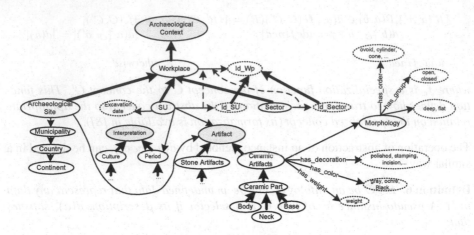

Fig. 3. Fragment of an archeology ontology. The concepts are represented in ellipses, the shady ones correspond to root concepts of different hierarchies of the ontology and the dashed-line ones correspond to datatypes. The thick lines represent the hierarchical relations between concepts, the thin lines aggregation relations and the dashed lines represent hierarchies between relations. The name of relations used in examples has been drawn.

$$c_{\Phi D}(\{d, d'\}), \qquad \qquad \text{if } \{d, d'\} \subseteq \Phi,$$

$$d' \uplus_t d, \qquad \qquad \text{if } d \in_* C, d' \in_* C', C \sqsubseteq C'$$

$$d(d \uparrow_{C'} d') \cup d(d') \setminus \{R(d', b_1), ..., R(d', b_n)| \qquad \text{if } d \in_* C, d' \in_* C',$$
$$b_1, ..., b_n \in_* C''\} \cup \qquad C' \sqsubseteq C \text{ and during the process}$$
$$\cup \{R(d', b_1 \uplus_t ... \uplus_t b_n)|\{R(d', b_1), ..., R(d', b_n)| \qquad \text{no indefinitions are obtained}$$
$$b_1, ..., b_n \in_* C''\}$$

$$\qquad \qquad \qquad \qquad otherwise$$
$$undefined,$$

In particular, two types of functions for data combination can be identified:

- *Unification Functions* which map data from 2^{Φ} to Φ. They can be oriented to statistical indexes, such as means or deviations. Of special interest is the function *restrictive unification* defined for all types of data as:

$$f(\{d_1, ..., d_n\}) = \begin{cases} d_1, & \text{if } d_1 = ... = d_n \\ undefined, & \text{otherwise} \end{cases}$$

- *Generalization Functions*: which map data from 2^{Φ} to a *compact notation*.

Example 3. Let d be an instance represented by the set of axioms {*has_color(d, "black"), has_decoration(d, "incisions"), has_weight(d, 20g)*} and d' an instance represented by the set {*has_color(d', "ochre red"), has_decoration(d', "incisions"), has_weight(d', 12g)*}. The outcome of the combination of d and d', d'', by using the union of sets as combination function for the data associated to *has_color* and *has_decoration* relations and the maximum of values as combination function for the data associated to has_weight is represented by $\{has\_color(d", \{"black", "ochre\ red"\}),\ has\_decoration(d", \{"incisions"\}),$ $has\_weight(d", 20g)\}$. However, if all relations have to be combined using unification functions, the result is undefined, because d and d' have different values for the same relations.

5 Multidimensional Conceptual Spaces

The definitions of multidimensional conceptual schema and multidimensional conceptual space are given below. The former can be seen as the structure which defines how to analyze the information. The latter is the container where the analyzed instances are described according to the specifications of the schema.

Definition 9. *Let O be an ontology, $C \in N_C$ and $CC = path_1, ..., path_n$ a set of paths from concept C toward concepts $C_1^*, ..., C_n^*$, respectively, (i.e., $path_i = \oplus_{1 \leq j \leq n_{i-1}}$ $\left\langle RS_{i_j}, C_{i_j}^* \right\rangle \oplus \langle R_{in_i}, C_i^* \rangle$). The tuple $E = (D_1, ..., D_n, c_{\phi_1}, ..., c_{\phi_n})$ is an n-dimensional (or simply multidimensional) conceptual schema of O associated to C using paths CC, where it is satisfied that $\forall D_i = (Cd_i, \overset{\sqsubseteq}{\rightarrow}), \forall C_{i_j} \in Cd_i, C \overset{\sqsubseteq^*}{\rightarrow} C_{i_j}^* \overset{\sqsubseteq^*}{\rightarrow} C_i^*$ and c_{ϕ_i} is a function which assigns a combination function to each simple type of data that can be reached from a concept in Cd_i. If no pair of paths in CC have common suffix the resulting schema has orthogonal dimensions.*

The concepts C_i^* are used to identify a unique base concept for dimension D_i, avoiding the double counting problem. Moreover, the multidimensional schema of orthogonal dimensions are useful to avoid multiple processing of common parts.

Definition 10. *Let O be a concrete ontology with Abox A. Let E be an m-dimensional conceptual schema of O associated to C using the paths in CC. The set of tuples $\{t_1, ..., t_n\}$, $t_i = (d_{i_1}, ..., d_{i_m})$, $t_i \neq t_j, \forall i, j \in \{1, ..., n\}$ is called m-dimensional conceptual space of O with respect to E, if for each data:*

1. *$d_{i_k} \in C_{i_s}^I, C_{i_s} \in Cd_i$, or*
2. *$\exists d \in C_{i_s}^I$ such that d_{i_k} is an instance selected with respect to a selector instance of class C_{i_s}, or*
3. *d_{i_k} is a generalized data of a dataset selected with respect to a selector instance of class $C_{i_s}, C_{i_s} \in Cd_i$.*

Each t_i represents a generalized instance of a set of instances of A selected with respect to a selector instance of class C which contains all paths in CC.

Example 4. The multidimensional schema shown in Table 1 has been constructed by using the following multidimensional conceptual space associated to $CeramicArtifact$ concept, $E = (D_1, D_2, D_3, c_{\phi_{union}}, c_{\phi_{union}}, c_{\phi_{max}})$, where:

$$D_1 = (\{Decoration\}, \emptyset), \qquad c_{\phi_{union}} = \{\langle \Phi, \hat{\cup} \rangle \,|\, \Phi \in \Phi^D\},$$
$$D_2 = (\{Morphology\}, \emptyset), \qquad c_{\phi_{max}} = \{\langle \Re, Max \rangle\}$$
$$D_3 = (\{Weight\}, \emptyset),$$

Table 1. Analysis and generalization of ceramic artifacts. Each row maintains the number of instances represented by the associated description (values between parenthesis).

Decoration	Morphology	Weight [g]
incisions	group:{open} order:{ovoid, spheroid}	20 (3)
polished	group:{open, closed} order:{cone, ovoid}	500 (3)
stamping	group:{open} order:{cone}	5 (2)
without dec.	group:{closed} order:{spheroid}	12 (2)

$\hat{\cup}$ is defined by $\hat{\cup}(d_1, ..., d_n) = \{d_1\} \cup ... \cup \{d_n\}$ and Max represent the maximum function for real numbers (in this case to compute the maximum weight).

Algorithm for the Generation of Multidimensional Conceptual Spaces. Algorithm 1 describes how an m-dimensional conceptual space is obtained from a given ontology $\mathcal{O}$ and a conceptual schema of $\mathcal{O}$ associated to the class C, E. The technique of

---

Algorithm 1. Generation of multidimensional conceptual spaces

---

Require: $\mathcal{O}, E, C, \beta_m, \beta_M$
{ $\mathcal{O}$, instantiated ontology with Abox $\mathcal{A}$,
C, reference class to generate multidimensional spaces,
E, multidimensional schema,
β_m, β_M, minimum and maximum percentage of different values in each dimension}
Ensure: E'
{E', multidimensional space associated to the schema E and the ontology $\mathcal{O}$ }

First part: Eliminate the irrelevant dimensions and create the generalization mappings between the data.
$\mathcal{A}_C = \{a | a \in C^I\}$.
Iterate $\mathcal{A}_C$ and group different data associated to each dimension of E.
if $\beta_m \geq \frac{|D_i|}{|\mathcal{A}_C|} \geq \beta_M$ (D_i must be removed) **then**
$\quad E = (D_1, ..., D_{i-1}, D_{i+1}, ..., D_n, c_{\phi_1}, ..., c_{\phi_{i-1}}, c_{\phi_{i+1}}, ..., c_{\phi_n})$
else
$\quad$ Generate mappings (d, d') for each value d collected in D_i, $d(d \downarrow_{C_i^{sup}} d')$ with C_i^{sup} being one of the classes direct ancestors of C_i, $d \in C_i^I$.
Second part: Creation of the multidimensional space
$E' = \{\}$
for all $a \in \mathcal{A}_C$ **do**
$\quad$ Let $t = (d_1, ..., d_n)$ be the tuple of data associated to a, according to E.
$\quad t' = (d_1', ..., d_n')$, computed from the mapping generated in the previous step, being C_i' the type of data d_i.
$\quad$ **if** $\exists t'' = (d_1'', ..., d_n'') \in E'$, being C_i'' the type of data d_i'' such that $\forall i \in \{1, .., n\}$ **then**
$\quad\quad t'' = (d_1'' \hat{\cup}_t d_1', ..., d_n'' \hat{\cup}_t d_n')$
$\quad$ **else**
$\quad\quad E' = E' \cup \{t'\}$

Algorithm 2. Generation of interesting schemata of multidimensional conceptual spaces

---

Require: $\mathcal{O}, C, \gamma, \alpha$
  $\{\mathcal{O}$, ontology with Abox A
  C, reference class for generating a schema of multidimensional spaces,
  γ, minimum allowed information gain
  α, minimum number of objects in a description$\}$
Ensure: SP
  $\{SP$, set of paths associated to C whose subsets can be used to form interesting multidimensional schemata$\}$

Let $Paths_C$ be the dictionary of paths starting from C with key in the destination concept.
$S = \{S_i | S_i = \{a \in C_i^I\}, \forall C_i, i \in \{1, ..., n\}, C \sqsubseteq C_i, C_i \neq C_j, j \in \{1, ..., n\}, i \neq j\}$
$SP = ComputePseudoSchemata(Paths_C, S, \gamma)$

function $ComputePseudoSchemata(Paths_C, S, \gamma)$:
  $\{$Outputs a pair, in which the first element is a set of paths and the second a real value indicating the importance of the set of paths for the generation of interesting schemata$\}$

First phase: Compute the importance of the current clustering
$ve = \sum_{S_i \in S} imp(S_i)$, where

$$imp(S_i) = \begin{cases} 1, & \text{if } |S_i| > \alpha \\ 1 - |S_i|/\alpha, & \text{otherwise} \end{cases}$$

Second phase: Retrieve the subsets with highest information gains
if $|S| = 1$ **then**
  Output $(\emptyset, ve)$
else
  $SP = \emptyset$
  Compute information gain, G, for each destination concept $Paths_C$ according to the classification in S
  Let $\{path_1, ..., path_m\}$ be the set of paths which allow a high discrimination between objects ordered according to the gain value: $G(path_1) \geq ... \geq G(path_m) > \gamma$ and C_k the destination concept associated to path $path_k$, $k \in \{1, ..., m\}$.
  for all $k \in \{1, ..., m\}$ **do**
    $CD = \{C' | C' \sqsubseteq C_k\}$
    for all $C' \in CD$ **do**
      $S = \{S_{Cl_{C'}} = \{a \in C_i^I | C_i \in \{1, ..., n\}, C \sqsubseteq C_i;$ a is related to some data d
      according to $path_i \wedge d \in C'^I\}\}$
      $SP = SP \cup_{(sp,v) \in ComputePseudoSchemata(Paths_C \backslash \{path_i\}, S, \gamma)} \{\langle C_i \cup sp, v + ve \rangle\}$
  Output SP

---

attribute-oriented induction [10,11] was taken as inspiration for its simplicity and flexibility. One remarkable common feature between that technique and this one is that no restrictions are put on the data. The first step of the algorithm is to define a mapping between each data of each dimension and its generalized value, according to the conceptual schema for instances of type C. Then, the generalized instances are formed, substituting each m-tuple with a generalized m-tuple that constitutes the generalization of the instances of the same type for each dimension.

6   Interesting Conceptual Spaces

A conceptual space as previously defined allows users to freely browse the conglomerate of objects and review the aspects they consider more interesting. If the data analyst were not informed about the features of the object distribution in the domain, or if the number of such features were too high, hcr analysis capabilities would be strongly affected. Nevertheless, this problem can be overcome with an analysis tool able to suggest to the user some interesting analysis dimensions. This can be obtained with a customized *feature selection* process. The concept of feature selection was introduced for the task of dimensionality reduction originally defined in Statistics and widely studied in Machine Learning.

Anyway, it is necessary to define a way of assessing the importance of a given feature subset. The measures widely used in the literature are the information gain, the Gini index, the uncertainty and the correlation coefficients. Nevertheless, the large number of studies that argue in favor of decision trees and information gain (like ID3 and C4.5), made us decide to choose such a combination for our feature selection process. More exactly, in this work we propose to compute interesting conceptual multidimensional schemata associated to a concept C by way of Algorithm 2, an adaptation of the one proposed by [12]. The purpose of such a customized algorithm is that of using the distributions of a set of objects in relation to a set of concepts, in order to select the compositions of relations (paths) that ensure the highest information gain with respect to the distribution. The main block of the algorithm is procedure $ComputePseudoSchemata$ which selects, as a first step, the paths with highest information gain. Then, for each path, the initial distribution is subdivided according to the possible values of the data associated to the objects through such a path, and the process of subdivision of the clusters is repeated while the information gain is maintained in a desirable range. A parallel task performed during this process is the computation of the weights which indicate the interest estimation for each conceptual schema that may be generated for each path set.

A further filtering step can be done for each dimension taking into account the relation between the quantity of objects associated to a concept and to its ancestor concept. In this way, an uninteresting ancestor concept can be removed from the dimension, following the rule described in Figure 4.

Let Cp be an ancestor concept in a dimension, $\{C_1, ...C_n\}$ concepts directly specialized of Cp, and $objs$ a function which associates each concept with its objects set. Let β be the percentage of maximum correlation between the number of objects of descendant and ancestor concepts.
If $\exists C_i$ such that $|objs(C_i)| \geq \beta|objs(C_p)|$
    Promote concepts $C_1, ..., C_n$ to level of concept C_p
    Delete C_p from the dimension.

Fig. 4. Rules for filtering out uninteresting concepts

7 Using a Multidimensional Conceptual Space

As explained in the introduction, the major advantage of a multidimensional space is that a user can see her data from different points of view. Tabular models in 3D, function graphs, histograms and relational graphs are the most friendly tools to use for the analysis of results. The possibility of realizing generalizations and selections at each level also represents a powerful analysis skill. In this way, it is possible to characterize object classes in relation to others, allowing for the comparison and discovering of concept features.

Although these are the analysis methods that have traditionally been used, an analyst may be interested in other more complex insights about the behavior of her data. Various pattern analysis tools have been described in the literature, especially with the development of data mining research. It is thus plausible to create new algorithms for the extraction of interesting patterns in the multidimensional conceptual environment. Some of the most interesting patterns to extract are:

- *Characterization Patterns*: they represent rules for characterizing a class of objects according to the values of a subset of its dimensions. They can be expressed by: $class\ X \Rightarrow Condition[p_c]$, where $p_c = \frac{100 \times count(Condition)}{count(classX)}$, which means that, in class X, $Condition$ occurs in a p_c percentage of the cases, $Count$ is a function that counts the number of times a certain condition occurs, and n is the total numbers of analyzed objects. p_c is known as characterization coefficient.
- *Discrimination Patterns*: they represent rules for characterizing a class of objects for which a given pattern is not observable with a certain frequency in any other class. They can be expressed by: $class\ X \Leftarrow Condition[p_d]$, where $p_d = \frac{100 \times count(Condition \land classX)}{count(Condition)}$. p_d is known as discrimination coefficient.
- *Association Rules at Different Levels*: they represent rules for characterizing the multidimensional space by using co-occurrence of data found in the features of the multidimensional space. They can be expressed by: $Condition_1 \Rightarrow Condition_2[s, c]$, where $s = \frac{100 \times count(Condition_1 \land Condition_2)}{n}, c = \frac{100 \times count(Condition_1)}{count(Condition_1 \land Condition_2)}$, which means that in a $s\%$ of the objects both $Condition_1$ and $Condition_2$ are observed and that $c\%$ of objects satisfying $Condition_1$ also satisfy $Condition_2$. s is known as the support of the rule and c as its confidence.

In order to customize these results to the model we presented, the multidimensional conceptual model must take into account how many objects in the concrete ontology are characterized by the description of each cell.

8 Conclusions

The proposal of this paper is the formalization of a data warehouse tool for the Semantic Web. The tool is based on the theoretical foundations of Description Logic and on the current developments of information data generalization.

Besides, an algorithm to generate interesting conceptual spaces according to the data distribution is proposed. Ideas for adapting Data Mining techniques in order to allow users a better knowledge discovering from the Semantic Web have also been exposed. Implementation of the proposal framework, on which we are now working, consists of two main components: 1) a ontology reasoner (which works in an off-line way) that retrieve instance models and abstraction/specialization functions between concepts; and 2) a data warehouse processor that use such models and functions in order to perform all the necessary generalizations. This second module is being optimized considering some of the OLAP solutions.

Acknowledgements. This research has been partially supported by the project TIN2005-09098-C05-04 (2006-2008).

References

1. Hacid, M.S., Sattler, U.: Modeling multidimensional databases: A formal object-centered approach. In: Proceedings of the Sixth European Conference on Information Systems (1998)
2. Baader, F., Sattler, U.: Description logics with aggregates and concrete domains. Inf. Syst. 28, 979–1004 (2003)
3. Buzydlowski, J.W., Song, I.Y., Hassell, L.: A framework for object-oriented on-line analytic processing. In: DOLAP 1998: Proceedings of the 1st ACM international workshop on Data warehousing and OLAP, pp. 10–15 (1998)
4. Trujillo, J., Palomar, M., Gómez, J., Song, I.Y.: Designing data warehouses with OO conceptual models. Computer 34, 66–75 (2001)
5. Nguyen, T.B., Tjoa, A.M., Wagner, R.: An object oriented multidimensional data model for OLAP. In: Web-Age Information Management, pp. 69–82 (2000)
6. Binh, N.T., Tjoa, A.M.: Conceptual multidimensional data model based on object-oriented metacube. In: Vaudenay, S., Youssef, A.M. (eds.) SAC 2001. LNCS, vol. 2259, pp. 295–300. Springer, Heidelberg (2001)
7. Abelló, A.: YAM^2: A Multidimensional Conceptual Model. PhD thesis, Universitat Politécnica de Catalunya (2002)
8. Danger, R.: Extracción y análisis de información desde la perspectiva de la Web Semántica (Information extraction and analysis from the viewpoint of Semantic Web, in spanish). PhD thesis, Universitat Jaime I (2007)
9. Baader, F., Kusters, R., Wolter, F.: Extensions to description logics. In: The description logic handbook: theory, implementation, and applications, pp. 219–261. Cambridge University Press, Cambridge (2003)
10. Carter, C.L., Hamilton, H.J.: Efficient attribute-oriented generalization for knowledge discovery from large databases. IEEE Transactions on Knowledge and Data Engineering 10, 193–208 (1998)
11. Han, J., Nishio, S., Kawano, H., Wang, W.: Generalization-based data mining in object-oriented databases using an object cube model. Data Knowledge Engineering 25, 55–97 (1998)
12. Han, J., Kamber, M.: Data Mining: Concepts and Techniques. Morgan Kaufmann, San Francisco (2001)

Aspects Based Modeling of Web Applications to Support Co-evolution

Buddhima De Silva and Athula Ginige

School of Computing and Mathematics, University of Western Sydney
Locked Bag 1797, Penrith South DC, NSW 1797, Australia
bdesilva@scm.uws.edu.au, a.ginige@uws.edu.au

Abstract. When an information system is introduced to an organisation it changes the original business environment thus changing the original requirements. This can lead to changes to processes that are supported by the information system. Also when users get familiar with the system they ask for more functionality. This gives rise to a cycle of changes known as co-evolution. One way to facilitate co-evolution is to empower end-users to make changes to the web application to accommodate the required changes while using that web application. This can be achieved through meta-design paradigm. We model web applications using high level abstract concepts such as user, hypertext, process, data and presentation. We use set of smart tools to generate the application based on this high-level specification. We developed a hierarchical meta-model where an instance represent a web application. High level aspects are used to populate the attribute values of a meta-model instance. End-user can create or change a web application by specifying or changing the high level concepts. This paper discusses these high level aspects of web information systems. We also conducted a study to find out how end-users conceptualise a web application using these aspects. We found that end-users think naturally in terms of some of the aspects but not all. Therefore, in meta-model approach we provided default values for the model attributes which users can overwrite. This approach based on meta-design paradigm will help to realise the end-user development to support co-evolution.

Keywords: Conceptual Model, Web Information Systems, Aspects, co-Evolution.

1 Introduction

In traditional approach to develop web information systems, when an organization finds the need for new information system, the requirements are analysed. Then if they go for a custom made information system the design specification is produced and the system gets developed tested and deployed. Otherwise organization may buy a product to match their requirements and adapt their processes accordingly. The system then needs to be maintained until decommissioned. However, we have found that many such systems that were deployed with the university and client organizations after a period of time no longer meet the user requirements. The failure is due to 3 reasons:

J. Filipe et al. (Eds.): ICSOFT/ENASE 2007, CCIS 22, pp. 283–293, 2008.
© Springer-Verlag Berlin Heidelberg 2008

1) When an information system is introduced to an organisation it changes the original business environment thus changing the original requirements.
2) The processes that information system is supporting may change.
3) When users get familiar with the system they ask for more functionality.

This gives rise to a cycle of changes known as co-evolution [1]. Meta-design is proposed as a solution to the co-evolution [1, 2]. Meta-design paradigm characterises objectives, techniques, and processes for creating new media and environments allowing 'owners of problems' (that is, end-users) to act as designers [3, 4]. A fundamental objective of meta-design paradigm is to create socio-technical environments that empower users to engage actively in the continuous development of systems rather than being restricted to the use of existing systems. If we are to empower end users to actively participate in development tasks then, they should be provided with a suitable web application development environment. Rode's study on mental model of end-user developers reveals that end-users do have no or very little concerns about the critical issues in web applications such as authentication, session management, etc.[5]. Therefore we identify the need to bridge the gap between end-user mental model and developer's mental model for the success of meta-design paradigm. This gives rise to following 2 requirements:

1) End-users need to model a web application using high level aspects adequate to specify a web application.
2) Then we have to provide a set of tools and a framework that they can use to develop and change the application by specifying and changing these aspects.

Thus rather than developing "the application" for end-users to use we need to develop a meta-model to represent various aspects of applications and a set of tools which the end-users can use to create the applications that they want by populating the instance values in the meta-model. Then end-users can populate or change the values relating to various aspects according to their requirements to instantiate or change the web application. We have created a meta-model for web based information systems to support the meta-design paradigm [6].

We analysed a set of requirement specifications end-users written for different information systems within the University to identify how they specify web applications. In that study we found that end-users tend to specify aspects of information systems at a conceptual level. Therefore, we reviewed the literature on conceptual modelling of web applications to find different aspects required to model web applications at a conceptual model. We found that many conceptual models are proposed to model a specific type of web applications such as data-intensive web applications, process intensive web applications. From these, we derived a set of aspects required to define web conceptual model for any type of web application. Then we refined the set of aspects by modelling different types of web applications. These systems varied from simple web sites to e-commerce web sites. We have excluded the special types of web applications such as e-mail or chat applications which are optimised for specific functionality. We found that there are 2 broader categories of web applications; information centric, and process centric.

- Information centric: simple web sites with unstructured information. The focus is on effective presentation of information.

- Process Centric: web sites that support business processes by enabling users to perform actions such as filling a form, approving a form etc. These can further be divided into 2 types;
  - Data intensive: The focus is on efficient presentation on structured data such as product catalogue.
  - Workflow intensive: The focus is on efficient automation of business process consisting of sequence of steps such as order processing system.

In this paper, we discuss the 'holistic' web conceptual model to define different types of web information systems. This will answer the research question; how can we specify Web Information Systems? We use WebML, Web modelling language with extensions to model these aspects. Then for end-users we provide techniques such as visual tool, program by example to model these aspects. In section 2, we review the existing web conceptual models to identify the required aspects to model web applications. Section 3 presents the different aspects of web information systems. In section 4, we analyse end-user specifications to find out the concepts they use to define these aspects. Section 5 concludes the paper.

2 Related Work

Most of the web modeling languages developed at early days such as HDM-Lite [7], UWE [8], WebRE [9], W2000 [10], OOHDM [10] and initially WebML [11] emphasized aspects required to model data intensive web applications such as data and hypertext. All these approaches provided data models to define the entities used and relationships between them, navigation & composition models to organize and to access pages and presentation model to define look and feel. WebML consists of models such as structural model (i.e. data model), hypertext model (composition navigation models), presentation models, personalised models and operational models. The operational model is providing a way to model functions in a data intensive web application. Modelling in WebML is based on visual notations which are simple and complete.

Later some of these languages such as WebML [11], UWE [8] and OOHDM [10] were extended to model process intensive web applications. WebML was integrated with Business Process Modellign Language (BPML) to model workflows in web applications. In OOHDM [12], the conceptual model is derived based on Object Oriented Modelling (OMT) principles. It is defined using classes and relations. The navigation model is also modelled as two classes, the navigation class schema and navigation context schema. State charts are used to define the browsing semantics. The interfaces are defined using configuration diagrams.

HDM-Lite [7] uses structure schema, navigation schema and presentation schema to model web applications.

Two types of use cases exist in UWE [8] approach called navigation and functional. Use cases are explained using behavioural diagrams or textual format.

Recently, Koch et.al have modelled the conceptual model of web application based on the web application behaviour and structure [9, 13]. Their Requirement metamodel is focused on navigation and presentation of information. Therefore, business process is modelled as a kind of navigation. This type of generalization leads to complex models of web applications.

Jakob, et al. [14] and Schimid et.al [15] have identified the problem with most of these web application models as neglecting the business processes at the early stages of modeling. Jakob, et al. [14] use "Operational model" to model the business logic aspect of data intensive applications. However, their operational model is in the logical level, not in the conceptual level. They are more focused on application generation not on requirement elicitation. In Kobti et al.'s Conceptual framework [16], they use WebML to model the hypertext aspects of web applications and Web Service Choreography Description Language (WS-CDL) to model the processes. In this approach they model the business process independent of hypertext modeling language. But the transaction nature of WS-CDL won't be sufficient to define complex business processes. Also it doesn't provide a mechanism to model the access control.

Oliveira et.al [17] have modeled the process-centric web applications using state models. They have modelled Access control as conditions associated with states and navigation as state transition. They have associated views with states. Even though this is an attempt to model the web application at logical level we see the state chart can be used to model the state dependant behaviour of use cases.

However, most of these modelling approaches have embedded the aspects such as process model and user model within other aspects rather than defining these separately during the analysis phase of web applications. Also we have identify the possibility of abstracting composition model at a higher level than the granular elements such as label, text, list box for process intensive web applications.

3 Aspects of Web Conceptual Model

We have analysed modelling of different web applications from information-centric and process-centric categories. Based on our finding we have revised WebML conceptual modelling aspects to enhance the naturalness and completeness. We believe these modifications will help end-user developers to efficiently develop or modify their web applications. We have identified 5 high level aspects required to completely define any type of web application called Hypertext model, User model, Process model, Data model and Presentation model.

- Hypertext model: Hypertext model consists of two sub models: composition and navigation. Navigation model defines how to get in to a specific page and how to go through a group of pages in the web application. Thus navigation defines the sequence of views presented to the user when user is interacting with the application. For example possible sequence of views present in the e-commerce application is customer details view, order detail view, payment detail view. Composition model defines the collection of elements in the view which help users to interact with web application. For example the composition model of a "log in" page consists of log in instruction provided at the top of the page, user name and password elements submit and cancel buttons.

- User Model: The underline structure in user model is that users belong to groups. User model consists of two sub models called access control and personalisation models. Access control model defines what functions user group can access in an information system. For example prospective buyer can access the shopping cart function on an e-commerce application. But only users who are waiting for delivery of an order can access their order data. Personalisation model defines the attributes for the user that can be used to personalise the web application at group level and user level. When a user logged on in the same e-commerce application the list of products a user see on a personalized home page can be based on user profile.

- Data Model: Data model defines relevant objects and relationships among objects used in the web application. Data defines the domain objects stored and retrieved through the web application. For example in the above e-commerce application we have to store product information such as name, type, price, availability, available units, etc. We may also want to store order information such as customer name, product quantity, payment amount, etc. We may want to keep relationships between order and product to maintain stock management.

- Process Model: Process model defines high level abstraction of tasks and workflows to be performed in a web application. When modelling web applications sometimes it is necessary to model some routing tasks that should happen when a user interact with the application. Typically this is expressed in a use case such as use case description of "request for more information" in an e-commerce application. These functionalities can be executed independent of the status of any other processes. Details of these functionalities are captured in a Task Model. On the other hand there may be workflows in web applications to sequence series of tasks to carry out a business process. Some examples are order processing and stock management. In a workflow, when a task needs to be performed may depend on the status of some other tasks. The order in which tasks need to be performed is captured in the Workflow model.

- Presentation: Presentation model captures the layout and graphic appearance of all generic elements appearing on the views of the web application. This is modelled at two levels in a web information system: site level, and page level. At site level we define the look and feel of the web application. For example in the e-commerce site we may define the size and style of the fonts used in labels and style, size and color for the instructions etc. At page level we define the same attributes only applicable to a specific page.

We find that some of these aspects can be specified independently. For example data aspect can be specified using objects and relationships among the objects. However, other aspects such as composition model and process model depend on data model to find out the objects or object attributes which associate with user interfaces or tasks. The process model also depends on composition model for actions which trigger states and data model for data updates. Presentation aspect depends on personalisation sub model of the user model. Navigation model depends on access control sub model of the user model to find out access rights of the user when authorisation is required. These dependencies are shown in Fig. 1.

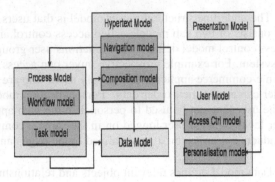

Fig. 1. Dependencies between aspects

3.1 Hypertext Model

Hypertext model consist of composition and navigation models. Navigation can be defined explicitly using menus or implicitly with in the composition model in a web information system. We have to define navigation for two different types of tasks: state dependant, and state independent. In an e-commerce application the tasks such as "view product catalogue", "order" are available for all users. Also views such as "home page" or "contact us" pages are available for all users. Then there can be tasks such as "view order summary" which is available for authorised users. These tasks which are always available for users are categorised as state independent tasks. These tasks can be executed independent of the state of any processes. Once a user log in, user will be provided with a menu to access state independent tasks. On the other hand, the "view order status is an example for state dependant functions. These tasks are available to users only if an order waiting for processing for that particular user at that time. Navigation model can identify views to sequence from the access control model.

Composition model defines the user interfaces of a web application. There are three different types of Composition models called unstructured, form, and table. Unstructured page can contain content such as images, links and, text. Form or table model consists of UI guide, UIElementGroup, and or UIElements, and UI Actions. UI elements can be WebML data units, which store or present data. UI elements can be in input mode or output mode. In a form we have UI elements in Input mode, i.e. called entry units in WebML. In a table we can have UI elements in output mode. It can be WebML index, single data, multi data, scroll data or search data unit. UIGuide provide the guidelines to use the particular interface. For example, in a form interface, the guidelines can help the users to understand the purpose of the form. In a table interface, the guidelines can help the users to interpret the report properly. UI is associated with a primary business object. For a UI element in input mode, we can also have associated help tip. Help tips help users to enter the values of the form UI element correctly. We extend WebML entry unit by adding help information for each field of entry unit. Sometimes, it is required to logically group UIElements. For example a product order may include more than one product. The data of the each product order can consist of quantity of product, price of product. Thus, product details can be in a UIElement group of the order UI. UI Actions such as add product, amend product actions can be associated with that UIElement group. Order UI Model can have UI Action to process the order.

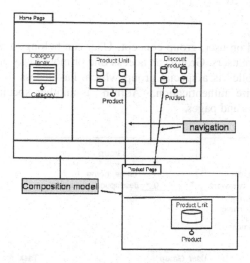

Fig. 2. Example Hypertext model from Amazon

Fig. 2 shows the partial hypertext model for amazon site. The composition model consists of table of products. We model table with WebML multi data unit as shown in Fig. 2.

3.2 Presentation Model

Presentation model specifies the look and feel of a web application. It dictates the layout and graphic appearance of all generic elements appeared on a page. We specify the presentation model at two levels: site level and page level. The template and style settings of the presentation model at site level apply to all the web pages of the application. If we want to model a custom page we have to define a template and style settings at page level.

If we consider personalisation then presentation model has to get the values of the presentation model for the user from the user model. Fig. 3 shows the site level template for amazon. Then for some pages it may have page level templates.

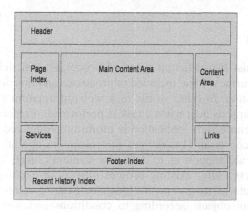

Fig. 3. Site level Template for Amazon

3.3 User Model

User model is based on user, group concept. User can belong to many groups. Group can have one or more users. Group can access one or more tasks or pages using access control model. "Public" is a special group which, has access to all tasks and pages which doesn't require authentication. "Admin" is another special group which has access to all the tasks and pages.

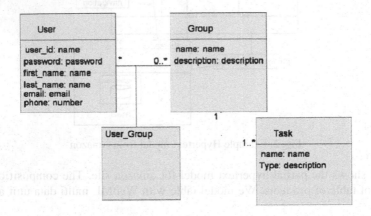

Fig. 4. User model

We can store values for presentation model at user level or group level. These personalised values are used by presentation model to generate views for given user. We have extended WebML conceptual model by adding a user model. We identify the need to define access control explicitly at conceptual level by end-user developers since this is required to generate the applications. We also identify possibility for expressing personalisation model with in the user model. Since user has attributes and relations similar to object we use class diagram to model users. User model is a special data model because it is common in any web information system. Fig. 4. shows the class diagram for user model to define the access control and personalisation sub model.

3.4 Process Model

Process model consist two different types of processes: task related and workflow related. We define business logic included in processes and this determines the behaviour of web application. In other words in a web information system, the business rules govern what happens next when a task is performed. Presence of process aspect in an Information-centric web application is minimum compared to a process-centric web application. Jacob et al. [14] defines the minimum operations a data intensive web application should support as to Add new content objects or new relations between content objects, alter existing content objects or existing relations between content objects, delete existing content objects or existing relations between content objects, filter content objects according to conditions and sort content objects by specified criteria. There can also be tasks such as send notification associate with data-intensive web applications.

In the e-commerce example when a customer submits an order, order processing officer should be able to see the order information. Conditional rules can also be associated to the flow. We define the flow of information in the workflow model. Workflow model defines the state, entry condition, exit condition, transition, transition activity, etc. Transition also associated with a task and UI action. A simplified process model for order processing system is shown in Fig. 5. User can place an order. Then order processing officer process order.

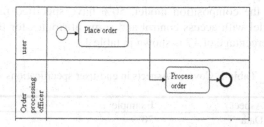

Fig. 5. Simplified workflow model for order processing

3.5 Data Model

Data Model defines the relevant objects and relationships among objects used in the web information system. Data model is the key concept in data-intensive web applications. Objects have attributes. We define high level attribute types such as photo, e-mail ,etc as defined in Smart Business Object Modelling Language (SBOML) [18]. This high level abstract attribute types will help to improve the naturalness in modelling data aspect. The partial data model for amazon is shown in Fig. 6. There are 3 different kinds of products such as book, software, electronic. Product can has discount during a given time frame. Product can have 0 or more reviews.

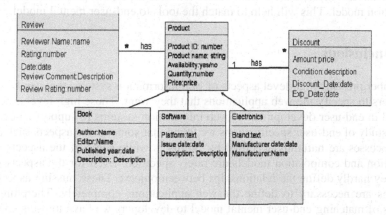

Fig. 6. Partial Data model

4 Study on End-User Specifications

We analysed 32 proposals submitted by users in administration positions of an academic organisation to develop or modify existing information systems. From these 32 proposals 17 were for new Information Systems or enhancements to existing Information Systems. Rest were for new hardware or hardware upgrades. We further analysed the software proposals to identify the concepts they have used. 12 out of 17 users have specified objects and some attributes for the objects. In most cases users had a manual system. Therefore they have an understanding of data that needs to be managed. No one has specified navigation personalisation or presentation models. Only one has specified the composition model. 50% have specified processes. 80% has specified user model with access control and some attributes for user groups. Usage of concepts as a percentage of 17 is shown in table 1.

Table 1. Covered Aspects in end-user specifications

Aspect	Example
Data	70%
Hypertext	4%
Presentation	0%
Process	50%
User	80%

This is only a sample of end-users. However, this sample of end-users also confirms Rode's conclusion on end-user mental model. They want systems to store data and then manipulate data. They are not concern about other aspects such as presentation and hypertext. All of that will come as usability issues when they start to use the system. However, when they become the owners of the system they need to create or modify the system. Then they also need to consider the other aspects such as hypertext model. When designing the tools for end-users to develop web information systems we have to incorporate default values for these aspects in presentation model and navigation model. This will help to match the tools to end-user mental model.

5 Conclusions

This paper presents high-level aspects of web information systems that can be used by end-users to specify the web applications that they want. These high level aspects can be used in end-user development of web information systems to support co-evolution. In our study of end-user specifications we found that some of the aspects such as data, and processes are naturally specified by end-users while some of the aspects such as navigation and composition models are rarely specified. Even for the aspects such as data they hardly define the relationships between objects. These missing aspect specifications are necessary to define the web applications completely. Therefore in the process of matching end-user mental model to developers we have to use some strategies and techniques to capture minimum aspects necessary to define a web application.

References

1. Costabile, M.F., Fogli, D., Marcante, A.: Supporting Interaction and Co-evolution of Users and Systems. In: Advanced Visual Interfaces -AVI (2006)
2. Costabile, M.F., et al.: A meta-design approach to End-User Development. In: VL/HCC (2005)
3. Fischer, G., Giaccardi, E.: A framework for the future of end user development. In: Lieberman, H., Paterno, F., Wulf, V. (eds.) End User Development: Empowering People to flexibly Employ Advanced Information and Communication Technology. Kluwer Academic Publishers, Dordrecht (2004)
4. Fischer, G., et al.: Meta Design: A Manifesto for End -User Development. Communications of the ACM 47(9), 33–37 (2004)
5. Rode, J., Rosson, M.B.: Programming at Runtime: Requirements and Paradigms for Non-programmer Web Application Development. In: IEEE Symposium on Human Centric Computing Languages and Environments-2003, Auckland, New Zealand (2003)
6. Ginige, A., De Silva, B.: CBEADS©: A framework to support Meta-Design Paradigm. In: 12th International Conference on Human-Computer Interaction (HCII 2007), Bejing, P.R. China. Springer, Heidelberg (2007)
7. Fratenali, P., Paolini, P.: A conceptual model and a tool environment for developing more scalable and dynamic Web applications. In: Schek, H.-J., Saltor, F., Ramos, I., Alonso, G. (eds.) EDBT 1998. LNCS, vol. 1377. Springer, Heidelberg (1998)
8. Koch, N., Kraus, A.: Power of UML-based Web Engineering. In: Second International Workshop on Web-oriented Software Technology (IWWOST 2002), Málaga, Spain (2002)
9. Escalona, M.J., Koch, N.: Metamodeling Requirements of Web Systems. In: Filipe, J., Cordeiro, J., Pedrosa, V. (eds.) WEBIST 2006. LNBIP 1, pp. 267–280. Springer, Heidelberg (2006)
10. Baresi, L., Garzotto, F., Paolini, P.: Extending UML for Modelling Web Applications. In: Annual Hawaii Int.Conf. on System Sciences, Miami, USA (2001)
11. Ceri, S., Fratenali, P., Bongio, A.: Web Modelling Language (WebML): a modelling language for designing Web sites. In: WWW9 Conference, Amsterdam (2000)
12. Schwabe, D., Rossi, G., Barbosa, S.D.J.: Systematic hypermedia application design with OOHDM. In: Seventh ACM conference on Hypertext, Bethesda, Maryland, United States. ACM Press, New York (1996)
13. Koch, N., Zhang, G., Escalona, M.J.: Model Transformation from Requirements to Web System Design. In: International Conference on Web Engineering (ICWE 2006), Palo Alto, California, USA. ACM Press, New York (2006)
14. Jakob, M., et al.: Modeling and Generating Application Logic for Data-Intensive Web Applications. In: Internationa Conference on Web Engineering (ICWE 2006). ACM Press, New York (2006)
15. Schmid, H.A., Rossi, G.: Modeling and designing processes in e-commenrce applications. Internet Computing IEEE 8(1), 19–27 (2004)
16. Kobti, Z., Sundaravadanam, M.: An enhanced conceptual framework to better handle business rules in process oriented applications. In: International Conference on Web Engineering (ICWE 2006), Palo Alto, California, USA. ACM Press, New York (2006)
17. De Oliveira, M.C.F., Turine, M.A.S., Masiero, P.C.: A Statechart-Based Model for Hypermedia Applications. ACM Transactions on Information Systems (TOIS) 19(1), 28–52 (2001)
18. Liang, X., Ginige, A.: Smart Business Objects: A new Approach to Model Business Objects for Web Applications. In: 1st International Conference on Software and Data Technologies, Setubal, Portugal (2006)

Part V
Knowledge Engineering

Recommending Trustworthy Knowledge in KMS by Using Agents

Juan Pablo Soto, Aurora Vizcaíno, Javier Portillo-Rodríguez, and Mario Piattini

University of Castilla–La Mancha
Alarcos Research Group – Institute of Information Technologies & Systems
Department of Information Technologies & Systems – Escuela Superior de Informática
Ciudad Real, Spain
JuanPablo.Soto@inf-cr.uclm.es,
{aurora.vizcaino,javier.portillo,mario.piattini}@uclm.es

Abstract. Knowledge Management is a critical factor for companies worried about increasing their competitive advantage. Because of this, companies are acquiring knowledge management tools that help them manage and reuse their knowledge. One of the mechanisms most commonly used with this goal is that of Knowledge Management Systems. However, sometimes Knowledge Management Systems are not very used by the employees, who consider that the knowledge stored is not very valuable. In order to avoid it, in this paper we propose a three-level multi-agent architecture based on the concept of communities of practice, with the idea of providing the most trustworthy knowledge to each person according to the reputation of the knowledge source. Moreover, a prototype that demostrates the feasibility of our ideas is described.

Keywords: Knowledge Management systems, Multi-agent architecture, Software Agents.

1 Introduction

Knowledge Management (KM) is an emerging discipline that is considered a key part in the strategy of using expertise to create a sustainable competitive advantage in today's business environment. Having a healthy corporate culture is imperative for success in KM. Zand in [1] claims that bureaucratic cultures suffer from a lack of trust and a failure to reward and promote cooperation and collaboration. Without a trusting and properly motivated workforce, knowledge is rarely shared or applied, and organizational cooperation and alignment are nonexistent.

Certain systems have been designed to assist organizations to manage their knowledge. These are called Knowledge Management Systems (KMS), defined by Alavi and Leidner [2], as an IT-based system developed to support/enhance the processes of knowledge creation, storage/retrieval, transfer and application. An advantage of KMS is that staff may also be informed about the location of information. Sometimes the organization itself is not aware of the location of the pockets of knowledge or expertise [3]. Moreover, a KMS is able to provide process improvements: it is better at

J. Filipe et al. (Eds.): ICSOFT/ENASE 2007, CCIS 22, pp. 297–309, 2008.

serving the clients, and provides better measurement and accountability along with an automatic knowledge management.

However, developing KMS is not a simple task since knowledge per se is intensively domain dependant whereas KMS are often context specific applications. KMS have received certain criticism as they are often installed in the company, thus overloading employees with extra work, since employees have to introduce information into the KMS and then worry about updating this information. Moreover, the employees often do not have time to introduce or search for knowledge, or they do not want to give away their own knowledge, and or to reuse someone else's knowledge [4]. As is claimed in [5] "employees resist being labeled as experts" and "they do not want their expertise in a particular topic to stunt their intellectual growth". Because of this resistance towards sharing knowledge, companies are using incentives to encourage employees to contribute to the knowledge growth of their companies [6]. Some of these incentives are organizational rewards and allocate people to projects not only to work but also to learn and to share experiences. These strategies are sometimes useful. However, they are not a 'silver bullet', since an employee may introduce information that is not very useful with the only objective of trying to simulate that s/he is collaborating with the system in order to generate points and benefits to get incentives or rewards. Generally, when this happens, the stored information is not very valuable and it will probably never be used. Based on this idea we have studied how the people obtain and increase their knowledge in their daily work. One of the most important developments concerning the nature of tacit, collective knowledge in the contemporary workplace has been the deployment of the concept 'Communities of Practice (CoPs)', by which we mean groups of people with a common interest where each member contributes knowledge about a common domain [7]. CoPs is necessarily bound to a technology, a set of techniques or an organization, that is to a common referent from which all members evaluate the authority or skill and reputation of their peers and the organization. A key factor for CoPs is to provide an environment of confidence where their members can to share the information and best practices.

In order to provide to companies the conditions to develop trustworthy KMS, we propose a multi-agent system that simulates the member's behaviors of CoPs to detect trustworthy knowledge sources. Thus in Section 2, we explain why agents are a suitable technology with which to manage knowledge. Then, in Section 3 we describe our proposal. After that, in Section 4 we illustrate how the multi-agent architecture has been used to implement a prototype which detects and suggests trustworthy knowledge sources for members in CoPs. In Section 5 previous work are outlined. Finally, in Section 6 the evaluation and future work are presented.

2 Why Intelligent Agents?

Due to the fundamentally social nature of knowledge management applications, different techniques have been used to implement KMS. One of them, which is proving to be quite useful is the agent paradigm [8]. Different definitions of intelligent agents can be found in literature. For instance, in [9] agents are defined as computer programs that assist users with their tasks. One way of distinguishing agents from other types of software applications and to characterize them is to describe their main properties [10]:

- Autonomy: agents operate without the direct intervention of humans or other agents, and have some kind of control over their actions and internal states
- Social ability: agents interact with other agents (and possibly humans) via some kind of agent communication language.
- Reactivity: agents perceive their environment and respond in a timely fashion.
- Pro-activeness: in the sense that the agents can take the initiative and achieve their own goals.

In addition, intelligent agents' specific characteristics turn them into promising candidates in providing a KMS solution [11]. Moreover, software agent technology can monitor and coordinate events, meetings and disseminate information [12], automation of complex processes [13], building and maintaining organizational memories [14]. Another important issue is that agents can learn from their own experience. Most agents today employ some type of artificial intelligence technique to assist the users with their computer-related tasks, such as reading e-mails, maintaining a calendar, and filtering information. Agents can exhibit flexible behaviour, providing knowledge both *"reactively"*, on user request, and *"pro-actively"*, anticipating the user's knowledge needs. They can also serve as personal assistants, maintaining the user's profile and preferences. The advantages that agent technology has shown in the area of information management have encouraged us to consider agents as a suitable technique by which to develop an architecture with the goal of helping develop trustworthy KMS.

Therefore, we have chosen the agent paradigm because it constitutes a natural metaphor for systems with purposeful interacting agents, and this abstraction is close to the human way of thinking about their own activities [15]. This foundation has led to an increasing interest in social aspects such as motivation, leadership, culture and trust [16]. Our research is related to this last concept of "trust" since artificial agents can be made more robust, resilient and effective by providing them with trust-reasoning capabilities.

3   A Three-Level Multi-agent Architecture

Before defining our architecture it is necessary to explain the conceptual model of an agent, which in our case is based on two related concepts: trust and reputation. The former can be defined as confidence in the ability and intention of an information source to deliver correct information [17], and the latter as the amount of trust an agent has in an information source, created through interactions with information sources. There are other definitions of these concepts [18, 19]. However, we have presented the most appropriate for our research since the level of confidence in a source is, in our case, based upon previous experience.

The reputation of an information source not only serves as a means of belief revision in a situation of uncertainty, but also serves as a social law that obliges us to remain trustworthy to other people. Therefore, people, in real life in general and in companies in particular, prefer to exchange knowledge with "trustworthy people".

People with a consistently low reputation will eventually be isolated from the community since others will rarely accept their justifications or arguments and will limit their interaction with them. It is for this reason that the remainder of this paper deals mainly with reputation.

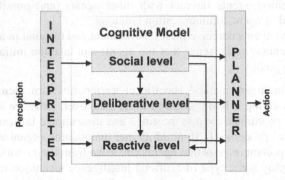

Fig. 1. General architecture

Taking these concepts into account we designed a multi-agent architecture which is composed of three levels (see Figure 1): reactive, deliberative and social. The reactive and deliberative levels are considered by other authors as typical levels that a multi-agent system must have [20]. The first level is frequently used in areas related to robotics where agents react to changes in the environment, without considering other processes generated in the same environment. In addition, the deliberative level uses a reasoning model in order to decide what action to perform.

On the other hand, the last level (social) is not frequently considered in an explicit way, despite the fact that these systems (multi-agent systems) are composed of several individuals, interactions between them and plans constructed by them. The social level is only considered in those systems that try to simulate social behavior. Since we wish to emulate human feelings such as trust, reputation and even intuition we have added a social level that considers the social aspects of a community. This takes into account the opinions and behavior of each of the members of that community. Other previous works have also added a social level. For instance in [21], the authors try to emulate human emotions such as fear, thirst, bravery, and also uses an architecture of three levels.

In the following paragraphs we will explain each of these levels in detail.

Reactive Level. This is the agent's capacity to perceive changes in its environment and to respond to these changes at the precise moment at which they happen. It is in this level when an agent will execute the request of another agent without any type of reasoning. That is to say, the agent must act quickly in the face of critical situations.

Deliberative Level. The agent may also have a behavior which is oriented towards objectives, that is, it takes the initiative in order to plan its performance with the purpose of attaining its goals. In this level the agent would use the information that it receives from the environment, and from its beliefs and intuitions, to decide which is the best plan of action to follow in order to fulfill its objectives.

Social Level. This level is very important as our agents are situated within communities and they exchange information with other agents. Thanks to this level they can cooperate with other agents. This level represents the actual situation of the community, and also considers the goals and interests of each community member in order to solve conflicts and problems which may arise between them. In addition, this level provides the support necessary to measure and stimulate the level of participation of the members of the community.

Two further important components of our architecture are the *Interpreter* and the *Planner* (see Figure 2). The former is used to perceive the changes that take place in the environment. The planner indicates how the actions should be executed.

In the following subsections we will describe each of the levels of which our architecture is composed in more detail.

3.1 Reactive Architecture

This architecture was designed to the reactive level of the agent. The architecture must respond at the precise moment in which an event has been perceived. For instance, when an agent is consulted about its position within the organization. This architecture is formed of the following modules:

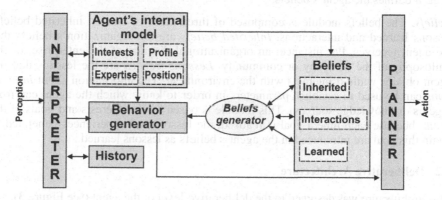

Fig. 2. Reactive architecture

Agent's Internal Model. As a software agent represents a person in a community this model stores the user's features. Therefore, this module stores the following parts:

- The *interests*. This part is included in the internal model in order to make the process of distributing knowledge as fast as possible. That is, the agents are able to search for knowledge automatically, checking whether there is stored knowledge that matches with its own interests. This behavior fosters knowledge-sharing and reduces the amount of work employees have to do because they receive knowledge without having to search for it.
- *Expertise.* This term can be briefly defined as the skill or knowledge of a person who knows a great deal about a specific thing. Since we are emulating communities of practice it is important to know the degree of expertise that each member of the community has in order to decide how trustworthy a piece of knowledge is, since people often trust in experts more than in novice employees.

- *Position.* Employees often consider information that comes from a boss as being more reliable than that which comes from another employee in the same (or lower) position as him/her [22]. Such different positions inevitably influence the way in which knowledge is acquired, diffused and eventually transformed in the local area. Because of this, these factors will be calculated in our research by taking into account a weight that can strengthen this factor to a greater or lesser degree.
- *Profile.* This part is included in the internal model to describe the profile of the person on whose behalf the agent is acting. Therefore, a person's preferences are stored here.

Behaviour Generator. This component is necessary for the development of this architecture since it has to select the agent's behaviour. This behaviour is defined on the basis of the agent's beliefs. Moreover, this component finds an immediate response to the perceptions received of the environment.

History. This component stores the interactions of the agents with the environment.

Belief Generation. This component is one of the most important in the cognitive model because it is in charge of creating and storing the agent's knowledge. Moreover, it defines the agent's beliefs.

Beliefs. The beliefs module is composed of three kinds of beliefs: inherited beliefs, lessons learned and interactions. *Inherited beliefs* are the organization's beliefs that the agent receives. For instance: an organizational diagram of the enterprise, or the philosophy of the company or community. *Lessons learned* are the lessons that the agent obtains while it interacts with the environment. The information about *interactions* can be used to establish parameters in order to know which the agent can trust (agents or knowledge sources). This module is based on the interests and goals of the agent, because each time a goal is realised, the lessons and experiences generated to attain this goal are introduced in the agent's beliefs as lessons learned.

3.2 Deliberative Architecture

This architecture was designed to the deliberative level of the agent (see Figure 3).

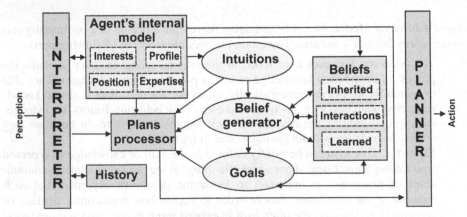

Fig. 3. Deliberative architecture

Its components are:

Agent's Internal Model. this module is the same as that which is described in the reactive architecture. It is composed of the interests, profile, position and expertise of the agent.

Plans Processor. This module is the most important of this architecture as it is in charge of evaluating the beliefs and goals to determine which plans have to be included in the Planner to be executed.

Belief Generator. This component, as in the previous architecture, is in charge of creating, storing and retaining the agent's knowledge. In addition, it is also in charge of establishing the agent's beliefs. The belief creation process is a continuous process that is initiated at the moment at which the agent is created and which continues during its entire effective life.

Intuitions. Intuitions are beliefs that have not been verified but which it thinks may be true. According to [23] intuition has not yet been modeled by agent systems. In this work we have tried to adapt this concept because we consider that in real communities people are influenced by their intuitions when they have to make a decision or believe in something. This concept is emulated by comparing the agents' profiles to obtain an initial value of intuition that can be used to form a belief about an agent.

History. This component stores the interactions of the agents with the environment.

3.3 Social Architecture

This architecture (see Figure 4) is quite similar to the deliberative architecture. The main differences are the social model and social behavior processor, which are explained in the following paragraphs.

Social Model. This module represents the actual state of the community, the community's interests and the members' identifiers.

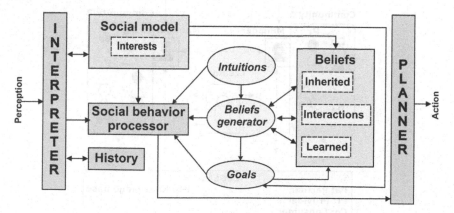

Fig. 4. Social architecture

Social Behavior Processor. This component processes the beliefs of the community's members. To do this, this module needs to manage the goals, intuitions and beliefs of the community in order to make a decision.

The social focus that this architecture provides permits us to give the agents the social behavior necessary to emulate the work relationships in an organization. In addition, this layer permits the decentralization of decision-making, that is, it provides methods by which to process or make decisions based on the opinions of the members of a community.

4 Implementation of the Architecture

To evaluate the feasibility of the implementation of the architecture, we have developed a prototype into which people can introduce documents and where these documents can also be consulted by other people. The goal of this prototype is to allow software agents to help employees to discover the information that may be useful to them thus decreasing the overload of information that employees often have and strengthening the use of knowledge bases in enterprises. In addition, we try to avoid the situation of employees storing valueless information in the knowledge base.

A feature of this system is that when a person searches for knowledge in a community, and after having used the knowledge obtained, that person then has to evaluate the knowledge in order to indicate whether:

- The knowledge was useful.
- How it was related to the topic of the search (for instance a lot, not too much, not at all).

One type of agent in our prototype (see Figure 5) is the User Agent which is in charge of representing each person that may consult or introduce knowledge in a knowledge

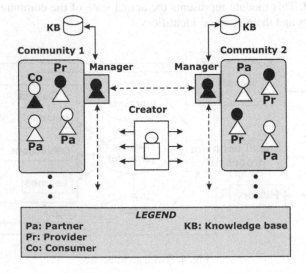

Fig. 5. Agents' distribution

base. The *User Agent* can assume three types of behavior or roles similar to the tasks that a person may carry out in a knowledge base. Therefore, the User Agent plays one role or another depending upon whether the person that it represents carries out one of the following actions:

- The person contributes new knowledge to the communities in which s/he is registered. In this case the User Agent plays the role of **Provider**.
- The person uses knowledge previously stored in the community. Then, the User Agent will be considered as a **Consumer**.
- The person helps other users to achieve their goals, for instance by giving an evaluation of certain knowledge. In this case the role is that of a **Partner**. So, Figure 5 shows that in Community 1 there are two User Agents playing the role of Partner (Pa), one User Agent playing the role of Consumer (Co) and another being a Provider (Pr).

The second type of agent within a community is called the *Manager Agent* (represented in black in Figure 5) which is in charge of managing and controlling its community. In order to approach this type of agent the following tasks are carried out:

- Registering an agent in its community. It thus controls how many agents there are and how long the stay of each agent in that community is.
- Registering the frequency of contribution of each agent. This value is updated every time an agent makes a contribution to the community.
- Registering the number of times that an agent gives feedback about other agents' knowledge. For instance, when an agent "A" uses information from another agent "B", the agent A should evaluate this information. Monitoring how often an agent gives feedback about other agents' information helps to detect whether agents contribute to the creation of knowledge flows in the community since it is as important that an agent contributes new information as it is that another agent contributes by evaluating the relevance or importance of this information.
- Registering the interactions between agents. Every time an agent evaluates the contributions of another agent the Manager agent will register this interaction. But this interaction is only in one direction, which means, if the agent A consults information from agent B and evaluates it, the Manager records that A knows B. But this does not mean that B knows A because B does not obtain any information about A.

Moreover, when a user wants to join a community in which no member knows anything about him/her, the reputation value assigned to the user in the new community is calculated on the basis of the reputation assigned from others communities where the user is or was a member. For instance, a User Agent called *j*, will ask each community manager where he/she was previously a member to consult each agent that knows him/her with the goal of calculating the average value of his/her reputation (R_{Aj}). This is calculated as:

$$R_{Aj} = (\sum_{j=1}^{n} R_{sj})/n \tag{1}$$

where n is the number agents who know j and $\mathbf{R}_{sj}$ is the value of reputation of j in the eyes of s. In the case of being known in several communities the average of the values $\mathbf{R}_{Aj}$ will be calculated. Then, the User Agent j presents this reputation value (similar to when a person presents his/her curriculum vitae when s/he wishes to join a company) to the Manager Agent of the community to which it is "applying". This reputation value permits to assign a reputation value, taking into account the previous experiences and relations with others agents, generating a flow and exchange of information between the agents. This mechanism is similar to the "word-of-mouth" propagation of information for a human [24].

In addition, $\mathbf{R}_{sj}$ value is computed as follows:

$$R_{sj} = w_e * E_j + w_p * P_j + w_i * I_{sj} + (\sum_{j=1}^{n} QC_{sj})/n \qquad (2)$$

where R_{sj} denotes the reputation value that $agent_s$ has in $agent_j$ (each agent in the community has an opinion about each of the other agent members of the community which it has interacted with).

E_j is the value of expertise which is calculated according to the degree of experience that a person has in a domain and it is given by the company.

P_j is the value assigned to the position of a person. This position is defined by the organizational diagram of the enterprise. Therefore, a value that determines the hierarchic level within the organization can be assigned to each level of the diagram.

I_{sj} denotes the intuition value that $agent_s$ has in $agent_j$ which is calculated by comparing each of the users' profiles.

In addition, previous experience should also be calculated. We suppose that when an agent s consults information from another agent j, the agent s should evaluate how useful this information was. This value is called QC_{sj} (Quality of j's Contribution). To attain the average value of an agent's contribution, we calculate the sum of all the values assigned to their contributions and we divide it between their total. In the expression n represents the total number of evaluated contributions.

Finally, w_e, w_p and w_i are weights with which the Reputation value can be adjusted to the needs of the organizations. For instance, if an enterprise considers that all their employees have the same category, then $w_p = 0$. The same could occur when the organization does not take its employees' intuitions or expertise into account.

In this way, an agent can obtain a value related to the reputation of another agent and decide to what degree it is going to consider the information obtained from this agent. The formulas (1) and (2) are processed in the social and deliberative architecture, respectively.

5 Previous Work in the Field

This research can be compared with other proposals that use agents and trust in knowledge exchange. For instance, in [24], the authors propose a model that allows agents to decide which agents' opinions they trust more and propose a protocol based on recommendations. This model is based on a reputation or word-of-mouth mechanism. The main problem with this approach is that every agent must keep rather complex data structures that represent a kind of global knowledge about the whole network. In [25], the authors propose a framework for exchanging knowledge in a mobile environment. They use delegate agents to be spread out into the network of a mobile community and use trust information to serve as the virtual presence of a mobile user. Another interesting work is in [26], where the authors describe a trust and reputation mechanism that allows peers to discover partners who meet their individual requirements through individual experience and by sharing experiences with peers that have similar preferences. This work is focused on *peer-to-peer* environments.

Barber and Kim present a multi-agent belief revision algorithm based on belief networks [17]. In their model the agent is able to evaluate incoming information, to generate a consistent knowledge base, and to avoid fraudulent information from unreliable or deceptive information sources or agents. This work has a similar goal to ours. However, the means of attaining it are different. In Barber and Kim's case they define reputation as a probability measure, since the information source is assigned a reputation value of between 0 and 1. Moreover, every time a source sends knowledge, the source should indicate the certainty factor that the source has of that knowledge. In our case, the focus is very different since it is the receiver who evaluates the relevance of a piece of knowledge rather than the provider as in Barber and Kim's proposal.

6 Evaluation and Future Work

Once the prototype is finished we will evaluate it. To do this, different approaches can be followed, from a multi-agent point of view or from a social one. First of all we have focused on the former and we are testing the most suitable number of agents advisable for a community. Therefore, several simulations have been performed. As result of them we found that:

- The maximum number of agents supported by the Community Manager Agent when it receives User Agents' evaluations is approximately 800. When we tried to work with 1000 agents for instance, the messages were not managed conveniently. However, we could see that the Manager Agent could support a high number of petitions, at least, using simpler behavior.
- On the other hand, if we have around 10 User Agents launched, they need about 20 or more interactions to know all agents of the community. If a User Agent has between 10 and 20 interactions with other members it is likely that it interacts with 90% of members of its community, which means that the agent is going to know almost all the members of the community. Therefore, after several trials we detected that the most suitable number of agents for one community was around 10 agents and they needed a average of 20 interactions to know (to have a contact with) all the members of the community, which is quite convenient in order to obtain its own value of reputation about other agent.

All these results are being used to detect whether the exchange of messages between the agents is suitable, and to see if the information that we propose to be taken into account to obtain a trustworthy value of the reputation of each agent is enough, or if more parameters should be considered. Once this validation is finished we need to carry out further research to answer one important question, which is how the usage of this prototype affects the performance of a community. This is the social approach that we mentioned at the beginning of this section. As claimed in [27], to measure the performance of communities is a challenge since communities only have an indirect impact on business results. In order to do this we are going to take some ideas of the performance measurement framework for communities proposed by [28] where the performance of communities is measured in terms of output and values such as: personal knowledge, strength of relationships (this could be one of the most important values for our research) and access to information. This research will be critical to find how our proposal affects communities of practice.

Acknowledgements. This work is partially supported by the MECENAS (PBI06-0024), ESFINGE (TIN2006-15175-C05-05) and MELISA project (PAC08-0142-3315), Junta de Comunidades de Castilla-La Mancha, Consejería de Educación y Ciencia, in Spain and CONACYT (México) under grant of the scholarship 206147 provided to the first author.

References

1. Zand, D.: The Leadership Triad: Knowledge, Trust and Power. Oxford University Press, Oxford (1997)
2. Alavi, M., Leidner, D.E.: Knowledge Management and Knowledge Management Systems: Conceptual Foundations and Research Issues. MIS Quarterly 25(1), 107–136 (2001)
3. Nebus, J.: Framing the Knowledge Search Problem: Whom Do We Contact, and Why Do We Contact Them? In: Academy of Management Best Papers Proceedings, pp. 1–7 (2001)
4. Lawton, G.: Knowledge Management: Ready for Prime Time? Computer 34(2), 12–14 (2001)
5. Desouza, K., Awazu, Y., Baloh, P.: Managing Knowledge in Global Software Development Efforts: Issues and Practices. IEEE Software, 30–37 (2006)
6. Huysman, M., Wit, D.: Knowledge Sharing in Practice. Information Science and Knowledge Management. Springer, Heidelberg (2002)
7. Wenger, E.: Communities of Practice: Learning Meaning, and Identity. Cambridge University Press, New York (1998)
8. van Elst, L., Dignum, V., Abecker, A. (eds.): Agent-Mediated Knowledge Management. LNCS, vol. 2926. Springer, Heidelberg (2004)
9. Mohammadian, M.: Computational Intelligence Techniques Driven Intelligent Agents for Web Data Mining and Information Retrieval. In: Intelligent Agents for Data Mining and Information Retrieval, pp. 15–29. IDEA Group Publishing (2004)
10. Wooldridge, M., Jennings, N.: Intelligent Agents: Theory and Practice. Knowledge Engineering Review 10(2), 115–152 (1995)
11. Mercer, S., Greenwood, S.: A Multi-Agent Architecture for Knowledge Sharing. In: Proceedings of the Sixteenth European Meeting on Cybernetics and Systems Research (2001)
12. Balasubramanian, S., Brennan, R., Norrie, D.: An Architecture for Metamorphic Control of Holonic Manufacturing Systems. Computers in Industry 46(1), 13–31 (2001)
13. Encinas, J.C., García, A., Abarca, A.: Application of a RFID enhanced Multi-Agent Systems for the Control of a Distribution Center. In: Proceedings of 4th International Conference on Digital Enterprise Technology, pp. 454–460. Bath, UK (2007)

14. Abecker, A., Bernardi, A., van-Elst, L.: Agent Technology for Distributed Organizational Memories: The Frodo Project. In: Proceedings of 5th International Conference on Enterprise Information Systems (ICEIS), Angers France, vol. 2, pp. 3–10 (2003)
15. Wooldridge, M., Ciancarini, P.: Agent-Oriented Software Engineering: The State of the Art. In: Wooldridge, M., Ciancarini, P. (eds.) Agent Oriented Software Engineering. LNCS (LNAI), vol. 1975. Springer, Heidelberg (2001)
16. Fuentes, R., Gómez-Sanz, J., Pavón, J.: A Social Framework for Multi-agent Systems Validation and Verification. In: Wang, S., Tanaka, K., Zhou, S., Ling, T.-W., Guan, J., Yang, D.-q., Grandi, F., Mangina, E.E., Song, I.-Y., Mayr, H.C. (eds.) ER Workshops 2004. LNCS, vol. 3289, pp. 458–469. Springer, Heidelberg (2004)
17. Abdul-Rahman, A., Hailes, S.: Supporting Trust in Virtual Communities. In: Proceedings of the 33rd Hawaii International Conference on Systems Sciences (HICSS), vol. 6. IEEE Computer Society, Washington (2000)
18. Schulz, S., Herrmann, K., Kalcklosch, R., Schowotzer, T.: Trust-Based Agent -Mediated Knowledge Exchange for Ubiquitous Peer Networks (AMKM). LNCS (LNAI), vol. 2926, pp. 89–106 (2003)
19. Wang, Y., Vassileva, J.: Trust and Reputation Model in Peer-to-Peer Networks. In: Third International Conference on Peer-to-Peer Computing (P2P 2003), p. 150 (2003)
20. Barber, K., Kim, J.: Belief Revision Process Based on Trust: Simulation Experiments. In: 4th Workshop on Deception, Fraud and Trust in Agent Societies, Montreal Canada (2004)
21. Gambetta, D.: Can We Trust Trust? In: Gambetta, D. (ed.) Trust: Making and Breaking Cooperative Relations, pp. 213–237 (1988)
22. Marsh, S.: Formalizing Trust as a Computational Concept. PhD Thesis, University of Stirling (1994)
23. Ushida, H., Hirayama, Y., Nakajima, H.: Emotion Model for Life like Agent and its Evaluation. In: Proceedings of the Fifteenth National Conference on Artificial Intelligence and Tenth Innovative Applications of Artificial Intelligence Conference (AAAI 1998 / IAAI 1998), Madison, Wisconsin, USA, pp. 8–37 (1998)
24. Imbert, R., de Antonio, A.: When emotion does not mean loss of control. In: Panayiotopoulos, T., Gratch, J., Aylett, R.S., Ballin, D., Olivier, P., Rist, T. (eds.) IVA 2005. LNCS, vol. 3661, pp. 152–165. Springer, Heidelberg (2005)
25. Wasserman, S., Glaskiewics, J.: Advances in Social Networks Analysis. Sage Publications, Thousand Oaks (1994)
26. Mui, L., Halberstadt, A., Mohtashemi, M.: Notions of Reputation in Multi-Agents Systems: A Review. In: International Conference on Autonomous Agents and Multi-Agents Systems (AAMAS), pp. 280–287 (2002)
27. Geib, M., Braun, C., Kolbe, L., Brenner, W.: Measuring the Utilization of Collaboration Technology for Knowledge Development and Exchange in Virtual Communities. In: Proceedings of the 37th Annual Hawaii International Conference on Systems Sciences (HICSS), vol. 1 (2004)
28. McDermott, R.: Measuring the Impact of Communities. Knowledge Management 5(2), 26–29 (2002)

Recent Developments in Automated Inferencing of Emotional State from Face Images

Ioanna-Ourania Stathopoulou and George A. Tsihrintzis

Department of Informatics
University of Piraeus
Piraeus 185 34
Greece
{iostath,geoatsi}@unipi.gr

Abstract. Automated facial expression classification is very important in the design of new human-computer interaction modes and multimedia interactive services and arises as a difficult, yet crucial, pattern recognition problem. Recently, we have been building such a system, called NEU-FACES, which processes multiple camera images of computer user faces with the ultimate goal of determining their affective state. In here, we present results from an empirical study we conducted on how humans classify facial expressions, corresponding error rates, and to which degree a face image can provide emotion recognition from the perspective of a human observer. This study lays related system design requirements, quantifies statistical expression recognition performance of humans, and identifies quantitative facial features of high expression discrimination and classification power.

Keywords: Facial Expression Classification, Human Emotion, Knowledge Representation, Human-Computer Interaction.

1 Introduction

Facial expressions are particularly significant in communicating information in human-to-human interaction and interpersonal relations, as they reveal information about the affective state, cognitive activity, personality, intention and psychological state of a person and this information may, in fact, be difficult to mask.

When mimicking communication between humans, human-computer interaction systems must determine the psychological state of a person, so that the computer can react accordingly. Indeed, images that contain faces are instrumental in the development of more effective and friendlier methods in multimedia interactive services and human computer interaction systems. Vision-based human-computer interactive systems assume that information about a user's identity, state and intent can be extracted from images, and that computers can then react accordingly. Similar information can also be used in security control systems or in criminology to uncover possible criminals. Studies have concluded to six facial expressions which arise very commonly during a typical human-computer interaction session and, thus, vision-based human-computer

J. Filipe et al. (Eds.): ICSOFT/ENASE 2007, CCIS 22, pp. 310–319 , 2008.
© Springer-Verlag Berlin Heidelberg 2008

interaction systems that recognize them could guide the computer to "react" accordingly and attempt to better satisfy its user needs. Specifically, these expressions are: "neutral", "happy", "sad", "surprised", "angry", "disgusted" and "bored-sleepy".

It is common experience that the variety in facial expressions of humans is large and, furthermore, the mapping from psychological state to facial expression varies significantly from human to human and is complicated further by the problem of pretence, i.e. the case of someone's facial expression not corresponding to his/her true psychological state. These two facts make the analysis of the facial expressions of another person difficult and often ambiguous. This problem is even more severe in automated facial expression classification, as face images are non-rigid, have a high degree of variability in size, shape, color and texture and variations in pose, facial expression, image orientation and conditions add to the level of difficulty of the problem.

Towards achieving the automated facial image processing goal, we have been developing an automated facial expression classification system [19–26], called NEU-FACES, in which features extracted as deviations from the neutral to other common expressions are fed into neural network-based classifiers. Specifically, NEU-FACES is a two-module system, which automates both the face detection and the facial expression process.

To start specifying requirements and building NEU_FACES, we needed to conduct an empirical study first on how humans classify facial expressions, corresponding error rates, and to which degree a face image can provide emotion recognition from the perspective of a human observer. This study lays related system design requirements, quantifies statistical expression recognition performance of humans, and identifies quantitative facial features of high expression discrimination and classification power. The present work is the outcome of the participants' responses to our questionnaires.

An extensive search of the literature revealed a relative shortage of empirical stud ies of human ability to recognize someone else's emotion from his/her face image. The most significant of these studies are summarized next. Ekman and Friesen first defined a set of universal rules to "manage the appearance of particular emotions in particular situations" [3-7]. Unrestrained expressions of anger or grief are strongly discouraged in most cultures and may be replaced by an attempted smile rather than a neutral expression; detecting those emotions depends on recognizing signs other than the universally recognized archetypal expressions. Reeves and Nass [15] have already shown that people's interactions with computers, TV and similar machines/media are fundamentally social and natural, just like interactions in real life. Picard in her work in the area of affective computing states that "emotions play an essential role in rational decision-making, perception, learning, and a variety of other cognitive functions" [12, 13]. De Silva et al. [2] also performed an empirical study and reported results on human subjects' ability to recognize emotions. Video clips of facial expressions and corresponding synchronised emotional speech clips were shown to human subjects not familiar with the languages used in the video clips (Spanish and Sinhala). Then, human recognition results were compared in three tests: video only, audio only, and combined audio and video. Finally, M. Pantic et al. performed a survey of the past work in solving emotion recognition problems by a computer and provided a set of recommendations for developing the first part of an intelligent multimodal HCI [9–11].

In this paper, we present our empirical study on identifying those face parts that may lead to correct facial expression classification and on determining the facial features that are more significant in recognizing each expression. Specifically, in Section 2, we present emotion perception principles from the psychologist's perspective. In Section 3, we describe the questionnaire we used in our study. In Section 4, we show statistical results of our study. Finally, we summarize and draw conclusions in Section 5 and point to future work in Section 6.

2 Emotion Perception

The question of how to best characterize perception of facial expressions has clearly become an important concern for many researchers in affective computing. Ironically, this growing interest is coming at a time when the established knowledge on human facial affect is being strongly challenged in the basic psychology research literature [8]. In particular, recent studies have thrown suspicion on a large body of long-accepted data, even on studies previously conducted by the same people.

In the past, two main studies regarding facial expression perception have appeared in the literature. The first study is the classic research by psychologist Paul Ekman and colleagues [3-7] in the early 1960s, which resulted in the identification of a small number of so-called "basic" emotions, namely *anger, disgust, fear, happiness, sadness and surprise (contempt was added only recently)*. In Ekman's theory, the basic emotions were considered to be the building blocks of more complex feeling states [3], although in newer studies he is sceptical about the possibility of two basic emotions occurring simultaneously [7]. Following these studies, Ekman and Friesen [4] developed the, so-called, "facial action coding system (FACS)," which quantifies facial movement in terms of component muscle actions. Recently automated, the FACS remains the one of the most comprehensive and commonly accepted methods for measuring emotion from the visual observation of faces.

In the past few years, a second study by psychologist James Russell and colleagues summarizes previous works on human emotion perception [18] and challenges strongly the classic data [17], largely on methodological grounds. Russell argues that emotion in general (and facial expression of emotion in particular) can be best characterized in terms of a multidimensional affect space, rather than discrete emotion categories. More specifically, Russell claims that two dimensions, namely "pleasure" and "arousal," are sufficient to characterize facial affect space.

Despite the fact that divergent studies have appeared in the literature, most scientists agree that:

- Human experience emotions in subjective ways.
- The "basic emotions" deal with fundamental life tasks.
- The "basic emotions" mostly occur during interpersonal relationships, but this does not exclude the possibility of their occurring in the absence of other humans.
- Facial expressions are important in revealing emotions and informing other people about a person's emotional state. Indeed, studies have shown that people with congenital (Mobius Syndrome) or other (e.g. from a stroke) facial paralysis report great difficulty in maintaining and developing interpersonal relationships.

- Each time an emotion occurs, a signal will not necessarily be present. Emotions may occur without any evident signal, because humans are, to a very large extent, capable of suppressing such signals. Also, a threshold may need to be exceeded to bring about an expressive signal and this threshold may vary across individuals.
- Usually, emotions are influenced by two factors, namely *social learning* and *evolution*. Thus, similarities across different cultures arise in the way emotions are expressed because of past evolution of the human species, but differences also arise which are due to culture and social learning.
- Facial expressions are emotional signals that result into movements of facial skin and connective tissue caused by the contraction of one or more of the forty four bilaterally symmetrical facial muscles. These striated muscles fall into two groups:
  - four of these muscles, innervated by the trigeminal (5th cranial) nerve, are attached to and move skeletal structures (e.g., the jaw) in mastication
  - forty of these muscles, innervated by the facial (7th cranial) nerve, are attached to bone, facial skin, or fascia and do not operate directly by moving skeletal structures but rather arrange facial features in meaningful configurations.

Based on these studies and by observing human reactions, we identified differences between the "neutral" expression of a model and its deformation into other expressions. We quantified these differences into measurements of the face (such as size ratio, distance ratio, texture, or orientation), so as to convert pixel data into a higher-level representation of shape, motion, color, texture and spatial configuration of the face and its components. Specifically, we locate and extract the corner points of specific regions of the face, such as the eyes, the mouth and the brows, and compute their variations in size, orientation or texture between the neutral and some other expression. This constitutes the feature extraction process and reduces the dimensionality of the input space significantly, while retaining essential information of high discrimination power and stability.

3 The Questionnaire

In order to validate these *facial features* and decide whether these features are used by humans when attempting to recognize someone else's emotion from his/her facial expression, we developed a questionnaire where the participants were asked to determine which *facial features* helped them in the classification task. In the questionnaire, we used images of subjects of a facial expression database which we had developed at the University of Piraeus [25]. Our aim was to identify the facial features that help humans in classifying a facial expression. Moreover, we wanted to know if it is possible to map a facial expression into an emotion. Finally, another goal was to determine if a human observer can recognize a facial expression from isolated parts of a face, as we expect computer-classifiers to do.

3.1 The Questionnaire Structure

In order to understand how a human classifies someone else's facial expression and set a target error rate for automated systems, we developed a questionnaire in which each we asked 300 participants to state their thoughts on a number of facial expression-related questions and images.

Specifically, the questionnaire consisted of three different parts:

1. In the first part, the observer was asked to identify an emotion from the facial expressions that appeared in 14 images. Each participant could choose from the 7 of the most common emotions that we pointed out earlier, such as: "anger", "happiness", "neutral", "surprise", "sadness", "disgust", "boredom–sleepiness", or specify any other emotion that he/she thought appropriate. Next, the participant had to state the degree of certainty (from 0-100%) of his/her answer. Finally, he/she had to state which features (such as the eyes, the nose, the mouth, the cheeks etc.), had helped him/her make that decision. A typical question of the first part of the questionnaire is depicted in Figure 1.

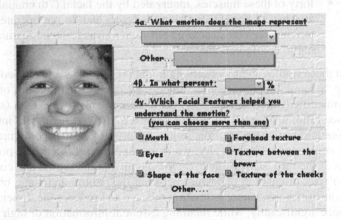

Fig. 1. The first part of the questionnaire

2. When filling the second part of the questionnaire, each participant had to identify an emotion from parts of a face. Specifically, we showed them the "neutral" facial image of a subject and the corresponding image of some other expression. In this latter image pieces were cut out, leaving only certain parts of the face, namely the "eyes", the "mouth", the "forehead", the "cheeks", the "chin" and the "brows." This is typically shown in Figure 2. Again, each participant could choose from the 7 of the most common emotions "anger", "happiness", "neutral", "surprise", "sadness", "disgust", "boredom–sleepiness", or specify any other emotion that he/she thought appropriate. Next, the participant had to state the degree of certainty (from 0-100%) of his/her answer. Finally, the participant had to specify which features had helped him/her make that decision.

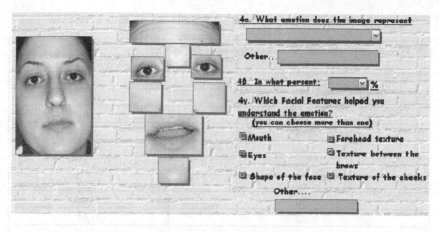

Fig. 2. The second part of the questionnaire

3. In the final (third) part of our study, we asked the participants to supply information about their background (e.g. age, interests, etc.). Additionally, each participant was asked to provide information about:

    - The level of difficulty of the questionnaire with regards to the task of emotion recognition from face images
    - Which emotion he/she though was the most difficult to classify
    - Which emotion he/she though was the easiest to classify
    - The percentage to which a facial expression maps into an emotion (0-100%)

3.2 The Participant and Subject Backgrounds

There were 300 participants in our study. All the participants were Greek, thus familiar with the greek culture and the greek ways of expressing emotions. They were mostly undergraduate or graduate students and faculty in our university and there age varied between 19 and 45 years.

4 Statistical Results

4.1 Test Data Acquisition

Most users agreed that a facial expression represents the equivalent emotion with a percentage of 70% or higher. The results are shown in Table 1.

Based on the participants' answers in the second part of our questionnaire it was observed that smaller error rates could be achieved if parts rather than the entire face image were displayed. The differences in error rates are quite significant and show that the extracted facial parts are well chosen. An exception to this observation

Table 1. Percentage to which a facial expression represents an emotion

Percentage to which an expression represents an emotion (%)	Percentage of user answers (%)
0	0,00
10	0,00
20	0,76
30	2,27
40	1,52
50	9,85
60	14,39
70	31,06
80	21,97
90	15,91
100	2,27

occurred with the "angry" and "disgusted" emotions where we observed a 6,44% and 5,10% increase in the error rate in the second part of our questionnaire. This is expected to be observed in the performance of automated expression classification systems when shown a face forming an expression of anger of disgust. More specifically, these differences in the error rates are shown in Table 2. As shown in the last column (P-value) these results are statistically significant.

Table 2. Error rates in the two parts of the questionnaire

Emotion	Error rates		Difference	P-value
	1st Part	**2nd Part**		
Neutral	61,74	----------	**61,74**	----------
Happiness	31,06	3,79	**27,27**	0,000000 003747
Sadness	65,91	17,42	**48,48**	0,000000 000035
Disgust	81,26	86,36	**-5,10**	0,02932 4580032
Boredom	49,24	21,97	**27,27**	0,000012 193203
Angry	23,86	30,30	**-6,44**	0,026319 945845
Surprise	10,23	4,55	**5,68**	0,001390 518291
Other	9,47	18,18	**-8,71**	

The facial features that helped the users to understand the emotions are mostly the eyes, the mouth, and the cheeks. In some expressions, e.g. the "angry", there were some other features very important, for example the texture between the brows in this case. The most important facial features are shown in Table 3.

Table 3. Important features for each facial expression

	A	B	C	D	E	F	G
1	66,3	81,6	63,6	82,6	77,3	55,7	83,7
2	84,5	67,8	76,1	81,1	79,9	81,4	88,8
3	10,2	22,7	4,2	6,1	4,9	30,9	46,4
4	20,8	14,4	31,1	7,6	14,4	10,0	11,4
5	18,2	59,5	8,7	3,0	4,2	8,9	23,7
6	46,6	8,1	30,7	28,8	60,6	21,4	5,1
7	0,0	2,5	3,0	3,0	3,0	2,3	1,5

1	Eyes
2	Mouth
3	Texture of the Forehead
4	Shape of the Face
5	Texture between the brows
6	Texture of the cheeks
7	Other

A	Neutral
B	Angry
C	Bored-Sleepy
D	Disgusted
E	Happy
F	Sad
G	Surprised

5 Summary and Conclusions

Automated expression classification in face images is a prerequisite to the development of novel human-computer interaction and multimedia interactive service systems. However, the development of integrated, fully operational such automated systems is non-trivial. Towards building such systems, we have been developing a novel automated facial expression classification system [26], called NEU-FACES, in which features extracted as deviations from the neutral to other common expressions are fed into neural network-based expression classifiers. In order to establish the correct feature selection, in this paper, we conducted an empirical study of the facial expression classification problem in images, from the human's perspective. This study allows us to identify those face parts of the face that may lead to correct facial expression classification. Moreover, the study determines those facial features that are more significant in recognizing each expression. We found that the isolation of parts of the face resulted to better expression recognition than looking at the entire face image.

6 Future Work

In the future, we will extend this work in the following directions: (1) we will improve our NEU-FACES system by applying techniques based on multi-criteria decision theory for the facial expression classification task, (2) we will investigate the application of quality enhancement techniques to our image dataset and seek to extract additional classification features from them, and (3) we will extend our database so as to contain *sequences of images of facial expression formation* rather than simple static images of formed expressions and seek in them additional features of high classification power.

Acknowledgements. Support for this work was provided by the General Secretariat of Research and Technology, Greek Ministry of Development, under the auspices of the PENED-2003 basic research program.

References

1. Csikszentmihalyi, M.: Flow: The Psychology of Optimal Experience, Harper and Row, New York (1994)
2. De Silva, L.C., Miyasato, T., Nakatsu, R.: Facial Emotion Recognition Using Multimodal Information. In: Proc. IEEE Int. Conf. on Information, Communications and Signal Processing - ICICS, Singapore, pp. 397–401 (September 1997)
3. Ekman, P.: The Handbook of Cognition and Emotion. In: Dalgleish, T., Power, T. (eds.), Sussex, U.K, pp. 45–60. John Wiley & Sons, Ltd, Chichester (1999)
4. Ekman, P., Friesen, W.: Unmasking the Face. Prentice-Hall, Englewood Cliffs (1975)
5. Ekman, P.: Emotion In the Human Face. Cambridge University Press, Cambridge (1982)
6. Ekman, P., Campos, J., Davidson, R.J., De Waals, F.: Darwin, Deception, and Facial Expression. In: Emotions Inside Out, vol. 1000. Annals of the New York Academy of Sciences, New York (2003)
7. Ekman, P., Rosenberg, E.L.: What the Face Reveals: Basic and applied studies of spontaneous expression using the Facial Action Coding System (FACS). Oxford University Press, New York
8. Ortony, A., Clore, G.L., Collins, A.: The Cognitive Structure of Emotions. Cambridge University Press, Cambridge (1988)
9. Pantic, M., Rothkrantz, L.J.M.: Automatic Analysis of Facial Expressions: The State of the Art. IEEE Transactions on Pattern Analysis and Machine Intelligence 22(12), 1424–1445 (2000)
10. Pantic, M., Rothkrantz, L.J.M.: Toward an affect-sensitive multimodal HCI. Proceedings of the IEEE 91(9), 1370–1390 (2003)
11. Pantic, M., Valstar, M.F., Rademaker, R., Maat, L.: Web-based Database for Facial Expression Analysis. In: Proc. IEEE Int'l Conf. Multmedia and Expo. (ICME 2005), Amsterdam, The Netherlands (July 2005)
12. Picard, R.W.: Affective Computing. MIT Press, Cambridge (1997)
13. Picard, R.W.: Affective Computing: Challenges. International Journal of Human-Computer Studies 59(1-2), 55–64 (2003)
14. Reeves, B., Nass, C.: Social and Natural Interfaces: Theory and Design. CHI Extended Abstracts, pp. 192–193 (1997)

15. Reeves, B., Nass, C.: The Media Equation: How People Treat Computers, Television, and New Media Like Real People and Places. Cambridge University Press and CSLI, New York

16. Rosenberg, M.: Conceiving the Self. Basic Books, New York (1979)

17. Russell, J.A.: Core affect and the psychological construction of emotion. Psychological Review 110, 145–172 (2003)

18. Russell, J.A.: Is there universal recognition of emotion from facial expression?: A review of the cross-cultural studies. Psychological Bulletin 115, 102–114 (1994)

19. Stathopoulou, I.-O., Tsihrintzis, G.A.: A neural network-based facial analysis system. In: 5th International Workshop on Image Analysis for Multimedia Interactive Services, Lisboa, Portugal, April 21-23 (2004)

20. Stathopoulou, I.-O., Tsihrintzis, G.A.: An Improved Neural Network-Based Face Detection and Facial Expression Classification System. In: IEEE International Conference on Systems, Man, and Cybernetics 2004, The Hague, Netherlands, October 10-13 (2004)

21. Stathopoulou, I.-O., Tsihrintzis, G.A.: Pre-processing and expression classification in low quality face images. In: 5th EURASIP Conference on Speech and Image Processing, Multimedia Communications and Services, Smolenice, Slovak Republic, June 29 – July 2 (2005)

22. Stathopoulou, I.-O., Tsihrintzis, G.A.: Evaluation of the Discrimination Power of Features Extracted from 2-D and 3-D Facial Images for Facial Expression Analysis. In: 13th European Signal Processing Conference, Antalya, Turkey, September 4-8 (2005)

23. Stathopoulou, I.-O., Tsihrintzis, G.A.: Detection and Expression Classification Systems for Face Images (FADECS). In: 2005 IEEE Workshop on Signal Processing Systems (SiPS 2005), Athens, Greece, November 2 – 4 (2005)

24. Stathopoulou, I.-O., Tsihrintzis, G.A.: An Accurate Method for eye detection and feature extraction in face color images. In: 13th International Conference on Signals, Systems, and Image Processing, Budapest, Hungary, September 21-23 (2006)

25. Stathopoulou, I.-O., Tsihrintzis, G.A.: Facial Expression Classification: Specifying Requirements for an Automated System. In: 10th International Conference on Knowledge-Based & Intelligent Information & Engineering Systems, Bournemouth, United Kingdom, October 9-11 (2006)

26. Stathopoulou, I.-O., Tsihrintzis, G.A.: NEU-FACES: A Neural Network-based Face Image Analysis System. In: 8th International Conference on Adaptive and Natural Computing Systems, Warsaw, Poland, April 11-14 (2007)

Inconsistency-Tolerant Integrity Checking
for Knowledge Assimilation*

Hendrik Decker

Instituto Tecnológico de Informática,
Campus de Vera 8G, E-46071 Valencia, Spain
hendrik@iti.upv.es

Abstract. A recently introduced notion of inconsistency tolerance for integrity checking is revisited. Two conditions that facilitate verification or falsification of inconsistency tolerance are discussed. Based on a definition of updates that do not cause integrity violation but may be executed in the presence of inconsistency, this notion is then extended to several knowledge assimilation tasks: integrity maintenance, view updating and repairing of integrity violation. Many knowledge assimilation approaches turn out to be inconsistency-tolerant without needing any knowledge about the status of integrity of the underlying database.

1 Introduction

Knowledge assimilation (abbr.: *KA*) [18,23,6,15] is the process of integrating new data into a body of information such that the latter's integrity remains satisfied. For instance, KA takes place in data warehousing, decision support, diagnosis, quality assurance, content management, machine learning, robotics, vision, natural language understanding etc.

Also in more commonplace systems, KA is a very important issue. In databases, for example, a common instance of KA is *integrity maintenance*, i.e., when updates of relational tables are rejected or modified in order to preserve integrity. For example, the deletion of a row r in table T_1 may not be possible without further ado if the primary key of T_1 is referenced by a foreign key constraint of table T_2. For maintaining the postulated referential integrity, the delete request for r either necessitates the deletion of each row in T_2 that references r or the insertion of a new or modified row r' into T_1 with the same primary key values as in the referenced row r.

A somewhat more involved task of KA which subsumes integrity maintenance is *view updating*, i.e., the translation of a request for updating a virtual table of rows derivable from the view's defining query, to changes in the queried base tables. The goal of KA for view updating is to compute integrity-preserving translations for realizing update requests. Translations that would violate integrity are filtered out by KA. However, if integrity is violated, KA is called for to *repair* the violated constraints. Repairs that would violate other constraints are not valid. For instance, the deletion of referenced rows and the insertion of r' in the preceding example are possible repairs for maintaining integrity.

---

* Supported partially by FEDER and the Spanish MEC grant TIN2006-14738-C02-01.

J. Filipe et al. (Eds.): ICSOFT/ENASE 2007, CCIS 22, pp. 320–331, 2008.

All approaches to consistency-preserving KA employ some integrity checking mechanism, for making sure that the assimilation of a new piece of knowledge will not violate any constraint, i.e., that integrity satisfaction is an invariant of database state transitions. Usually, KA methods require that each constraint be satisfied by the underlying database before assimilating new knowledge, i.e., that integrity satisfaction is total. As shown in [8], many methods ensure that all consistent parts of the database remain consistent even when the strict requirement of total integrity satisfaction is waived. So, it comes as no surprise that this requirement can be abandoned for KA in general.

In section 2, we revisit definitions and results for inconsistency-tolerant integrity checking. In section 3, we discuss two conditions that ensure inconsistency tolerance. In section 4, we generalize the definitions and results of sections 2 and 3 to integrity maintenance, view updating and inconsistency repair. In the concluding section 5, we also address related work and look out to future research. Throughout, we use terms and notations of standard database logic.

2 Integrity Checking

2.1 Basic Notions

Recall that integrity constraints are well-formed sentences of first-order predicate calculus. W.l.o.g., we assume they are represented in prenex form (i.e., roughly, all quantifiers explicitly or implicitly appear outermost), which subsumes prenex normal form (i.e., prenex form with all negations innermost) and denial form (i.e., clauses with empty conclusion). In each database state, they are required to be *satisfied*, i.e. true in, or at least consistent with that state. Otherwise, they are said to be *violated* in D.

An *integrity theory* is a finite set of integrity constraints. It is *satisfied* if each of its members is satisfied, and *violated* otherwise. Let I optionally stand for an integrity constraint or an integrity theory, and D be a database. With $D(I) = sat$, we denote that I is satisfied in D, and $D(I) = vio$ that it is violated. Moreover, for an update U, let D^U denote the database obtained from executing U on D; D and D^U also are referred to as *old* and *new* state, respectively.

Different integrity checking methods use different notions to define and determine integrity satisfaction and violation. Abstracting away from such differences, each integrity checking method $\mathcal{M}$ can be formalized as a function that takes as input a database, an integrity theory and an update (i.e., a bipartite finite set of database clauses to be deleted and, resp., inserted). It outputs the value *sat* if it has concluded that integrity will remain satisfied in the new state, and outputs *vio* if it has concluded that integrity will be violated in the new state. Thus, the soundness and completeness of $\mathcal{M}$ can be stated as follows.

Definition 1. *An integrity checking method $\mathcal{M}$ is* sound *if, for each database D, each integrity theory I such that $D(I) =$ sat *and each update U, the following holds.*

$$\text{If } \mathcal{M}(D, I, U) = sat \text{ then } D^U(I) = sat. \tag{1}$$

Completeness of $\mathcal{M}$ can be defined dually, by the only-if half of (1). Note that both definitions are impartial to the question whether *sat* and *vio* are the only values that could be output by $\mathcal{M}(D, I, U)$ (another value could be, e.g., *unknown*). However, for simplicity, we do not consider any semantics of integrity that would have values other than *sat* and *vio* for integrity.

Definition 1 and its dual apply to virtually any integrity checking method in the literature. Of course, each of them is defined for certain classes of databases, constraints and updates (e.g., relational or stratified databases, range-restricted constraints and transactions consisting of insertions and deletions of base facts). So, whenever we say "each" (database, integrity theory, update), we mean all those for which the respective methods are defined at all. From now on, each method $\mathcal{M}$ considered in this paper is assumed to be sound.

For significant classes of databases and integrity theories, soundness and completeness has been shown for methods in [24,4,20,25,3] and others; also their termination, as defined below, can be shown. Other methods, e.g., [14,19], are only sound, i.e., they provide sufficient conditions that guarantee the integrity of the updated database. If these conditions do not hold, further checks may be necessary. For later (theorem 4), also the termination of methods is of interest.

Definition 2. *An integrity checking method $\mathcal{M}$ is said to be* terminating *if, for each database D, each integrity theory I and each update U, the computation of $\mathcal{M}(D, I, U)$ halts and outputs either* sat *or* vio.

Note that by definition 2, the computation of $\mathcal{M}(D, I, U)$ terminates, no matter whether $D(I) = sat$ or not. Thus, such an $\mathcal{M}$ is complete, although it is clear that complete methods are not necessarily terminating.

Definitions 1 and 2 are independent of the diversity of criteria by which methods often are distinguished, e.g., to which classes of databases, constraints and updates they apply, how efficient they are, which parts of the data in (D, U, I) are actually accessed, whether they are complete or not, whether constraints are 'soft' or 'hard', whether integrity is checked in the old or the new state, or whether simplification steps are precompiled at schema specification time or taken at update time. Such distinctions are studied, e.g., in [22,9], but do not matter much in this paper, except when explicitly mentioned.

2.2 Inconsistency-Tolerant Integrity Checking

Common to all methods is that they require *total integrity satisfaction*, i.e., before each update, each constraint must be completely satisfied, without exception. In [8], we have shown how this requirement can be relaxed. Informally speaking, it is in fact possible to tolerate (i.e., live with) individual inconsistencies in the database, i.e., some cases of violated constraints, while trying to make sure that updates do not cause any new cases of integrity violation, i.e., that the cases of constraints that were satisfied in the old state remain satisfied in the new state. The following definitions revisit previous ones in [8] for formalizing what we mean by "case" and "inconsistency tolerance" of integrity checking.

Definition 3 (Case). *Let C be an integrity constraint. The variables in C that are $\forall$-quantified but not dominated by any $\exists$ quantifier (i.e., $\exists$ does not occur left of $\forall$) are* *called* global variables of C. *For a substitution σ of the global variables of C, $C\sigma$ is called a* case *of C.*

For convenience, let a case of some constraint in an integrity theory I also be shortly called a *case of I*.

Note that cases have themselves the form of integrity constraints, and need not be ground. In particular, each constraint is a case of itself.

We remark that the following definition is somewhat more succinct than its equivalent in [8].

Definition 4 (**Inconsistency Tolerance**). *An integrity checking method $\mathcal{M}$ is inconsistency-tolerant if, for each database D, each integrity theory I, each case C of I such that $D(C)=$ sat, and each update U, the following holds.*

$$\text{If } \mathcal{M}(D,I,U) = \text{sat then } D^U(C) = \text{sat.} \tag{2}$$

In general, inconsistent cases may be unknown or not efficiently recognizable. However, by definition 4, inconsistency-tolerant methods are able to blindly cope with any degree of inconsistency. They guarantee that all cases of constraints that were satisfied in the old state will remain satisfied in the new state. Running such a method $\mathcal{M}$ means to compute the very same function as if total satisfaction were required. Since $\mathcal{M}$ does not need to be aware of any particular case of violation, no efficiency is lost, whereas the gains are immense: transactions can continue to run even in the presence of (obvious or hidden, known or unknown) violations of integrity (which is rather the rule than the exception in practice), while maintaining the integrity of all satisfied cases. Running $\mathcal{M}$ means that no new cases of integrity violation will be introduced, while existing "bad" cases may disappear (intentionally or even accidentally) by committing updates that have successfully passed the integrity test.

As shown in [8], inconsistency tolerance is available off the shelve, since most, though not all known approaches to database integrity are inconsistency-tolerant. The following examples illustrate this.

Example 1. Let $C = \leftarrow b(x,y) \wedge b(x,z) \wedge y \neq z$ be the constraint that no two entries with the same ISBN x in the relation b about books must have different titles y and z. Suppose U is to insert $b(11111, logic)$. The simplification $C' = \leftarrow b(11111, y) \wedge y \neq logic$ is generated and evaluated by most methods, with output *sat* if the query C' returns the empty answer, and *vio* otherwise. With the traditional prerequisite of total integrity satisfaction, this output says that D^U satisfies or, resp., violates integrity. Now, suppose that $b(88888, t_1)$ and $b(88888, t_2)$ are in D, possibly together with many other facts in b. Clearly, the case $\leftarrow b(88888, t_1) \wedge b(88888, t_2) \wedge t_1 \neq t_2$ of C is violated in D, i.e., integrity is not totally satisfied. However, the insertion of $b(11111, logic)$ is guaranteed not to cause any additional violation as long as the evaluation of C' yields the output *sat*, i.e., as long as there is no other entry in b with ISBN 11111.

Before, or instead of, evaluating simplified instances of relevant constraints, some methods, e.g. [14], may reason on the integrity theory alone for detecting the possibleinvariance of integrity satisfaction by given updates. That, however, may fail to be inconsistency-tolerant, as illustrated below.

Example 2. Let $D = \{q(a)\}$, $I = \{\leftarrow q(x), \leftarrow q(a), r(b)\}$ and $U = insert\ r(b)$. The case $\leftarrow q(a)$ of $\leftarrow q(x)$ clearly is violated while all other ground cases of I are satisfied. Simplification of $\leftarrow q(a), r(b)$ (which, unlike $\leftarrow q(x)$, is relevant for U) yields $\leftarrow q(a)$; the conjunct $r(b)$ is dropped because U makes it true. Methods that reason with possible subsumptions of simplifications by the integrity theory then easily detect that the simplification above is subsumed by the constraint $\leftarrow q(x)$. Using the intolerant assumption of total integrity satisfaction in the old state then leads to the faulty output *sat*, by the following argument: The constraint $\leftarrow q(x)$, which is assumed to be satisfied in D, is not relevant wrt U. Thus, it can be assumed to remain satisfied in D^U. So, since this constraint subsumes the simplification $\leftarrow q(a)$, integrity will remain satisfied. This argument, which is correct if integrity is totally satisfied in the old state, fails to be inconsistency-tolerant since it fails to identify the violated case of $\leftarrow q(a), r(b)$ caused by U.

3   Verifying and Falsifying Inconsistency Tolerance

To verify or falsify condition (2) of definition 4 for a given method can be laborious. However, there are various sufficient conditions by which inconsistency tolerance can be verified much more easily. Two of them are presented below, in theorems 1 and 4. The first has been used in [7,8] to verify inconsistency tolerance of the methods in [24,4,20,25]. The second is new. It also is a necessary condition, i.e., it also serves to falsify inconsistency tolerance. It arguably is even more apt to show or disprove the inconsistency tolerance of the already mentioned and other methods. Theorem 1 states that inconsistency tolerance is entailed by the first condition, labeled (3) below.

Theorem 1. *A method $\mathcal{M}$ for integrity checking is inconsistency-tolerant if, for each database D, each integrity theory I, each case C in I such that $D(C) = sat$, and each update U, the following holds.*

$$\text{If } \mathcal{M}(D, I, U) = sat \text{ then } \mathcal{M}(D, \{C\}, U) = sat \qquad (3)$$

Proof. Clearly, (2) follows from the transitivity of (3) and (4):

$$\text{If } \mathcal{M}(D, \{C\}, U) = sat \text{ then } D^U(C) = sat \qquad (4)$$

where (4) obviously is a special case of (1).   □

The second condition for verifying inconsistency tolerance is based on definition 5 below. Part *a)* is interesting also in itself, because it provides a notion of inconsistency tolerance that is independent of any method.

Definition 5
a) For a database D and an integrity theory I, an update U *causes violation if there is a case C of I such that $D(C) = sat$ and $D^U(C) = vio$. If, for each case C of I, $D(C) = sat$ entails $D^U(C) = sat$, then U is said to* preserve integrity.
b) For an integrity checking method $\mathcal{M}$, we say that $\mathcal{M}$ recognizes violation *if, for each database D, each integrity theory I and each update U that causes violation, $\mathcal{M}(D, I, U) = vio$.*

Note that definition 5 does not require any constraint to be satisfied in D. Thus, integrity-preserving updates embody a method-independent notion of inconsistency tolerance, since they may run in the presence of inconsistency without violating any case of any constraint that has been satisfied before the update.

Theorem 2 below relates the two parts of definition 5a and is evident. Theorem 3 is a corollary of 5a and definition 4. It states that updates can be checked for inconsistency tolerance by inconsistency-tolerant integrity checking methods. Theorem 4 relates definition 5b to definition 4.

Theorem 2. *For a given database and a given integrity theory, an update U preserves integrity if and only if it does not cause violation.*   □

We remark that theorem 2 would not hold if the semantics of integrity were not two-valued.

Theorem 3. *For a database D, an integrity theory I, an update U and an inconsistency-tolerant integrity checking method $\mathcal{M}$, U preserves integrity if $\mathcal{M}(D, I, U) =$ sat.*   □

In general, the only-if half of theorem 3 does not hold. For example, consider a view p defined by $p(x, y) \leftarrow s(x, y, z)$ and $p(x, y) \leftarrow q(x), r(y)$ in a database D in which $q(a)$ and $r(a)$ are the only tuples that contribute to the natural join of relations q and r. Further, let I consist of the constraint $\leftarrow p(x, x)$, and U be the insertion of the tuple $s(a, a, b)$. Clearly, U does not cause violation, since the case $C = \leftarrow p(a, a)$ is already violated in D. Hence, by theorem 2, U preserves integrity. However, the inconsistency-tolerant methods in [20,25] and others compute and evaluate the simplification $\leftarrow p(a, a)$ of $\leftarrow p(x, x)$ and thus output *vio*. On the other hand, note that inconsistency-tolerant methods which check for idle updates (e.g., the one in [4]) identify $p(a, a)$ as idle (i.e., a consequence of the update that is already true in the old state) and hence output *sat*.

Theorem 4. *Let $\mathcal{M}$ be a terminating integrity checking method. Then, $\mathcal{M}$ is inconsistency-tolerant if and only if it recognizes violation.*

Proof

If: Let $\mathcal{M}$ be a method that recognizes violation, D a database, I an integrity theory and U an update such that $\mathcal{M}(D, I, U) = sat$, and C a case of I such that $D(C) = sat$. We have to show that $D^U(C) = sat$. Since $\mathcal{M}(D, I, U) = sat$, theorem 2 entails that U does not cause violation, i.e., there is no case of I that is satisfied in D and violated in D^U. Thus, $D(C) = sat$ implies $D^U(C) = sat$.   □

Only if: Let $\mathcal{M}$ be inconsistency-tolerant and suppose that U causes violation. So, we have to show that $\mathcal{M}(D, I, U) = vio$. Since U causes violation, there is a case C such that $D(C) = sat$ and $D^U(C) = vio$. Hence, the inconsistency tolerance of $\mathcal{M}$ entails by definition 4 that $\mathcal{M}(D, I, U) \neq sat$. Since $\mathcal{M}$ is terminating, it follows that $\mathcal{M}(D, I, U) = vio$.   □

We remark that termination of $\mathcal{M}$ is used only in the proof of the *only-if* half. However, the last steps in each half of the proof rely on the assumption that the semantics of integrity is two-valued.

4   Inconsistency-Tolerant Knowledge Assimilation

As already indicated, our focus is on the KA tasks of integrity maintenance across updates, satisfaction of view update requests, and reparation of violated integrity constraints. Common to each of them and also other KA tasks is that they generate updates as candidate solutions where the integrity of the state obtained by executing such an update is one of possibly several filter criteria for distinguishing valid candidates. Other criteria typically ask for minimality of (the effect of) updates, or use some additional preference ordering, to select among valid candidates. For instance, integrity maintenance may sanction a given update after having checked it successfully for integrity preservation, or otherwise either reject or modify it so that integrity remains invariant.

Since integrity checking is an integral part of KA, the requirement of total satisfaction of all constraints has traditionally been postulated also by all methods for tackling the mentioned tasks. However, this requirement appears as unrealistic for KA in general as for mere integrity checking. In fact, it can be abandoned just as well, as shown in theorem 5 below. The latter relies on the following definition, which in turn recurs on definition 5a.

Definition 6
A KA method $\mathcal{K}$ is inconsistency-tolerant *if each update generated for tackling the task of $\mathcal{K}$ preserves integrity.*

Similar to definitions 1 and 4, definition 6 is as abstract as to apply to virtually all methods in the literature. We repeat that such methods originally have not been meant to be applied in case the current database state is inconsistent with its constraints. Strictly speaking, they are not even defined for such situations. However, the clue of inconsistency-tolerant methods is that, by definition, they produce reliable results even when they are run in situations for which they originally have not been thought for. The justification for the definition of inconsistency tolerance is that many methods turn out to comply with it. Thus, definition 6 provides a basis for KA to be applicable also if the underlying database is not consistent with its integrity constraints. In particular, the generated updates still are going to achieve what they are supposed to achieve. More precisely, view updating methods compute updates that make update requests true, and repair methods turn violated cases of constraints into satisfied cases, while the overall state of integrity is not exacerbated by the respective updates.

We remark that theorem 3 does not readily provide a means to test a given KA method $\mathcal{K}$ for inconsistency tolerance, because definition 6 asks that *each* update that ever might be generated by $\mathcal{K}$ be integrity-preserving. It could be said that definition 6 should be relaxed to the extent that not all, but just one of the generated updates would have to be inconsistency-tolerant. Then, a further test by an inconsistency-tolerant integrity checking method could act as a filter for eliminating updates that would cause violation. However, that would in fact amount to the definition of a modified KA method, extended by an inconsistency-tolerant integrity checking method. The following theorem reflects the usefulness of such integrity checking methods for KA.

Theorem 5. *Each KA method that uses an inconsistency-tolerant method to check updates for not causing violation is inconsistency-tolerant.*

Proof. Straightforward from definition 6 and theorems 2 and 3.   □

4.1 Inconsistency-Tolerant View Updating

With theorem 5, it is possible to identify several known view update methods as inconsistency-tolerant, due to their use of suitable integrity checking methods. Among them are the view updating methods in [5] and [12,13], as stated in theorem 6 below. For convenience, let us name them $\mathcal{D}ec$ and $\mathcal{GL}$, respectively.

Theorem 6

a) The view update method $\mathcal{D}ec$ is inconsistency-tolerant.
b) The view update method $\mathcal{GL}$ is inconsistency-tolerant.

Proof
a) $\mathcal{D}ec$ uses the inconsistency-tolerant integrity checking method in [4] for filtering out generated update candidates that would cause violation.   □

b) $\mathcal{GL}$ uses the inconsistency-tolerant integrity checking method in [20] for filtering out generated update candidates that would cause violation.   □

A related method by Kakas and Mancarella is described in [16,17]. For convenience, let us name it $\mathcal{KM}$. It does not use any integrity checking method as a separate module, as do the methods in [5,12,13], hence theorem 5 is not applicable. However, the inconsistency tolerance of $\mathcal{KM}$ can be tracked down as outlined in the remainder of this subsection.

For satisfying a given view update request, $\mathcal{KM}$ explores a possibly nested search space of 'abductive' derivations and 'consistency' derivations. Roughly, the goal of abductive derivations is to find successful deductions of a requested update, by which base table updates that satisfy the request are obtained; consistency derivations check these updates for integrity. Each update obtained that way consists of a set of positive and a set of negative literals that are all ground. Positive literals correspond to insertions, negative ones to deletions of rows in base relations. For more details, we refer the reader to the original papers as cited above. It may suffice here to mention that, for $\mathcal{KM}$, all constraints are assumed to be represented by denial clauses, so that they can be used as candidate input in consistency derivations.

It is easy to verify that, for an update request R, each update U computed by $\mathcal{KM}$ satisfies R, i.e., R is true in D^U even if some constraint is violated in D. What is at stake is the preservation of integrity in D^U, for each case that is satisfied in D, while unknown or irrelevant cases that are violated in D may remain to be violated in D^U. The following theorem states that satisfied cases are preserved by $\mathcal{KM}$.

Theorem 7. *The view update method $\mathcal{KM}$ is inconsistency-tolerant.*

Proof. By theorems 2 and 4, it suffices to show that each update computed by $\mathcal{KM}$ does not cause violation. To initiate a *reductio ad absurdum* argument, suppose that, for some update request in some database with some integrity theory I, $\mathcal{KM}$ computes an update U that causes violation. Then, by definitions 4 and 5a, there is a case D' of some constraint C in I such that such that $D(C') = sat$ and $D^U(C') = vio$. Thus, *a fortiori*, $D(C) = sat$ and $D^U(C) = vio$. Hence, by the definition of $\mathcal{KM}$, there is a consistency derivation δ rooted at one of the base literals in U, that uses C as input clause in its first step and terminates by deducing the empty clause. However, termination of any consistency derivation with the empty clause signals inconsistency, i.e., constraint violation. Hence, by definition, $\mathcal{KM}$ rejects U, because δ indicates that its root causes violation of C. Thus, $\mathcal{KM}$ never computes updates that would cause violation. □

4.2 Inconsistency-Tolerant Repairing

Repairing a database that is inconsistent with its integrity constraints can be difficult, for several reasons. For instance, there may be (too) many alternatives of possible repairs, even if a lot of options are filtered out by minimality or other selection criteria. To choose suitable filtering criteria can be a significant problem on its own already. Also, repairs can be prohibitively costly, due to the complexity of constraints and intransparent interactions between them and the stored data; cf., e.g., [21]. And, worse, the existence of unknown inconsistencies (which is common in practice) may completely foreclose the repair of known constraint violations, under the traditional inconsistency-intolerant semantics of classical first-order logic.

To see this, suppose that, for a database D, C_0 is a case of some constraint C, the violation of which is unknown, i.e., both $D \cup \{C_0\}$ and $D \cup \{C\}$ are inconsistent. Further, C_1 be a known violated case of the same or some other constraint, which is to be repaired. In general, all integrity constraints need to be taken into account for repairing violations, due to possible interdependencies between them. However, classical logic does not sanction any result of reasoning in an inconsistent theory, since anything (and thus nothing reliable at all) may follow from inconsistency. Thus, no repair of any known inconsistency can be trusted, unless it can be ensured that there is no unknown inconsistency. So, since it is hard to know about the unknown, repair may seem to be a hopeless task, in general.

Fortunately, inconsistency tolerance comes to the rescue. In the preceding example, an update U_1 such that $D^{U_1}(C_1) = sat$ can be obtained by running any inconsistency-tolerant view update method on the request to make C_1 true. Each terminating method will produce such an update U_1, independent of the integrity status of C_0, while all other cases of constraints that are satisfied in D remain satisfied in D^{U_1}.

For a database D, inconsistency-tolerant view updating can in general be used either for repairing all violated constraints in one go, or, if that task is too big, for repairing violated (cases of) constraints incrementally, as follows. W.l.o.g., suppose that all constraints $C_1,..., C_n$ ($n > 1$) are represented by the denial-like clauses *violated* $\leftarrow B_i$ ($1 \le i \le n$), where *violated* be a distinguished view predicate that is not used for any relation in D, and B_i is an existentially closed formula with predicates defined in D. A constraint of that form is satisfied if and only if B_i is not true in the given database state. So, to repair all violated constraints in one go, $\sim$*violated* can be be requested as a view update in the database $D \cup \{C_1,..., C_n\}$, asking that *violated* be not true (cf. [10]).

It is easy to see that any terminating inconsistency-tolerant view update method will return the required repair.

Otherwise, the following incremental approach may be tried. For each i at a time, the update request $\sim B_i$ be processed and satisfied, if possible, by an integrity-preserving update, computed by an inconsistency-tolerant view update method. Clearly, the end result will in general depend on the sequential ordering of the C_i. Here, as with any policy for choosing among several candidate updates for satisfying a request, application-specific considerations may help.

For instance, suppose the management of some enterprise has decided to dissolve their research department. In the database of that enterprise, let a foreign key constraint of the *works-in(EMP,DEPT)* relation ask for the occurrence of the second attribute's value of each tuple of *works-in* in the primary key's value of some tuple in the *dept* relation. To repair the cases of this constraint that have become violated by the deletion of the tuple *dept(research)*, the following updates can be performed.

First, a downsized new research-oriented department is established by inserting the fact *dept(investigation)*. No violation of any key constraint is caused by that. Then, for each employee e of the defunct research department, the tuple *works-in(e, research)* either is dropped (i.e., e is fired) or replaced by *works-in(e, investigation)*, or replaced by *works-in(e, development)*, for some already existing department *development*.

As an aside, we remark that the last two of the three candidate repairs of this example, which is quite typical for reorganizing enterprise departments, may also serve to criticize the adequacy of the usual minimality criteria in the literature, since they comply with none of them.

More importantly, note that each such repair is not acceptable by any inconsistency-intolerant method that would insist on total integrity satisfaction, because some violated cases of constraints are likely to survive across updates. However, each repair that does not cause violation of any of the mentioned constraints is sanctioned by inconsistency-tolerant methods that check the preservation of all satisfied cases.

5   Conclusions

The semantic consistency of data is a major concern of knowledge engineering. Consistency requirements usually are expressed by integrity constraints. Knowledge assimilation methods are employed for preserving constraint satisfaction across changes. To require total satisfaction, as most known approaches do, is unrealistic. To relax that, we have revisited and extended a notion of inconsistency tolerance. We have shown that it is possible to use existing KA methods for checking, preserving and maintaining integrity upon updates, for satisfying view update requests and for repairing violated constraints, even if the knowledge suffers from inconsistencies.

Arguably, our concept of inconsistency tolerance is less complicated and more effective than the one associated to the field of consistent query answering (CQA) [1] and others, as documented in [2]. The latter of course have several other merits of their own that are not questioned by inconsistency tolerance as discussed in this paper. In fact, we expect that our work, and in particular our notion of inconsistency-tolerant repair, can be beneficial for the further development of CQA. We intend to look into this in future research. We also intend to investigate the capacity of inconsistency tolerance of advanced procedures such as in [11].

References

1. Arenas, M., Bertossi, L., Chomicki, J.: Consistent query answers in inconsistent databases. In: Proc. 18th PODS, pp. 68–79. ACM Press, New York (1999)
2. Bertossi, L., Hunter, A., Schaub, T.: Inconsistency Tolerance. LNCS, vol. 3300. Springer, Heidelberg (2005)
3. Christiansen, H., Martinenghi, D.: On simplification of database integrity constraints. Fundam. Inform. 71(4), 371–417 (2006)
4. Decker, H.: Integrity enforcement on deductive databases. In: Proc. 1st EDS, pp. 381–395. Benjamin/Cummings (1987)
5. Decker, H.: Drawing updates from derivations. In: Kanellakis, P.C., Abiteboul, S. (eds.) ICDT 1990. LNCS, vol. 470, pp. 437–451. Springer, Heidelberg (1990)
6. Decker, H.: Some notes on knowledge assimilation in deductive databases. In: Kifer, M., Voronkov, A., Freitag, B., Decker, H. (eds.) Dagstuhl Seminar 1997, DYNAMICS 1997, and ILPS-WS 1997. LNCS, vol. 1472, pp. 249–286. Springer, Heidelberg (1998)
7. Decker, H., Martinenghi, D.: Checking violation tolerance of approaches to database integrity. In: Yakhno, T., Neuhold, E.J. (eds.) ADVIS 2006. LNCS, vol. 4243, pp. 139–148. Springer, Heidelberg (2006)
8. Decker, H., Martinenghi, D.: A relaxed approach to integrity and inconsistency in databases. In: Hermann, M., Voronkov, A. (eds.) LPAR 2006. LNCS (LNAI), vol. 4246, pp. 287–301. Springer, Heidelberg (2006)
9. Decker, H., Martinenghi, D.: Getting rid of straitjackets for flexible integrity checking. In: Proc. DEXA 2007 Workshop FlexDBIST, pp. 360–364. IEEE Computer Soceity, Los Alamitos (2007)
10. Decker, H., Teniente, E., Urpí, T.: How to Tackle Schema Validation by View Updating. In: Apers, P.M.G., Bouzeghoub, M., Gardarin, G. (eds.) EDBT 1996. LNCS, vol. 1057, pp. 535–549. Springer, Heidelberg (1996)
11. Dung, P.M., Kowalski, R., Toni, F.: Dialectic proof procedures for assumption-based admissible argumentation. Artif. Intell. 170(2), 114–159 (2006)
12. Guessoum, A., Lloyd, J.: Updating knowledge bases. New Generation Computing 8(1), 71–89 (1990)
13. Guessoum, A., Lloyd, J.: Updating knowledge bases II. New Generation Computing 10(1), 73–100 (1991)
14. Gupta, A., Sagiv, Y., Ullman, J., Widom, J.: Constraint checking with partial information. In: Proc. 13th PODS, pp. 45–55. ACM Press, New York (1994)
15. Kakas, A., Kowalski, R.A., Toni, F.: The role of abduction in logic programming. In: Gabbay, D., Hogger, C., Robinson, J. (eds.) Handbook of Logic in Artificial Intelligence and Logic Programming, vol. 5, pp. 235–324. Oxford University Press, Oxford (1998)
16. Kakas, A., Mancarella, P.: Database updates through abduction. In: Proc. 16th VLDB, pp. 650–661. Morgan Kaufmann, San Francisco (1990)
17. Kakas, A., Mancarella, P.: Knowledge assimilation and abduction. In: Martins, J.P., Reinfrank, M. (eds.) ECAI-WS 1990. LNCS, vol. 515, pp. 54–70. Springer, Heidelberg (1991)
18. Kowalski, R.A.: Logic for Problem Solving. Elsevier, Amsterdam (1979)
19. Lee, S.Y., Ling, T.W.: Further improvements on integrity constraint checking for stratifiable deductive databases. In: Proc. 22nd VLDB, pp. 495–505. Morgan Kaufmann, San Francisco (1996)

20. Lloyd, J.W., Sonenberg, L., Topor, R.W.: Integrity constraint checking in stratified databases. JLP 4(4), 331–343 (1987)
21. Lopatenko, A., Bertossi, L.: Complexity of consistent query answering in databases under cardinality-based and incremental repair semantics. In: Schwentick, T., Suciu, D. (eds.) ICDT 2007. LNCS, vol. 4353, pp. 179–193. Springer, Heidelberg (2007)
22. Martinenghi, D., Christiansen, H., Decker, H.: Integrity checking and maintenance in relational and deductive databases and beyond. In: Ma, Z. (ed.) Intelligent Databases: Technologies and Applications, pp. 238–285. Idea Group (2006)
23. Miyachi, T., Kunifuji, S., Kitakami, H., Furukawa, K., Takeuchi, A., Yokota, H.: A knowledge assimilation method for logic databases. New Generation Comput. 2(4), 385–404 (1984)
24. Nicolas, J.-M.: Logic for improving integrity checking in relational data bases. Acta Informatica 18, 227–253 (1982)
25. Sadri, F., Kowalski, R.: A theorem-proving approach to database integrity. In: Minker, J. (ed.) Foundations of Deductive Databases and Logic Programming, pp. 313–362. Morgan Kaufmann, San Francisco (1988)

Improving Cutting-Stock Plans with Multi-objective Genetic Algorithm

Ramiro Varela[1], Cesar Muñoz[1], María Sierra[1], and Inés González-Rodríguez[2]

[1] University of Oviedo. Department of Computing. Artificial Intelligence Centre
Campus of Viesques, 33271 Gijón, Spain
{ramiro,UO0166793}@uniovi.es, maríasierra@aic.uniovi.es
[2] University of Cantabria. Department of Mathematics, Statistics and Computing, Los Castros
s/n,39005, Santander, Spain
ines.gonzalez@unican.es

Abstract. In this paper, we confront a variant of the cutting-stock problem with multiple objectives. It is an actual problem of an industry that manufactures plastic rolls under customers' demands. The starting point is a solution calculated by a heuristic algorithm, termed SHRP that aims mainly at optimizing the two main objectives, i.e. the number of cuts and the number of different patterns; then the proposed multi-objective genetic algorithm tries to optimize other secondary objectives such as changeovers, completion times of orders weighted by priorities and open stacks. We report experimental results showing that the multi-objective genetic algorithm is able to improve the solutions obtained by SHRP on the secondary objectives and also that it offers a number of non dominated solutions, so that the expert can chose one of them according to his preferences at the time of cutting the orders of a set of customers.

Keywords: Multi-objective optimization, Genetic Algorithms, Cutting Stock, Meta-heuristics.

1 Introduction

This paper deals with a real Cutting-Stock Problem (CSP) in manufacturing plastic rolls. The problem is a variant of the classic CSP, as it is usually considered in the literature, with additional constraints and objective functions. We have solved this problem in [6,8] by means of a GRASP algorithm [7] termed Sequential Heuristic Randomized Procedure (SHRP), which is similar to other approaches such as the SVC algorithm proposed in [1]. Even though SHRP tries to optimize all objective functions, in practice it is mainly effective in optimizing the main two ones: the number of cuts and the number of patterns. It is due to SHRP considering all objective functions in a hierarchical way that it pays much more attention to the first two ones than to the remaining.

In this work we propose a Multi-Objective Genetic Algorithm (MOGA) that starts from a solution computed by SHRP algorithm and tries to improve it regarding three secondary objectives: the cost due to changeovers or setups, the orders' completion time weighted by priorities and the maximum number of open stacks.

J. Filipe et al. (Eds.): ICSOFT/ENASE 2007, CCIS 22, pp. 332–344, 2008.
© Springer-Verlag Berlin Heidelberg 2008

The paper is organized as follows. Next section is devoted to briefly describe the production process of plastic rolls. In section 3, the problem formulation is given. As this formulation is rather complex, in section 4 we have introduced an example to clarify both the formulation and the whole process of obtaining a solution. In section 3, the main characteristics of the proposed MOGA are described. In section 4, we report results from a experimental study. Finally, in section 5, we summarize the main conclusions and some ideas for future work.

2   The Production Process

Figure 1 shows the schema of the cutting machine. A number of rolls are cut at the same time from a big roll according to a cutting pattern. Each roll is supported by a set of cutting knives and a pressure roller of the appropriate size. At each of the borders, a small amount of product should be discarded; therefore there is a maximum width that can be used from the big roll. There is also a minimum width, due to the capability of the machine to manage trim loss. Moreover, a maximum number of rolls can be cut at the same time. When the next cut requires a different cutting pattern, the process incurs in a setup cost due to changing cutting knives and pressure rollers.

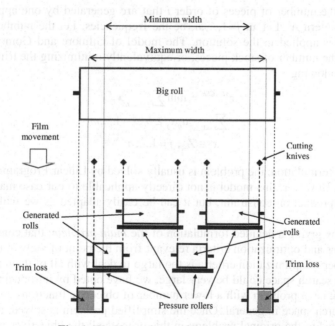

Fig. 1. Working schema of the cutting-machine

The problem has also a number of constraints and optimization objectives that make it different from conventional formulations. For example underproduction is not allowed and the only possibility for overproduction is a stock declared by the expert. Once a cut is completed, the rolls are packed into stacks. The stack size is fixed for each roll width, so a given order is composed by a number of stacks, maybe the last one being uncompleted. Naturally, only when a stack is completed it is taken away from the proximity of the cutting machine. So, minimizing the number of open stacks is also convenient in order to facilitate the production process. Moreover, some orders have more priority than others. Consequently the delivery time of orders pondered by the client priority is an important criterion as well.

3 Problem Formulation

The problem is a variant of the *One Dimensional Cutting-Stock Problem*, also denoted 1D-CSP. In [2] the first model is proposed for this problem by of Gilmore and Gomory. It is defined by the following data: $(m, L, l=(l_1,...,l_m), b = (b_1,...,b_m))$, where L denotes the length of each stock piece (here the width of the big roll), m denotes the number of piece types (orders) and for each type $i=1,...,m$, l_i is the piece length (roll width), and b_i is the order demand. A *cutting pattern* describes how many items of each type are cut from a stock length. Let column vectors $A^j=(a_{1j},...,a_{mj}) \in \mathbf{Z}_+^m$, $j=1,...,n$, represent all possible valid cutting patterns, i.e. those satisfying

$$\sum_{i=1,...,m} a_{ij} l_i \le L$$

where a_{ij} is the number of pieces of order i that are generated by one application of the cutting pattern A^j. Let x_j, $j=1,...,n$, be the frequencies, i.e. the number of times each pattern is applied in the solution. The model of Gilmore and Gomory aims at minimizing the number of stock pieces, or equivalently minimizing the trim-loss, and is stated as following

$$z^{1D-CSP} = \min \sum_{j=1,...,n} x_j$$
$$s.t. \quad \sum_{j=1,...,n} a_{ij} x_j \ge b_i, i = 1,...,m$$
$$x_j \in Z_+, j = 1,...,n$$

From this formulation, the problem is usually solved by Linear Programming based methods [9]. However, this model is not directly applicable to our case mainly due to the non-overproduction constraint, but it can be easily adapted as we will see in the sequel.

We start by giving a detailed formulation of the *main problem*; that considering all characteristics and optimization criteria relevant from the point of view of the experts. As the number of optimization criteria is too large to deal with all of them at the same time and the search space could be very large, we have opted by introducing a *simplified problem*; i.e. a problem with a lower number of objective functions and also with a smaller search space in general. Once the simplified problem is solved, the solution will be adapted to the original problem; in this process all the objectives will be considered.

3.1 The Main Problem

In order to clarify the problem definition, we present the data of the machine environment and the clients' orders, the form and semantics of a problem solution, the problem constraints and the optimization criteria in the hierarchical order in which they are usually considered by the expert.

Given

- The set of parameters of the cutting machine: the maximum width of a cut L_{max}, the minimum width of a cut L_{min}, the maximum number of rolls that can be generated in a cut C_{max}, the minimum and the maximum width of a single roll, W_{min} and W_{max} respectively, and the increment of width ΔW between two consecutive permitted roll widths.
- The setup costs. There is an elementary setup cost SC and some rules given by the expert that allows calculating the total setup cost from a configuration of the cutting machine to the next one. The setup cost is due to roller and cutter changes as follows. The cost of putting in or taking off a pressure-roller is SC; the cost of putting in an additional knife is $3SC$, and the cost of dismounting a knife is $2SC$.
- The types of pressure-rollers $PR = \{PR_1,...,PR_p\}$ and the mapping F_{PR} from roll widths to pressure-rollers.
- The mapping F_{ST} from roll widths to stack sizes or number of rolls in each stack unit.
- The orders *description* given by $(M=\{1,...,m\}$, $b = (b_1,...,b_m)$, $l=(l_1,...,l_m)$, $p=(p_1,...,p_m))$ where for each order $i = 1,...,m$, b_i denotes the number of rolls, l_i denotes the width of the rolls and p_i the order priority.
- The stock *allowed* for overproduction $(S =\{m+1,...,m+s\}$, $bs = (b_{m+1},...,b_{m+s})$, $ls = (l_{m+1},...,l_{m+s}))$ where for each $i=1,...,s$, b_{m+i} denotes the number of rolls of type $m+i$ allowed for overproduction and l_{m+i} denotes the width of these rolls.
- The set of feasible cutting patterns, for the orders and stock given, $\mathbf{A}$ where each $A^j \in \mathbf{A}$ is, $A^j=(a_{1j},...,a_{mj},a_{(m+1)j},...,a_{(m+s)j}) \in Z_+^{m+s}$ and denotes that, for each $i=1,...,m+s$, a_{ij} rolls of order i are cut each time the cutting pattern A^j is applied. A cutting pattern A^j is feasible if and only if both of the following conditions hold

$$L_{min} \le L_j = \sum_{i \in M \cup S} a_{ij}l_i \le L_{max}, \quad C_j = \sum_{i \in M \cup S} a_{ij} \le C_{max}$$

being L_j and C_j the total width and the number of rolls of pattern A^j respectively. $D_j = L_{max} - L_j$ denotes the trim-loss of the cutting pattern.

The objective is to obtain a *cutting plan* (Π, x), where $\Pi=(A^1,...,A^{|\Pi|}) \in \mathbf{A}^{|\Pi|}$ and $x = (x_1,...,x_{|\Pi|}) \in Z_+^{|\Pi|}$ denotes the pattern frequencies. The cutting patterns of Π are applied sequentially, each one the number of times indicated by its frequency. A^j_l, $0 \le j \le |\Pi|$, $0 \le l \le x_j$, denotes the lth cut corresponding to pattern A^j and $CI(A^j_l)$ is the cut index defined as

$$CI(A^j_l)= \sum_{k=1,...,j-1} x_k +l$$

Given an order $i \in M$ its first roll is generated in cut A^j_l such that A^j is the first pattern of Π with $a_{ij} \ne 0$, this cut is denoted $CU_{start}(i)$. Analogously, the last roll of order i is generated in cut A^k_{xk} so that A^k is the last pattern of Π with $a_{ik} \ne 0$, this cut is denoted $CU_{end}(i)$.

As we have considered feasible cutting patterns, the only constraint that should be required to a solution is the following

- The set of rolls generated by the application of the cutting plan (Π, x) should be composed by all rolls from the orders and, eventually, by a number of rolls from the stock. That is, let s_i is the number of rolls of stock $i \in S$ in the solution

$$\forall i \in S, \quad s_i = \sum_{A^j \in \Pi} a_{ij} x_j$$

Then, the constraint can be expressed as follows:

$$\forall i \in M \quad \sum_{A^j \in \Pi} a_{ij} x_j = b_i, \quad \forall i \in S, \quad 0 \le s_i \le b_i$$

Regarding objective functions, as we have remarked, we consider two main functions

1. Minimize the number of cuts, given by $\sum_{j=1,\dots,|\Pi|} x_j$. The optimum value is denoted $z^{1D\text{-}CSP}$.

2. Minimize the setup cost, given by $\sum_{j=1,\dots,|\Pi|} SU(A^{j-1}, A^j)$, where $SU(A^{j-1}, A^j)$ denotes the setup cost from pattern A^{j-1} to pattern A^j calculated as it is indicated above. Configuration A^0 refers to the situation of the cutting machine previous to the first cut.

And two secondary functions

3. Minimize the completion times of orders weighted by their priorities given by

$$\sum_{i \in M} P_i \, CI\big(CU_{end}(i)\big)$$

4. Minimize the maximum number of open stacks along the cut sequence. Let $R(i, A^j_l)$ denote the number of rolls of order i generated from the beginning up to completion of cut A^j_l.

$$R\big(i, A^j_l\big) = \sum_{k=1,\dots,j-1} a_{ik} x_k + a_{ij} l$$

and let $OS(i, A^j_l)$ be 1 if after cut A^j_l there is an open stack of order i and 0 otherwise. Then, the maximum number of open stacks along the cut sequence is given by

$$\max_{j=1,\dots,|\Pi|, l=0,\dots,x_j} \sum_{i \in M} OS\big(i, A^j_l\big)$$

3.2 The Simplified Problem

In the main problem, as formulated in the previous section, it is often the case that two or more orders have the same width or a stock has the same width as one of the orders. So, from the point of view of the cutting process, two cutting patterns A^i and A^j are equivalent if both patterns define the cutting of the same number of rolls of the same sizes, i.e. given the set of widths $L = \{l_e, e \in M \cup S\}$, with cardinal $|L| = ms$, $ms \le m + s$ we have

$$A^i \equiv A^j \Leftrightarrow \sum_{k=0,\dots,m+s, \, l_k = l} a_{ki} = \sum_{k=0,\dots,m+s, \, l_k = l} a_{kj}, \forall l \in L$$

Now the simplified problem can be stated as follows.

Given

- The set of parameters of the cutting machine, the setup costs, the types of pressure-rollers and mapping F_{PR}: as they are in the main problem and the mapping function F_{ST} as they are in the main problem.
- The simplified orders description given by $(M'=\{1,...,m'\}$, $b' = (b'_1,...,b'_{m'})$, $l'=(l'_1,...,l'_{m'}))$, where for each order $i = 1,...,m'$, b'_i denotes the number of rolls and $l'_i \in L$ denotes the width of the rolls. The simplified orders list b' are obtained from the original order list b so as

$$b'_i = \sum\nolimits_{k=1,...,m\ l_k=l'_i} b_k$$

- The stock allowed for overproduction $(S' =\{m'+1,...,m'+s\}$, $bs' = (b'_{m'+1},...,b'_{m'+s})$, $ls' = (l'_{m'+1},...,l'_{m'+s}))$ where for each $i=1,...,s$, $b'_{m'+i} = b_{m+i}$ denotes the number of rolls of type $m'+i$ allowed for overproduction and $l'_{m'+i} = l_{m+i} \in L$ denotes the width of these rolls (notice that two different stock orders cannot have the same width). Here both l' and ls' are lists with no repeated elements, so they can be seen as sets such that $l' \cup ls'=L$, although, it is possible that $l' \cap ls' \neq \emptyset$. In what follows, we assume L to be ordered, beginning with $l'_1,...,l'_{m'}$ followed by the elements from ls' that do not belong to l'. $L=(l'_1,......, l'_{ms})$, $ms \leq m'+s$.
- The set of simplified feasible cutting patterns for the simplified orders and stock given, E, obtained from the set of feasible cutting patterns for the original problem A, $|E| \leq |A|$, where every $E^j \in E$ is $E^j=(e_{1j},...,e_{msj}) \in Z_+^{ms}$ meaning that, for each $i=1,...,ms$, e_{ij} rolls of width l'_i are cut each time the cutting pattern E^j is applied. In other words, each element of E is an equivalence class of the quotient set of A with the above relation, so it is a simplified representation of a number of cutting patterns of A.

The objective is to obtain a *simplified cutting plan* (Π',x'), where $\Pi'=(E^1,...,E^{|\Pi'|}) \in E^{|\Pi'|}$ and $x'=(x'_1,...,x'_{|\Pi'|}) \in Z_+^{|\Pi'|}$ denotes the pattern frequencies.

As all the simplified cutting patterns are feasible, the only constraint that should be requited to a solution is the following

- The set of rolls generated by the application of the simplified cutting plan (Π',x') should be composed by all rolls from the orders and, eventually, by a number of rolls from the stock. That is, let s'_i the number of rolls of stock of width l'_i in the solution, being 0 if there is no $m'+k \in S'$ such that $l'_i=l'_{m'+k}$,

$$\forall i \in \{m'+1,..,ms\},\quad s'_i = \sum_{E^j \in \Pi'} e_{ij}x'_j$$

Then, the constraint can be expressed as follows:

$$\forall i \in M',\quad \sum_{E^j \in \Pi'} e_{ij}x'_j = b'_i+s'_i,$$

$$0 \leq s'_i \leq b'_{m'+k}$$

The objective functions are

1.   Minimize the number of cuts calculated by $\sum_{j=1,...,|\Pi'|} x'_j$.
2.   Minimize the number of simplified cutting patterns $|\Pi'|$.
3.   Maximize the amount of stock generated, that is $\sum_{i=1,...,ms} l'_i s'_i$, so the trim-loss is minimized for a given number of cuts.

Now let us to clarify how a solution of the simplified problem can be transformed in a solution to the main problem. To do so, we have to map each simple cut from a simplified pattern E^j to any of the cuts of pattern A^k of the equivalence class defined by E^j. In doing so, we can consider different orderings in the simplified cutting plan, and also different orderings between the single cuts derived from a simplified cutting pattern, in order to satisfy all the optimization criteria of the main problem. As we can observe, objectives 1 and 3 are the same in both problems, but objective 2 is different. The reason to consider objective 2 in the simplified problem is that minimizing the number of patterns $|\Pi'|$ it is expected that the setup cost of the main problem is to be minimized as well. This is because the setup cost between two consecutive cuts A^k and A^l of the main problem is null if both A^k and A^l belongs to the same equivalence class E^j.

To solve this simplified problem, in [6,8] we have proposed a GRASP algorithm. Then, the solution given by this algorithm is transformed into a solution to the main problem by a greedy algorithm that assigns items to actual orders so as to optimize objectives 2, 3, and 4 in hierarchical order, while keeping the values of the first two objectives. To be more precise, we clarify, in the next section, how a simplified solution is transformed into an actual solution by means of an example.

4   An Example

In this section we show an example to clarify the whole process of obtaining a cutting plan for the main problem and, in particular, how a simplified solution is transformed in a solution to the main problem by MOGA. The problem data and final results are displayed as they are by the application program.

Figure 2 shows an instance and the corresponding simplified problem. A real instance is given by a set of orders, each one defined by a client name, a client identification number, the number of rolls, the width of the rolls and the order priorities. Additionally, the maximum and minimum allowed width of a cut should be given, in this case 5500 and 5700 respectively and also a stock description to choose a number of rolls from if it is necessary to obtain valid cutting patterns. In this example up to 10 rolls of each width 1100, 450 and 1150 could be included in the cutting plan. Furthermore, some other parameters (not shown in Figures) are necessary, for instance, two additional data should be given to evaluate the number of open stacks and setup cost: the number of rolls that fit in a stack (mapping F_{ST}) and the correspondence between the size of pressure rollers and the width of the supported rolls (mapping F_{PR}).

Problem Data (Main Problem)							(Simplified Problem)	
ROLLS	WIDTH	ORDER	CLIENT	PRIORITY	Max. width	5700	ROLLS	WIDTH
20	600	20001	Client 1	2	Min. width	5500	30	600
10	600	20002	Client 2	2			28	850
15	850	20003	Client 3	1	Stock		15	950
13	850	20004	Client 4	1	1150	10	14	1350
15	950	20005	Client 5	1	550	10	20	550
14	1350	20006	Client 6	1	1500	10	33	900
20	550	20007	Client 7	1				
18	900	20008	Client 8	2				
15	900	20009	Client 9	1				

a) main instante

Cutting Plan for the Simplified Problem				
FREQUENCY	14	3	3	1
P	600	550	950	900
A	900	950	950	600
T	1350	950	900	600
T	850	550	900	1500
E	600	900	950	1500
R	850	900	900	550
N	550	900	-	-
PATTERN WIDTH	5700	5700	5550	5650
TRIM LOSS	0	0	150	50	500
			NUMBER OF PATTERNS	4
			NUMBER OF CUTS	21

b) simplified instance

Fig. 2. An example of problem data (main and simplified instance)

Here we have supposed that every stack contains 4 rolls and that the correspondence between pressure roller types and width rolls is the following: type 1 (0-645), type 2 (650-1045), type 3 (1050-1345), type 4 (1350-1695). All the allowed widths are multiples of 5 and the minimum width of a roll is 250 while the maximum is 1500. Finally, the maximum number of rolls in a pattern is 10.

As we can observe in Figure 1, the main instance with 10 orders is reduced to a simplified instance with only 6 orders. This is a conventional 1D-CSP instance with two additional constraints: the maximum number of rolls in a pattern and the minimum width of a pattern. Figure 2 shows a solution to the simplified problem with 21 cuts and 4 different patterns, where 3 stock rolls have been included in order that the last pattern to be valid.

Figure 3 shows the final solution to the main problem. The figure shows the order identifiers, where 0 represents to the stock. A solution is a sequence of cutting patterns, where each pattern represents not only a set of roll widths, but also the particular order the roll belongs to. The actual solution is obtained from a simplified solution by means of a greedy algorithm that firstly considers the whole set of individual cuts as they are expressed in the simplified solution. Then it assigns a customer order to each one of the roll widths in the simplified cuts, and finally considers all different actual patterns maintaining the order derived from the simplified solution. The MOGA proposed in this paper starts from this solution and tries to improve it by considering different ordering of the cutting patterns.

Cutting Plan for the Main Problem (roll widths and evaluation functions)							
FREQUENCY	8	2	4	3	3	1	
P	600	600	600	550	950	900	
A	900	900	900	950	950	600	
T	1350	1350	1350	950	900	600	
T	850	850	850	550	900	1500	
E	600	600	600	900	950	1500	
R	850	850	850	900	900	550	
N	550	550	550	900			
PATTERN WIDTH	5700	5700	5700	5700	5550	5650	
TRIM LOSS	0	0	0	0	150	50	500
CHANGEOVERS	28	0	0	4	5	10	47
OPEN STACKS	5-3-5-0-5-3-5-1	5-3	5-1-4-3	2-3-2	1-1-0	0	5
						WEIGHTED TIME	188
						NUMBER OF CUTS	21

Cutting Plan for the Main Problem (order identifiers)						
FREQUENCY	8	2	4	3	3	1
P	20001	20001	20002	20007	20005	20009
A	20008	20008	20008	20005	20005	20002
T	20006	20006	20006	20005	20009	20002
T	20003	20004	20004	20007	20009	0
E	20001	20001	20002	20008	20005	0
R	20003	20004	20004	20008	20009	0
N	20007	20007	20007	20008		

Fig. 3. A cutting plan for the problem of Figure 2

The changeover of each pattern refers to the cost of putting in and out cutting knives and pressure rollers from the previous pattern to the current one. As we can observe, the first pattern has a changeover cost of 28 because it is assumed that it is necessary to put in all the 7 cutting knives and 7 pressure rollers before this pattern. In practice, this is not often the case as a number of cutting knives and pressure rollers remain in the machine from previous cuts. Regarding open stacks, each column shows the number of them that remain incomplete in the proximity of the machine from a cut to the next one, i.e. when a stack gets full after a cut, or it is the last stack of an order, it is not considered.

5   Multi-objective Genetic Algorithm

According to the previous section, the encoding schema is a permutation of the set of patterns comprising a solution. So, each chromosome is a direct representation of a solution, which is an alternative to the initial solution produced by the greedy algorithm. The initial solution is the one of Figure 3 which is codified by chromosome (1 2 3 4 5 6 7 8 . . . 21), i.e. each gene represents a single cut. As objectives 2, 3 and 4 depend on the relative ordering of patterns and also on their absolute position in the chromosome sequence, we have used simple genetic order based operators [3] that maintain these characteristics from parents to offsprings.

The algorithm structure is quite similar to a conventional single GA: it uses generational replacement and roulette wheel selection. The main differences are due to its multi-objective nature. The MOGA maintains, apart from the current population, a set of non dominated chromosomes. This set is updated after each generation, so that it finally contains an approximation of the pareto frontier for the problem instance.

In order to assign a single fitness to each chromosome, the whole population is organized into dominant groups as it follows. The first group is comprised by the non dominated chromosomes. The second group is comprised by the non dominated chromosomes from the remaining population and so on. The individual fitness is assigned so that a chromosome in a group has a larger value than any chromosome in the subsequent groups. Moreover, inside each group, the fitness of a chromosome is adjusted by taking into account the number of chromosomes in its neighbourhood in the space defined by the three objective functions. The chromosomes' neighbors are those that are in the chromosome's *niche count*. The evaluation algorithm is as it follows.

Step 0. Set F to a value sufficiently large

Step 1. Determine all non-dominated chromosomes Pc from the current population and assign F to their fitness.

Step 2. Calculate each individual's *niche count* m_j:

$$m_j = \sum_{k \in P_c} sh(d_{jk})$$

where

$$sh(d_{jk}) = \begin{cases} 1 - (d_{jk}/\sigma_{share})^2 & \text{if } d_{jk} < \sigma_{share} \\ 0 & \text{otherwise} \end{cases}$$

and d_{jk} is the phenotypic distance between two individuals j and k in Pc and σ_{share} is the maximum phenotypic distance allowed between any two chromosomes of Pc to become members of a niche.

Step 3. Calculate the shared fitness value of each chromosome by dividing it fitness value by its niche count.

Step 4. Create the next non dominated group with the chromosomes of Pc, remove these chromosomes from the current population, set F to a value lower than the lowest fitness in Pc, go to step 1 and continue the process until the entire population is all sorted.

This evaluation algorithm is adapted from [10]. In this paper, G. Zhou and M. Gen propose a MOGA for the Multi-Criteria Minimum Spanning Tree (MCMSP). In the experimental study they consider only two objective functions.

In order to compute d_{jk} and σ_{share} values we normalize distances in each one of the three dimensions to take values in [0,1]; this requires calculating lower and upper bounds for each objective. The details of these calculations are given in [5]. Also, we have determined empirically that $\sigma_{share} = 0,5$ is a reasonable choice.

6 Experimental Study

In this section we present results from some runs of a prototype implemented in (Muñoz, 2006) for the problem instance of Figure 2. The program is coded in Builder C++

for Windows and the target machine was Pentium 4 at 3,2 Ghz. with HT and 1Gb of RAM.

In the first set of experiments, the MOGA starts from the solution of Figure 2. Table 1 summarizes the values of the three objective functions (changeovers, weighed time and maximum open stacks) for each of the solutions in the approximate pareto frontier obtained in three runs with different parameters. As we can observe, the quality of the solutions are in direct ratio with the processing time given to the MOGA. The values of objective functions for the initial solution of Figure 3 are 47/188/5, which is dominated by some of the solutions of Table 1. So, it is clear that it is possible to improve on secondary objectives in solutions obtained by procedure SHRP.

In the second set of experiments, we have taken 9 more simplified solutions, different from that of Figure 2b, and have applied MOGA to each of them with the same parameters as in the second run of Table 1. In these experiments, we have considered also the MOGA without fitness adjustment, i.e. by considering a niche count equal to 1 in all cases. The results are summarized in Table 2. As we can observe, in general, MOGA reaches better solutions with fitness adjustment, even though it takes a larger time (about 640 s. versus 600 s.). Only for instances 8 and 9 is the version without fitness adjustment equal or better.

On the contrary, for instance 1 the results without fitness adjustment are clearly much worse. These results show that MOGA is able to reach solutions better than the initial one. Here it is important to remark that the initial solution is not included into the initial population of MOGA and that this population is completely random; i.e. all cuts are randomly distributed, what usually translates into a very high changeover cost. In practice, good solutions tend to aggregate equal cuts consecutively in order to minimize changeovers. This fact could be exploited when generating the initial population in order to reduce the computation time required by MOGA.

Also, these results suggest that the neighborhood strategy should be reconsidered, in particular that a static value for parameter σ_{share} is not probably the best choice.

Table 1. Summary of results from three runs of MOGA starting from the solution of Figure 3 for the problem of Figure 1. Parameters of MOGA refer to /Population size/Number of generation/, the remaining Crossover probability/Mutation probability/σ_{share} are 0,9/0,1/0,5. Each cell shows the cost of /changeovers/weighed times/maximum open stacks.

Run	1	2	3
Pars.	/200/200/	/500/500/	/700/700/
Time(s.)	37	649	1930
Pareto frontier reached	49/188/6	47/176/6	39/172/6 (*)
	49/186/7	47/184/5 (*)	47/184/5 (*)
	44/196/5 (*)	39/179/7	
		45/184/6	

(*) These values represent solutions non-dominated by any other reached in all three runs

Table 2. Summary of results of MOGA starting from 9 different simplified solutions to the instance of Figure 2a with the same values of number of cuts (21) and patterns (4), except solution 6 which has 3 patterns, with different amount of stock generated. For each instance, two runs have been done with parameters /500/500/0,9/0,1/0,5, the first (Normal) in the same conditions as before; while in the second, the niche count is not computed but it is taken as 1 in all cases.

Inst.	Initial	Normal	Niche c. = 1
1	39/184/5	38/174/5 (*)	55/188/6
		38/197/4 (*)	42/212/8
			48/199/6
			50/197/5
			49/190/6
			52/190/5
			48/201/5
2	39/187/5	38/174/5 (*)	38/174/5 (*)
		38/197/4 (*)	
3	43/187/5	42/174/5 (*)	42/174/5 (*)
		42/197/4 (*)	
4	46/185/5	43/185/5 (*)	55/177/5 (*)
5	56/185/5	55/177/5 (*)	54/194/7 (*)
			63/182/6
6	36/186/5	38/177/5 (*)	38/177/5 (*)
7	56/192/5	50/188/5 (*)	50/203/5
			50/197/6
			50/195/7
8	55/213/5	61/193/5	56/181/5 (*)
		63/179/6 (*)	52/182/5 (*)
		70/188/5	67/173/5 (*)
		71/179/5	63/179/6 (*)
		60/180/6 (*)	62/180/6
		71/178/6	
		61/193/5	
9	42/201/6 (*)	54/180/6	51/177/7 (*)
		46/181/5 (*)	52/176/7 (*)
		55/180/5	44/181/6 (*)
		44/182/5 (*)	53/172/7 (*)
		60/179/6	54/169/7 (*)
		59/183/4 (*)	54/173/5 (*)

(*) These values represent solutions non-dominated by any other reached from the same simplified solution

7 Conclusions

In this paper we have proposed a multi-objective genetic algorithm (MOGA) which aims to improve solutions to a real cutting stock problem obtained previously by another heuristic algorithm. This heuristic algorithm, termed SHRP, focuses mainly on the two main objectives and considers them hierarchically. Then, the MOGA tries to improve other three secondary objectives at the same time, while keeping the values of the main objectives. We have presented some results over a real problem instance showing that the proposed MOGA is able to improve the secondary objective functions with respect to the initial solution, and that it offers the expert a variety of non-dominated solutions.

As future work, we plan reconsidering the MOGA strategy in order to make it more efficient and more flexible so that it can take into account the preferences of the experts with respect to each one of the objectives. In order to improve efficiency we will try to devise local search techniques and initialization strategies based on heuristic dispatching rules.

Also, we will consider alterative evolutionary strategies for multi-objective optimization [3] and other multi-objective search paradigms such as exact methods based on best first search [4]. In this way we could compare different strategies for this particular problem.

Acknowledgements. This research has been supported by contract CN-05-127 of the University of Oviedo and the company ERVISA, by the Principality of Asturias (FI-CYT) under research contract FC-06-BP04-021 and by the Spanish Ministry of Education and Science under project MEC-FEDER TIN2007-67466-C02-01.

References

1. Belov, G., Scheithauer, G.: Setup and Open Stacks Minimization in One-Dimensional Stock Cutting. INFORMS Journal of Computing (submitted, 2006)
2. Gilmore, P.C., Gomory, R.E.: A linear programming approach to the cutting stock problem. Operations Research 9, 849–859 (1961)
3. Goldberg, D.E.: Genetic Algorithms in Search, Optimization and Machine Learning. Addison-Wesley, Reading (1989)
4. Mandow, L., Pérez-de-la-Cruz, J.L.: A new approach to multiobjective A* search. In: IJCAI 2005 19th Int. Joint Conf. on Artificial Intelligence, pp. 218–223 (2005)
5. Muñoz, C.: A Multiobjective Evolutionary Algorithm to Compute Cutting Plans for Plastic Rolls. Degree project, University of Oviedo, School of Computing, Gijón (2006) (in Spanish)
6. Puente, J., Sierra, M., González-Rodríguez, I., Vela, C.R., Alonso, C., Varela, R.: An actual problem in optimizing plastic rolls cutting. Workshop on Planning, Scheuling and Temporal Reasoning. In: CAEPIA 2005, Santiago de Compostela (2005)
7. Resende, M.G.C., Ribeiro, G.C.: Greedy randomized adaptive search procedures. In: Glover, F., Kochenberg, G. (eds.) Handbook of Metaheuristics, pp. 219–249. Kluwer Academic Publishers, Dordrecht (2002)
8. Varela, R., Vela, C.R., Puente, J., Sierra, M.R., González-Rodríguez, I.: An effective solution for an actual cutting stock problem in manufacturing plastic rolls. In: Annals of Operations Research (submitted, 2007)
9. Umetani, S., Yagiura, M., Ibaraki, T.: One-dimensional cutting stock problem to minimize the number of different patterns. European Journal of Operational Research 146, 388–402 (2003)
10. Zhou, G., Gen, M.: Genetic algorithm approach on multi-criteria minimum spanning tree problem. European Journal of Operational Research 114, 141–152 (1999)

Knowledge Purpose and Visualization

Wladimir Bodrow and Vladimir Magalashvili

Department of Business Informatics
University of Applied Sciences Berlin
Treskowallee 8
10318 Berlin, Germany
w.bodrow@fhtw-berlin.de, vmagalashvili@googlemail.com

Abstract. Knowledge visualization is currently under investigation from different points of view especially because of its primary importance in research fields like Artificial Intelligence, Knowledge Management, Business Intelligence etc. and its growing value in the business practice. The concepts and technology of knowledge visualization in the presented contribution are considered from a purpose perspective which focuses on the interdependencies between different knowledge elements. This way the influence of these elements on each other in every particular situation can be visualized. This is crucial e.g. for decision making.

Keywords: Knowledge visualization, knowledge transfer, knowledge management.

1 Knowledge Transfer in Knowledge Management Process

Looking back onto more than ten years of knowledge management history we detected that there is still no universal definition of knowledge management. Some valuable attempts to define knowledge management were made by Probst [1], Davenport and Prusak [2], Nonaka and Takeuchi [3], Maier [4] etc. Most of today's accepted definitions describe knowledge management as creation, communication, and application of knowledge. The main goal of knowledge management is therefore to improve these processes. Outgoing from different perspectives and different aims correspondingly the descriptions of the activities in the knowledge management process vary significantly. But almost all of them [5, 4] emphasize knowledge transfer - which is also called sharing, diffusion, exchange, dissemination, or distribution, together with knowledge application as one of the most important activities in the knowledge management process. Below we will use the term transfer synonymously for sharing and exchanging of knowledge as well as for diffusion, dissemination and distribution of knowledge, knowing that sharing and exchange refer to bidirectional processes in opposition to dissemination, diffusion and distribution which represent the unidirectional (knowledge) flow. In respect to the presented research this difference is useless therefore only the transfer term will be applied.

The efficient transfer of knowledge has proven to be a difficult task. In this context the adequate visualization of knowledge can significantly improve its transfer. Therefore in the following analysis we will concentrate on this particular aspect of the knowledge transfer.

J. Filipe et al. (Eds.): ICSOFT/ENASE 2007, CCIS 22, pp. 345–356, 2008.
© Springer-Verlag Berlin Heidelberg 2008

2 Why Knowledge Visualization?

To answer this question we first take a quick look at some recently established changes within e-learning. A decade ago the re-orientation in e-learning to blended learning documents the inability of pure e-learning solutions to transfer knowledge to the students and to empower them to solve new problems without any assistance. Such an empowerment needs more then stupid memorizing of data and information, even if these are more or less appropriately visualized and submitted based on classical IT-tools. To understand the orientation on blended learning one has to understand, what is going on during the face-to-face teaching part in one blended learning course? Among other things the lecturer explains to the students why he and later they have to apply this particular approach or have to use a special interdependency between the elements in the task or solution under consideration. Above all he explains how such interdependencies are applicable to reach a particular purpose and how certain reasons may be involved in this process. Exactly this knowledge enables the students to select the appropriate approach or tool to solve the problem or task delegated to them.

A well-known saying, that a picture is worth 1000 words, leads us to the suggestion that the visualization of knowledge can increase the effectiveness of its representation, understanding and consequently of knowledge transfer. Visualization can be considered as a way of internalization [3] of knowledge (transformation from explicit to tacit knowledge). But what exactly do we mean by knowledge visualization? How does it differ from visualization of information or data? Knowledge visualization as opposed to information visualization is a rather new field within knowledge management research. It has received more attention recently due to the business's interests. There exist already some attempts to define knowledge visualization [6]. Two examples of such definitions are presented below.

3 Knowledge Visualization Today

Following Drosdol [7] knowledge visualization refers to "the result of transformation from information to knowledge, representation of connections and links, designing the space between information elements, development of meaning, creating meaningful structures fitting the contents, helping to generate new knowledge which can be used by people, staff, leaders, decision-makers".

Burkhard [8], [9], [10] defines knowledge visualization as "the use of visual representations to improve the transfer and creation of knowledge between at least two persons". Moreover he describes the difference between knowledge and information visualization. The latter is not trivially derivable from the presented definition. Information visualization also uses "visual representations to improve the transfer of knowledge", even if it's primary goal is to retrieve the information. According to Burkhard's definition it can be considered as knowledge visualization. The recipient (depending on their capabilities) can obtain or perhaps create new knowledge only by getting the visual information. There are many Software-tools that visualize a huge amount of data and information. Experience gathered in this field is very helpful i.e. for development of decision support systems. Obviously the presented definition is too general to be accepted as a definition for knowledge visualization.

Our approach compared to other definitions is not knowledge element-driven, but purpose-driven [11], [12]. It is based on an appropriate application of the different interrelations between the knowledge elements according to the selected purpose. We consider the knowledge element oriented approach as very similar to information or data visualization where the different declared attributes of the particular object can be visualized using graphics and other media. Alternatively we follow the idea, that the most important aspect in visualizing knowledge (especially!) is the multi-valence of the explanations of the interdependencies between knowledge elements. Consequential a context-dependent *explanation* of such interdependencies as well as a *definition* of their purposes from this perspective are the core characteristics of our approach.

4 How Do We Define Knowledge Visualization?

How can we define knowledge visualization from a knowledge management perspective (not from a view of cognitive psychology, pedagogy or graphic design)? Our aim is not to define a visualizing technique (like sketch, diagram, image etc.), but a general proper way for the representation of knowledge using visualization techniques.

If somebody tries to illustrate a solution of a complex (for example business) problem, they do not only visualize single elements of a transferred concept based on its attributes, but also the connections and/or interdependencies of these elements. However it is usually not enough for the recipient to understand the logic of the concept (and to accept the proposed solution). What the recipient needs is an explanation of those dependencies in respect to the task or problem to be solved. Why are the selected elements connected to each other in the considered case or in general? How do these visual dependencies help understand the knowledge to be transferred? Why does the knowledge have to be visualized based on a selected concept (motivation)? How is this visualization going to be helpful for the solution investigated and for other applications? Which role do the skills and preferences of both partners play in the particular knowledge transfer and its visualization?

From our point of view knowledge visualization has to answer these questions to be classified as such. Without explaining the meaning and purpose of the connections between the different visual elements, the visualization loses its value. It reduces to something like data or information visualization – visual representation of abstract data [13], [14], [15]. For instance according Card "information visualization is the use of computer-supported, interactive, visual representation of abstract data to amplify cognition" [13].

Summarizing the features and perspectives mentioned above the following definition of knowledge visualization can be derived:

Definition
Knowledge visualization is a set of graphical entities used to transfer knowledge from an expert to a person (or group of persons), which clarifies its complexity and explains the meaning and the purpose of the relevant interdependencies.

Firstly, according to the definition above the sender of knowledge can be both: human or artifact, whereas the recipient from today's perspective can be a single person or a group of persons.

Secondly, the visualization should represent a task or problem to be solved (e.g. business workflow process, structure of a business unit with its responsibilities etc.). This way it provides the answer for the question why the knowledge has to be transferred.

In this research we only consider the dependencies of the first order in the visualized structure (see Fig. 1.). That means we only analyze the connections (uni- or bidirectional) between different but single elements and not between groups of elements or indirect relations (n-way dependencies).

The connections can be considered from two perspectives:

Why this connection? – What is the purpose of this connection? Why does this connection have to be used? Which problems can be solved based on it?

Which particular dependency or influence is used in this connection? – It should explain the connection between two selected knowledge elements. Accordingly the dependency can be interpreted as a specialization for the more general or strategic formulated purpose of the single connection between two knowledge elements.

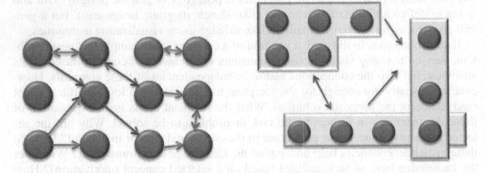

Fig. 1. Examples of first order (left) and second order dependencies between elements

According to the previous discussion we can define knowledge visualization formally as

KnowVis = F (E, D, P) where

F is a certain function of

E – a set of knowledge elements (different visual features as tables, charts, nodes of trees, circles etc.)

D – a set of dependencies/influences between knowledge elements

P – a purpose(s) of interdependencies.

From another perspective each dependency can be defined as

$D = f(e1, e2, s12, s21, p12, p21)$ where

$e1, e2$ are two knowledge elements from E

$p12, p21$ represent the corresponding purposes $s12, s21$ are the strengths of the influence of $e1$ on $e2$ and vice versa.

Sometimes one should only concentrate on the most important dependencies between knowledge elements in order to avoid extreme complexity in the visualized structure. Therefore it sounds reasonably that the connections receive such attributes as the "strength" of interdependency. Based on this attribute (strength) one can select for his analysis only connections with certain strength – for instance such with the most influence like in the Fig. 2.

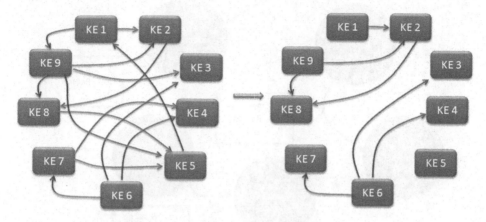

Fig. 2. Selection of connections with the strongest influence (right)

Our concept has something in common with the idea of Novak's concept maps [16]. Novak defines concept maps as tools for organizing and representing knowledge. They include concepts (enclosed in circles or boxes), and relationships between concepts or propositions. These relations are indicated by a connecting line and a linking word (often a verb).

But the key difference from Novak's to our concept is that each relation in knowledge visualization is provided by the explanation of its purpose. How does this explanation support the whole idea of knowledge transfer?

The choice of visualization technique certainly depends on the type of knowledge transferred and on the recipient's capabilities.

As just mentioned, knowledge visualization should clarify the purpose of the connections between visual entities. This does not mean that the recipient receives only one "right application". The given explanations will contain a description of how the sender would apply this knowledge. Those application suggestions will help the recipient to utilize the best practice by creating his own analogies and associations during his individual decision making. The way in which the obtained knowledge can be applied depends on the effectiveness of the visualization (choice of visual self-describing features, clear dependencies, etc.) and the intellectual (abstract thinking, logical conclusions, experience, etc.) capabilities of the recipient.

An example for purpose-driven knowledge visualizations is presented in Fig. 3. The explanations menus for all connections as shown above can overload the graphic. Therefore they should rather be implemented as context sensitive menu-boxes appearing after a mouse click on the connection to be clarified.

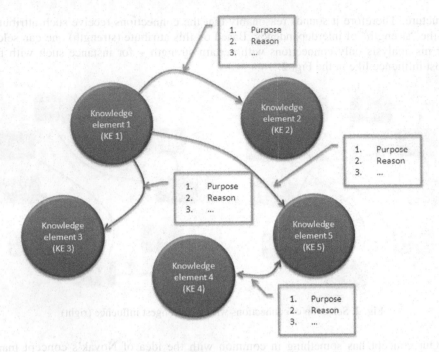

Fig. 3. The purpose-driven knowledge visualization metaphor

Which advantages can be expected from such visualizations?

Firstly, it is easier for the recipient to understand the knowledge transferred from the sender – presented elements and interdependencies between them allow the problem investigation from different perspectives and support this way new proposal for its solutions.

Secondly, described visualization makes the learning process easier and support better knowledge retention through integration of new knowledge into the already existing structure.

Moreover, this explanation of the dependencies and purposes of the relations will stimulate by recipients such creativity processes as associative, logical and analogical reasoning and consequently decision making.

5 Flexibility of the Approach and Creativity Promotion

In the proposed approach knowledge visualization, as mentioned above, consists of a set of knowledge elements and a number of connections between them with the explanation of their purposes. By allowing or disallowing to change the set of connections and the set of knowledge elements we can distinguish the following three levels of flexibility.

5.1 Rigid or Fixed Level of Visualization

Knowledge elements together with their connections and explanations are predefined and not editable. The visualized solution to a problem is to be accepted unconditionally without any possibility to change the connections between the knowledge elements.

This fixed visualization can be used, for example, in the learning approach based on the idea that learning should be placed in a problem-solving context (e.g. "The Adventures of Jaspers Woodbury" [17]).

The extension/improvement of instruction in e-learning solutions by utilizing our knowledge visualization tool, with the fixed visualization can be realised based on the following scenario:

The knowledge visualization tool displays knowledge elements related to the problem without any connection. As the recognition of interdependencies between presented elements is the most important task, they should be defined by an expert beforehand. By dragging different knowledge elements to each other the connections (if existing) will appear on the screen and the corresponding explanation (purposes and reasons) will be displayed according to the selection from the list [12]. In this concept the "right" solution is predefined by an expert and cannot be changed.

5.2 Connection Flexible Level of Visualization (Semi Flexible Level)

This level of flexibility is suitable in case more than one solution is possible for the problem under investigation. However, all of them are based on the same set of knowledge elements. Each solution shows a discrepancy to another one through utilizing different connections between the knowledge elements or/and their explanations. In such a case the following steps are essential:

The expert proposes his solution of a problem (Fig. 4.).

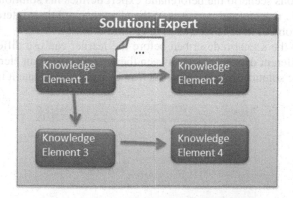

Fig. 4. Expert's solution of a problem

Using the same set of knowledge elements the learner creates his or her own solution (Fig. 5.). Based on the comparison of both solutions it is possible to examine the discovery/understanding of interdependencies between the defined elements and their utilization in the particular task.

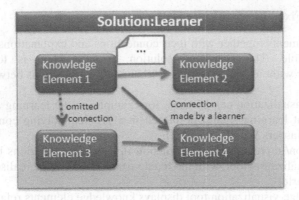

Fig. 5. Learner's Solution based on the same set of knowledge elements

The learner is enabled to discover/define his or her own view on the existing interdependencies and their influence on the problem.

5.3 Flexible Level of Visualization

This level allows building completely different solutions to a stated problem. Both the knowledge elements and the connections between them can vary from one solution to another.

This kind of knowledge visualization can be used e.g. in the cognitive apprenticeship instruction approach [18]. Within this approach an apprentice learns a trade (e.g. tailoring or timbering) under a master/teacher. A learner listens to the master explaining how the problem should be solved. After that the apprentice attempts to imitate the observed behaviour of the master in a real-world context.

Like in previous scenario the beforehand expert defines his solution of the problem and describes the reasons and purposes of the elements and their interdependencies in respect to the problem to be solved (Fig. 6. is similar to Fig. 4.).

In contrast to the scenario described before the learner can use different knowledge elements and different dependencies between them. If he misses an element, needed for his version of the solution, he can define such an element and situate it in the network of

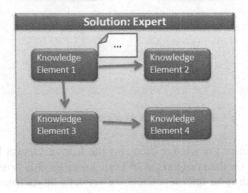

Fig. 6. Expert's solution of a problem

other elements (Fig. 7.). The placement of the new element will be described by its attributes but also by the interdependencies between them and the "rest of the community". Important aspects of this explanation are the purposes and reasons of the defined connections to the existing elements in respect to the various tasks and problems.

A similar case can be observed if many experts consider the investigated problem from different perspectives and use different knowledge elements in their solutions.

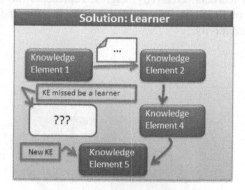

Fig. 7. Learner's solution of a problem

6   Incentive to Creativity

Analyzing all three described levels in Fig. 8. we can formulate that different levels of knowledge visualization require different creative capabilities. On the other side they can stimulate the learner's creativity in different manner and at different levels.

So far at fixed level the learners try to be approximately as good as their teacher/ expert, the learners at the flexible level develop their own solution regardless of the

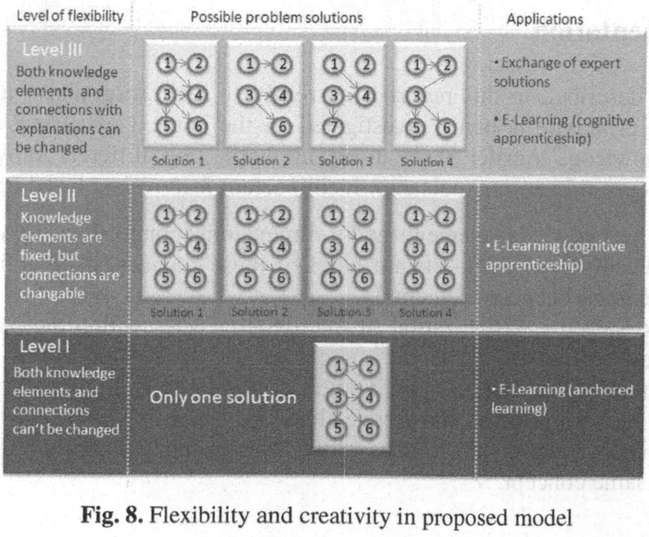

Fig. 8. Flexibility and creativity in proposed model

expert. The applicability of the levels described before depends upon the subject to be learned, the learning goals and the creativity of the learner.The higher levels of flexibility in knowledge visualization stimulate learners' creativity capabilities. If experts define all elements and interdependencies between them beforehand the learner will be placed in a kind of fixed network with its own rules and possibilities. The option to extend this network with their own contribution is the basis for innovation and consequently it stimulates learners' creativity. Fig. 9. represents the dependence between the discussed levels of flexibility in knowledge visualization and the required and/or stimulated learners' creativity.

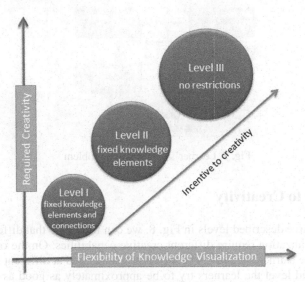

Fig. 9. Flexibility levels and creativity

7 Implementation

The concept described in this report is currently in realization and evaluation. The implemented prototype is being investigated in the context of various applications where the knowledge transfer plays an essential role (e.g. different knowledge management systems, e-Learning tools etc.) Its important features are listed below:

- Editor for knowledge elements and n-dimensional connections between them.
- Flexible edition of the facets/attributes of these connections to define the interdependencies between elements.
- Context sensitive visualization of interdependencies within the particular case analyses.
- Activation of the context sensitive pull-down menu with different interdependencies between selected knowledge elements
- Possibilities for generalization as well as for specialization of the solution based on the same concept.

8 Conclusions

The approach of knowledge visualization described in this paper provides a new basis for knowledge transfer. In contrast to other definitions, in this research knowledge visualization is investigated from the purpose perspective. Following presented purpose-driven approach it is important to extend the usual map of relations between different knowledge elements with explanation of their interdependencies. The implementation of this approach allows context sensitive visualizations of these interdependencies in respect to the purposes of knowledge transfer or tasks under investigation. The clarification of the purposes integrated into the visualization of interdependencies between knowledge elements significantly improves the recipient's understanding and acceptance of the knowledge transferred.

References

1. Probst, G., Raub, S., Romhardt, K.: Wissen managen, Wie Unternehmen ihre wertvollste Ressource optimal nutzen. Gabler, Wiesbaden (1997)
2. Davenport, T.H., Prusak, L.: Working Knowledg. Harward Business School Press, Cambridge (1998)
3. Nonaka, I., Takeuchi, H.: The Knowledge Creating Company. Oxford, New York (1995)
4. Maier, R.: Knowledge Management Systems: Information and Communication Technologies for Knowledge Management. Springer, Heidelberg (2004)
5. Bodrow, W., Fuchs-Kittowski, K.: Wissensmanagement in der Wissenschaft. In: Fuchs-Kittowski, K., Umstätter, W., Wagner-Döbler, R. (eds.) Wissenschaftsforschung Jahrbuch 2004, Berlin (2004)
6. Tergan, S., Keller, T. (eds.): Knowledge and Information Visualization: Searching for Synergies. Springer, Heidelberg (2005)
7. Drosdol, J., Frank, H.-J.: Information and Knowledge Visualization in Development and Use of a Management Information System (MIS) for Daimler Chrysler. In: Tergan, S., Keller, T. (eds.) Knowledge and Information Visualization: Searching for Synergies. Springer, Heidelberg (2005)
8. Burkhard, R., Meier, M.: Tube map: Evaluation of a visual metaphor for interfuncional communication of complex project. In: I-KNOW 2004, Austria. Springer, New York (2004)
9. Burkhard, R.: Towards a Framework and a Model for Knowledge Visualization: Synergies between Information and Knowledge Visualization. In: Tergan, S., Keller, T. (eds.) Knowledge and Information Visualization: Searching for Synergies. Springer, Heidelberg (2005)
10. Eppler, M., Burkhard, R.: Knowledge Visualization – Towards a New Discipline and its Fields of Application. Working Paper of NetAcademy on Knowledge Media, St.Gallen (2004)
11. Bodrow, W., Magalashvili, V.: IT-based Purpose-Driven Knowledge Visualization. In: Proceedings of ICSOFT 2007, Barcelona, pp. 194–197 (2007)
12. Bodrow, W., Magalashvili, V.: Using Knowledge Visualization in IT-based Discovery Learning. In: Proceedings of E-learning Conference 2007, Istanbul, pp. 125–129 (2007)
13. Card, S.K., Mackinlay, J.D., Scheiderman, B.: Readings in Information Visualization; Using Vision to Think. Morgan Kaufmann, San Francisco (1999)

14. Chen, C.: Information Visualization and Virtual Environments. Springer, London (1999)
15. Chen, C., Geroimenko, V.: Visualizing the Semantic Web: XML-Based Internet and Information Visualization. Springer, Heidelberg (2003)
16. Novak, J.D., Gowin, D.B.: Learning How to Learn. New York, Cambridge (1984)
17. The Adventures of Jasper Woodbury,
    http://peabody.vanderbilt.edu/projects/funded/jasper/preview
    /AdvJW.html
18. Bandura, A.: Self-efficacy: The exercise of control. Freeman, New York (1997)

Empirical Experimentation for Validating the Usability of Knowledge Packages in Transferring Innovations

Pasquale Ardimento, Maria Teresa Baldassarre, Marta Cimitile,
and Giuseppe Visaggio

Department of Informatics, University of Bari, Via Orabona 4, Bari, Italy
SER & Practices s.r.l. Spinoff of University of Bari
{ardimento,baldassarre,cimitile,visaggio}@di.uniba.it

Abstract. Transfer of research results following to technological innovation and to the experience collected in applying the innovation within an enterprise is a key success factor. A critical factor in transferring innovations to software processes concerns the knowledge transfer activity which requires the knowledge be explicit and understandable by stakeholders. As so many researchers have been studying alternative ways to conventional approaches i.e. books, papers, reports and other written communication means that favour knowledge acquisition on behalf of users. In this context, we propose the Knowledge Package (KP) structure as alternative. We have carried out an experiment which compared the usability of the proposed approach with conventional ones, along with the efficiency and the comprehensibility of the knowledge enclosed in a KP rather than in a set of Conventional Sources. The experiment has pointed out that knowledge packages are more efficient than conventional ones, for knowledge transfer. The experiment has been described according to guidelines that allow for replications. In this way other researchers can confirm or refute the results and enforce their validity.

Keywords: Knowledge packaging, empirical investigation.

1 Introduction

Transferring innovation in software processes is critical in that software development (production and maintenance) is man-centered and because the products of the development process are destined to be used by humans to improve their abilities in all application domains. For this reason, the most relevant concerns in transferring innovations in software processes concern knowledge transfer. This issue has two interesting perspectives:

- *usability*: intended as acquiring and institutionalizing explicit knowledge independently from the knowledge producer. The main sources of knowledge for an innovation remain research and experience collected on software processes and products. With respect to the first source, knowledge is expressed through papers, reports, books and any other conventional sources for socializing knowledge. The second type of knowledge is often embedded in processes and products. As so, it is difficult to acquire being scarcely or not at all explicit.

J. Filipe et al. (Eds.): ICSOFT/ENASE 2007, CCIS 22, pp. 357–370, 2008.
© Springer-Verlag Berlin Heidelberg 2008

- *exploitation*: requires that innovation, made usable by formalizing knowledge, must be enriched by technical and economical characteristics that allows to identify the best approach for introducing and institutionalizing new knowledge in processes together with the resources, risks and mitigation actions [3].

This paper focuses on the first aspect. Usability in this context relates to knowledge expression so that it is comprehensible and reusable by others that are not the authors of the knowledge. Conventional sources aim at convincing addressees of the validity of models, processes or techniques they refer to. As so, the authors do not look after content usability as specified above. Our conjecture is confirmed by the large amount of research results in literature that have not been transformed into innovations or transferable practices in business processes [3].

Given these premises, we have defined a knowledge packaging model for producing usable Knowledge Packages (KP). The model is based on experience verifiable in literature in terms of usability [4, 5, 6, 7]. Also, it has been refined following to our experience with industrial partners in the context of innovation management [30, 31, 32, 33, 34, 35]. The contents of a KP can be part of a process, method or technique that may be adopted to overcome software process or product problems. To summarize, the content of a KP describes a technology or, alternatively, a practice considering that the technology must be useful for practitioners. We have produced a set of KPs according to the knowledge packaging model using knowledge generated by our researchers and based on the experience acquired in our research work. The KPs are obtained using conventional sources retrieved in literature or other resources available on the web that express research results. We have used the same sources for packaging the contents. This has allowed us to better perceive the difficulties that arise in collecting, understanding and using research results through knowledge that researchers explicit. The experience gained in transforming knowledge from conventional sources into a KP has confirmed that much work is needed in producing the KP and that often a lot remains tacit [38].

Whatever the source, expressing knowledge according to our model requires more effort on behalf of the knowledge producer than what is spent to explicit the contents with conventional methods and instruments. As so, it is important for researchers to perceive that the additional effort assures more value to research results, i.e. usability and therefore, diffusion of research or experience results. In this paper usability is measured by two factors:

- *efficiency* in acquiring knowledge on behalf of a user: the faster it is to acquire new contents, the higher the probability that the acquisition process of research results or innovation is completed.
- *comprehensibility* of acquired knowledge: what has been correctly and completely acquired by a user so he can apply the contents to his work.

In order to be convincing we must prove that KPs improve usability with respect to conventional sources. This goal has been addressed through a *controlled experiment* that compares the two approaches for expressing knowledge and measures the usability factors in each one. As it arises from the next sections, results are encouraging. However, in order for experimental results to be generalizable, they must be externally valid [36]. As so, the experiment should be replicated many times in different contexts. Considering that our aim it to convince researchers to produce KPs, we must also allow them to replicate this study in order to confirm or refuse the results we have obtained in this paper. For this reason, the paper describes the experiment according to a model that allows for strict replications [37].

The rest of the paper is organized as follows: related works are described in section 2; section 3 illustrates the proposed approach for knowledge representation, section 4 illustrates the measurement model used; results of the study and lessons learned are presented in section 5; finally in section 6 conclusions are drawn.

2 Related Works

The problems related to knowledge transfer and valorization are investigated in industrial and academic contexts and sometimes it's not possible to distinguish the two because there is a convergence between industry and academia. Some companies have established internal organizations whose task is to acquire new knowledge [8, 9] to face knowledge transfer needs. For example, Shell Chemical has organized some groups with the aim at finding knowledge from outside sources, Hewlett Packard is commercializing not only its own ideas, but also innovations from other entities [8], Philips Research is participating to consortiums that direct one to one collaboration with innovative organizations [9].

There are also many studies that are focused on the use of Internet together with its Search Engines for knowledge diffusion and transfer. But in this direction our analysis shows that INTERNET, does not offer appropriate technologies for searching knowledge that is produced and published by a research organization nor by an enterprise, which is reusable in innovation projects by other research organizations or enterprises [10, 11, 12, 13, 14, 15]. A validation of this statement is proposed in [16], which proves that INTERNET retrieves information and not knowledge. The most accredited reason for this limitation is that usually general queries produce a large amount of documents and that there is not a natural language interface of the search engine. Furthermore, dedicated search engines, like digital libraries, improve the search precision although they do not overcome the problems described above.

There are also approaches based on the use of specialized search engines to find search results [17] and of systematic review to extract knowledge from conventional sources and use it in application domains. Such search engines and related methods do not add information, so they cannot be translated into a KP.

Another approach to knowledge search and transfer is based on the use of ontology [18, 19]. This approach is of interest for many studies which currently lack tools for creation and management. Much attention is being focused on these issues but the available experimental evidence is not yet sufficient for large-scale use. Ontologies can be used as support to a Knowledge Base (KB) containing many KPs. Indeed, ontology assures more accurate searches of KP contents within the KB with respect to user needs.

Knowledge packages are being studied by many research centers and companies. The existing knowledge bases sometimes have a semantically limited scope. This is the case of the IESE base [20] that collects lessons learned or mathematical prediction models or results of controlled experiments. In other cases the scope is wider but the knowledge is too general and therefore not very usable. This applies to the MIT knowledge base [5], that describes business processes but only at one or two levels of abstraction. There are probably other knowledge bases that cover wider fields with greater operational detail but we do not know much about them because they are private knowledge bases, for example the Daimler-Benz Base [4].

3 Proposed Approach

Our approach focuses on a knowledge-experience base, named PROMETHEUS [29], whose contents make it easier to achieve knowledge transfer among research centers; between research centers and production processes; among production processes. This knowledge-experience base is public to allow one or more interested communities to develop around it and exchange knowledge [21]. The knowledge-experience that is stored in the knowledge base must be formalized as KEP. Given the aim of this work, we are interested in the knowledge contents of the package. So, we will refer to KEP as KP.

3.1 Knowledge Package Structure

The proposed KP includes all the elements shown in Figure 1. A user can access one of the package components and then navigate along all the components of the same package according to her/his training or education needs. Search inside the package starting from any of its components is facilitated by the component's Attributes.

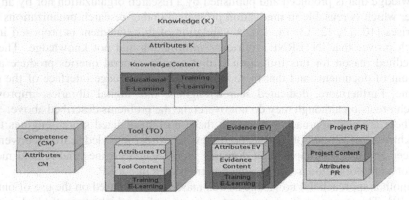

Fig. 1. Diagram of a Knowledge/Experience package

It can be seen in the figure that the Knowledge component (K) is the central one and its main element is the Knowledge Content (KC). It contains the knowledge package expressed in text form, with figures, graphs, formulas and whatever else may help to understand the content. The KC is organized as a tree. Starting from the root, (level 0) navigation to the lower levels (level 1, level 2, …) is possible through links. The higher the level of a node, the lower the abstraction of the content is which focuses more and more on operative elements. The root and each intermediate node contain the reasoned index of the underlying components. The KC consists of the following: research results for reference, analysis of how far the results on which the innovation should be built can be integrated into the system; analysis of the methods for transferring them into the business processes; details on the indicators listed in the attributes of the K inherent to the specific package, analyzing and generalizing the experimental data evinced from the evidence and associated projects; analysis of the results of any applications of the package in one or more projects, demonstrating the success of the application or any improvements required, made or in course; details on how to acquire the package.

Whenever knowledge is a prerequisite for understanding a node's content, the package points to an Educational E-learning course (EE). Instead, if use of a demonstrational prototype is required to become operative, the same package will point to a Training E-learning course (TE).

To integrate the knowledge package with the skills, K refers to a list of resources possessing the necessary knowledge, collected in the CoMpetence component (CM).

When a package also has support tools, rather than merely demonstration prototypes, K links the user to the available tool. For the sake of clarity, we point out that this is the case when the knowledge package has become an industrial practice, so that the demonstration prototypes included in the archetype they derived from have become industrial tools. The tools are collected in the TOol component (TO). Each tool available is associated to an educational course, again of a flexible nature, in the use of the correlated TE course.

A knowledge package is generally based on conjectures, hypotheses and principles. As they mature, their contents must all become principle-based. The transformation of a statement from conjecture through hypothesis to principle must be based on experimentation showing evidence of its validity. The experimentation, details of its execution and relative results, are collected in the Evidence component (EV), pointed to by the knowledge package.

Finally, a mature knowledge package is used in one or more projects, by one or more firms. At this stage the details describing the project and all the measurements made during its execution that express the efficacy of use of the package are collected in the Projects component (PR) associated with the package.

3.1.1 Knowledge Attributes

As shown in Figure 1, each component in the knowledge package has its own attributes. For all the components, these allow rapid selection of the respective elements in the knowledge base [21], Figure 2.

To facilitate the research, a set of selection classifiers and a set of descriptors summarizing the contents are used. The classifiers include: the key words and the problems the package is intended to solve. The descriptors mainly include: a brief summary of the content, prerequisite conditions for correct working and skills required to acquire the package, a history of the essential events occurring during the life cycle of the package, giving the reader an idea of how it has been applied, improved, and how mature it is, a description of the advantages that potential stakeholders could obtain applying the knowledge and finally a description of the risks concerning knowledge adoption, acquisition plans describing how to acquire the package and estimating the resources required for each activity. The history may also include information telling the reader that the content of all or some parts of the package are currently undergoing improvements.

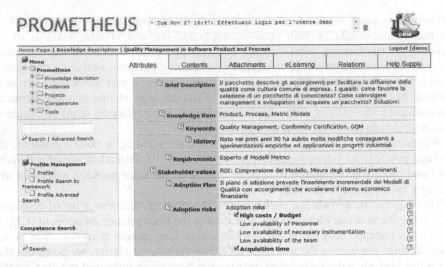

Fig. 2. Excerpt of a Knowledge package

The K descriptors also contain: the economic impact generated by applying the package; the impact on the processes, describing the impact that acquisition would have on the value of all the processes in the production cycle. There are also indicators estimating the costs. Thus, all these indicators allow a firm to answer the following questions: what specific changes need to be made? What would the benefits of these changes be? What costs and risks would be involved? How can successful acquisition be measured?

3.2 Experiment Details

The study has been described according to [37] which illustrates guidelines for reporting an experiment in order to assure its replication. In this paper we will not comment on the sections of the guidelines. Interested readers can refer to the report.

Motivation. The study aims to verify the approach of knowledge representation through knowledge packages collected in the Prometheus tool, compared with equivalent knowledge expressed in conventional sources and in sources extracted from the web and used for the production of a knowledge package. In particular, in accordance to the research questions, it investigates quality characteristics of efficiency and comprehensibility.

For efficiency we investigate whether the analysis and extraction of knowledge through an KP requires less effort than through conventional sources such as papers; for comprehensibility we investigate whether knowledge extraction is less error prone if KPs are used rather than conventional sources. Indeed, in the latter case, knowledge is most likely scattered in various parts of the paper or papers. With respect to his quality characteristic, we also investigate whether given a topic it is more difficult to understand the problem and extract knowledge from papers rather than an KP. The research goals listed below have been addressed through a controlled experiment.

Research Goals. Two research goals (RG) have been defined for the study according to the GQM template [39, 40].

RG1:
Analyze knowledge extraction using an Knowledge Package (KP)
With the aim of evaluating it
With respect to efficiency (compared to knowledge extracted from conventional sources)
From the view point of the knowledge user
In the context of a controlled experiment on a knowledge package tool called Prometheus.

RG2:
Analyze knowledge extraction using an Knowledge Package (KP)
With the aim of evaluating it
With respect to comprehensibility (compared to knowledge extracted from conventional sources)
From the view point of the knowledge user
In the context of a controlled experiment on a knowledge package tool called PROMETHEUS.

In accordance to the above goals, we have stated and investigated the following hypotheses:

EFFICACY: there are differences in terms of effort for extracting knowledge with KP rather than with CS
COMPRENSIBILITY: there are differences in terms of errors made in extracting knowledge through KP rather than through CS

Variables. The *dependent variables* of the study are Efficiency and Comprehensibility. Efficiency indicates to what point the Knowledge Representation criteria is effective (in terms of effort spent) for extracting knowledge and answering a specific set of questions. Comprehensibility indicates to what point the resources described in Prometheus or in Conventional Sources are easy to understand and to extract in order to answer a set of questions. The *independent variables* are the two treatments: the problems examined with KP and with known Conventional Sources. Two different types of problems were investigated: Balanced Scorecard and Reengineering Process. A set of 4 questions have been defined for each problem. This has been considered an appropriate number that balances the need for a sufficient amount of data without having to count on an excessive amount of effort and risk to bore and tire experimental subjects.

Experimental Subjects. The experimental subjects involved in the experimentation are first year students of a graduate course in Informatics from the University of Bari, with a common background in computer science or engineering. Also, they have carried out many projects, through individual and group work, in collaboration with local enterprises. As so, they are a proper representation of the enterprise population.

However, it is important to note that the selected set of experimental subjects, even if variegate, is not completely representative of the population of all software stakeholders such as managers, end users and so on. As consequence, at this first stage, it is not possible to generalize the results of the empirical investigation. Rather, results represent a first important step towards this direction. The sample was made up of 82 students randomly assigned to either of two groups (GROUP A and GROUP B). Each group was asked to answer questions assigned using, alternatively KP or Conventional Sources (CS) extracted from literature. All of the students have previous knowledge on the topic concerning Balanced Scorecard because it is part of their course curricula. They have no previous knowledge on the Reengineering Process topic.

Experiment Design. The experiment was organized in two experimental runs, RUN1 and RUN2, one per day in two consecutive days. During each run we changed the content of the KP and the content of the questions used to extract information from the source. So, in RUN1, the KP/ CS content, along with the questions for extracting information, related to Balanced Scorecard [22, 23, 24, 25]; and in RUN2 they referred to Reengineering [26, 27, 28]. Within a RUN, each group was assigned to either a KP or CS.

Study Instrumentation. At the beginning of each run, each experimental subject were handed a complete set of instrumentation. It contained the CS in digital version or KP according to the treatment and group. The KP was accessible through PROMETHEUS. The students examined the material and answered the questions reporting them on the data form. The start and end time were recorded by the researchers when handing in and collecting the forms. Comprehensibility was evaluated according to the number of errors made, while the effort, expression of efficiency, was reported on the data form.

4 Experiment Execution

At the beginning of each run, each experimental subject received a complete set of instrumentation (described above). It contained the papers or KP according to the treatment and group. The students examined the material and answered the questions reporting them on the data form. The start and end time were recorded by the researchers when handing in and collecting the forms.

The dependent and independent variables, presented in the previous section, have been formalized in metrics and collected through the empirical investigation on both types of knowledge extraction treatments (KP and CS) with respect to the quality characteristics.

Efficiency Factor has been measured through Effort (EF), defined as amount of time, measured in person/hrs, spent by each subject for carrying out their task and answer the questions: EF=t'-t where:

- t: Time when KP/CS and forms are given to an experimental subject.
- t': Time when an experimental subject hands in the data form complete with answers.

Comprehensibility was measured as the average of points Pij attributed for answering the i-th question of the j-th experimental subject. All answers were evaluated according to the interval scale reported in table 1.

Table 1. Details of comprehensibility quality factor

Evaluation of Question	P_{ij} score
Wrong Answer: the j-th subject gave a wrong answer to the i-th question.	0
Lacking Answer: the question was not answered by the j-th subject	2
Incomplete Answer: the j-th subject gave a partially correct answer to the i-th question	4
Complete Answer: the i-th question has received a correct answer by the j-th subject	6

The researchers, as domain experts involved in the investigation, corrected all the answers to the questions given by the experimental subjects.

5 Experimental Results

The data collected during the experimentation have been synthesized through descriptive statistics such as box and line plots. In the next sections, the results have been commented for both dependent variables: effort (used for measuring efficiency of the knowledge representation technique) and comprehensibility of the technique adopted.

5.1 Efficiency

In RUN1, the subject performances, as it appears in figure 3 are closer. The mean values are respectively 0.0643 for PROMETHEUS and 0.0657 for Conventional

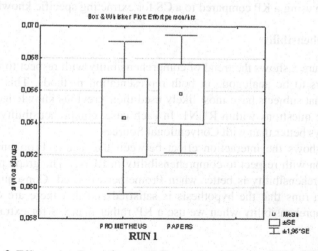

Fig. 3. Effort using Prometheus and Conventional Sources during RUN 1

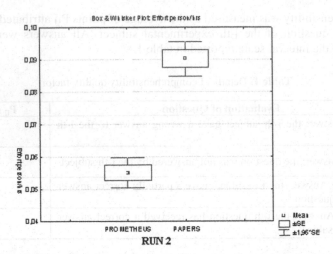

Fig. 4. Effort using Prometheus and Conventional Sources during RUN 2

Sources. Also, the dispersion of the results is very high for both knowledge representation methods. It seems as if the performances are independent from the technique used. Our explanation is that the experimental subjects were familiar with the topic (Balanced Scorecard) and so they used their previous experience and knowledge to answer the questions rather than strictly relate on the technique assigned (KP or CS).

Figure 4 illustrates the average effort in person/hrs spent by the experimental subjects in RUN2. It can be seen that there is less dispersion in the results for both knowledge representation techniques. Also, it can be seen how subjects using Conventional Sources spent, on average, a larger amount of time for answering the questions. This suggests that the structure of the packages promotes a more appropriate search of the knowledge contents for answering a question. As so, we can conclude in both cases that our hypothesis is satisfied and that there are differences in terms of effort spent in using a KP compared to a CS for extracting specific knowledge.

5.2 Comprehensibility

In RUN1, figure 5 shows the trend of comprehensibility with respect to the questions, which appears to be analogous in both representation methods. This confirms our assumption that subjects have most likely used their previous knowledge on the topic to answer the questions within RUN1. In each case, comprehensibility with Prometheus is always better than with Conventional Sources.

Figure 6 shows the interaction effect between the factors Problems*Knowledge Representation with respect to comprehensibility in RUN2. The graph points out that overall comprehensibility is better when Prometheus is used. Consequently, we can state for both runs that the hypothesis is satisfied and that there are differences in terms of comprehensibility when we use a KP rather than CS for extracting specific knowledge.

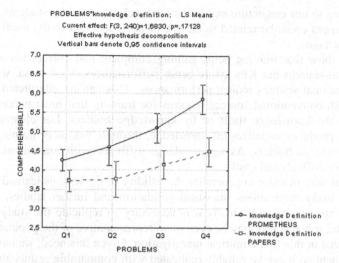

Fig. 5. Comprehensibility using Prometheus and CS for problem during RUN1

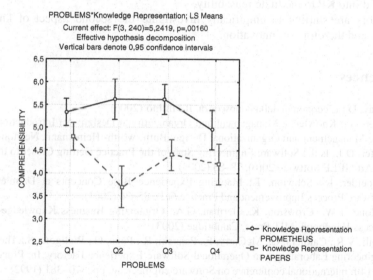

Fig. 6. Comprehensibility using Prometheus and CS for problem during RUN2

6 Conclusions and Future Works

For what concerns efficiency, the collected results provide some positive lessons learned about the use of Knowledge Packages. Indeed, the proposed approach with respect to the conventional ones:

- requires less effort for extracting information searched;
- represents explicit knowledge in a more comprehensible form.

According to our conjecture and to the feedback provided by students, the discovered differences could be related to the use of attributes and to the multi-level structure of a package.

Results show that training helps gaining complete and correct knowledge from conventional sources but KPs assure better performances. Vice versa, without training, conventional sources reduce performances. This can be interpreted as follows: transfer with conventional sources requires for training that must be carried out by authors of the knowledge itself or by knowledge holders. For clearness, when a knowledge producer socializes his knowledge, through lessons or training on job, he creates knowledge holders. As so, knowledge diffusion with conventional sources is longer, more difficult and costly.

It is clear that, in order to generalize the validity of the lessons learned proposed in this work, many replications, statistical validation and further studies, extended to other contexts, are needed. Finally it is necessary to replicate the study on a set of experimental subjects that may be even more representative of the population than the ones involved in this first empirical investigation. Given this need, we have described the experiment so it can be reliably replicated with comparable results able to extend external validity. Both [37] and [38] are accessible as technical reports and will be translated into KP to facilitate reusability.

Authors are starting to empirically investigate the second aspect of knowledge transfer, and therefore of innovation.

References

1. Foray, D.: L'economia della conoscenza. Il Mulino (2006)
2. Myers, P.: Knowledge Management and Organizational Design: An Introduction, Knowledge Management and Organizational Design. Butterworth- Heinemann, Newton (1996)
3. Reifer, D.J.: Is the Software Engineering State of the Practice Getting Closer to the State of the Art? IEEE Software 20(6), 78–83 (2003)
4. Schneider, K., Schwinn, T.: Maturing Experience Base Concepts at DaimlerChrysler. Software Process Improvement and Practice 6(2), 85–96 (2001)
5. Malone, T.W., Crowston, K., Herman, G.A.: Organizing Business Knowledge-The MIT Process Handbook. MIT Press, Cambridge (2003)
6. Basili, V.R., Caldiera, G., McGarry, F., Pajerski, R., Page, G., Waligora, S.: The Software Engineering Laboratory - an Operational Software Experience Factory. In: Proceedings of the 14th international conference on Software engineering, pp. 370–381 (1992)
7. Schneider, K., Hunnius, J.V.: Effective Experience Repositories for Software Engineering. In: Proceedings of the 25th International Conference on Software Engineering ICSE, pp. 534–539 (2003)
8. Halvorsen, P.K.: Adapting to changes in the (National) Research Infrastructure. Hewlett Packard Development Company, L.P (2004)
9. Hastbacka, M.A.: Open Innovation: What's mine it's mine. What if yours could be mine too. Technology Management Journal (December 2004)
10. Scoville, R.: Special Report: Find it on the Net. PC World (january 1996),
    http://www.pcworld.com/reprints/lycos.htm/

11. Leighton, H., Srivastava, J.: Precision among WWW search services (search engines): AltaVista, Excite, HotBot, Infoseek and Lycos (1997), http://www.winona.edu/library/

12. Ding, W., Marchionini, G.: A comparative study of the Web search service performance. In: Proceedings of the ASIS Annual Conference, vol. 33, pp. 136–142 (1996)

13. Leighton, H.: Performance of four WWW index services, Lycos, Infoseek, Webcrawler and WWW Worm (June 1996), http://www.winona.edu/library/

14. Chu, H., Rosenthal, M.: Search engines for the World Wide Web: a comparative study and evaluation methodology. In: Proceedings of the ASIS Annual Conference, vol. 33, pp. 127–135 (1996)

15. Clarke, S., Willett, P.: Estimating the recall performance of search engines. In: Proceedings ASLIB, vol. 49(7), pp. 184–189 (1997)

16. Ardimento, P., Caivàno, D., Cimitile, M., Visaggio, G.: Empirical Investigation of the efficacy and efficiency of tools for transferring Software Engineering Knowledge. JIKM Editor World Scientific Publishing, Singapore (submitted, 2007)

17. Kitchenham, B.: Procedures for Performing Systematic Reviews. Technical Report TR/SE-0401-ISSN (2004)

18. Zhang, Y.Y., Vasconcelos, W., Sleeman, D.: OntoSearch: An Ontology Search Engine. In: Proceedings of the twenty-fourth SGAI International Conference on Innovative Techniques and Applications of Artificial Intelligence (AI-2004), Cambridge, UK (2004)

19. Mingxia, G., Chunnian, L., Furong, C.: An Ontology Search Based on Semantic Analysis. In: Proceedings of the Third International Conference on Information Technology and Applications, vol. 1, pp. 256–259 (2005)

20. Althoff, K.D., Decker, B., Hartkopf, S., Jedlitschka, A., Nick, M., Rech, J.: Experience Management: The Fraunhofer IESE Experience Factory. In: Proceedings Industrial Conference Data Mining, Institute for Computer Vision and applied Computer Sciences, Leipzig, Germany. P. Perner (July 2001)

21. Ardimento, P., Cimitile, M., Visaggio, G.: Knowledge Management integrated with e-Learning in Open Innovation. Journal of e-Learning and Knowledge Society 2(3), Erickson Edn. (2006)

22. Becker, A.S., Bostelman, M.L.: Aligning Strategic and Project Measurement Systems. IEEE Software 16(3), 46–51 (1999)

23. Grembergen, W.V.: The Balanced Scorecard and IT Governance. Information Systems Control Journal 2, 1123–1124 (2000)

24. Abran, A., Buglione, L.: Balanced scorecards and GQM: What are the differences? In: FESMA-AEMES Software Measurement Conference (2000)

25. Mair, S.: A Balanced Scorecard for a Small Software Group. IEEE Software 19(6), 21–27 (2002)

26. Bianchi, A., Caivano, D., Visaggio, G.: Method and Process for Iterative Reengineering Data in a Legacy System. In: Proceedings of Working Conference on Reverse Engineering, pp. 86–96 (2000)

27. Bianchi, A., Caivano, D., Marengo, V., Visaggio, G.: Iterative Reengineering of Legacy Functions. In: Proceedings of IEEE International Conference Software Maintenance, pp. 632–641 (2001)

28. Bianchi, A., Caivano, D., Marengo, V., Visaggio, G.: Iterative Reengineering of Legacy Systems. IEEE Transactions on Software Engineering 29(3), 225–241 (2003)

29. SERLAB (Software Engineering Research LABoratory), http://193.204.187.180:8080/frame/controller_accessi_prometheus

30. Visaggio, G.: Assessing the Maintenance Process through Replicated, Controlled Experiment. The Journal of Systems and Software 44(3), 187–197 (1999)
31. Visaggio, G.: Value-based decision model for a renewal processes in software maintenance. In: Annals of Software Engineering, pp. 215–233. Kluwer Academic Publishers, Dordrecht (2000)
32. Visaggio, G.: Ageing of a Data-Intensive Legacy System: Symptoms and Remedies. Journal of Software Maintenance: Research and Practice 13(1), 281–308 (2001)
33. Visaggio, G.: Assessment of a Renewal Process Experimented on the Field. The Journal of Systems and Software 45(1), 3–17 (1999)
34. Baldassarre, M.T., Bianchi, A., Caivano, D., Visaggio, C.A., Stefanizzi, M.: Towards a Maintenance Process that Reduces Software Quality Degradation Thanks to Full Reuse. In: Proceedings of the 8th Workshop on Empirical Studies of Software Maintenance, WESS 2002, Montreal-Canada. IEEE Computer Society, Los Alamitos (2002)
35. Visaggio, G., Baldassarre, M.T., Bianchi, A., Caivano, D., Visaggio, C.A.: Full Reuse Maintenance Process for Reducing Software Degradation. In: Proceedings of the 8th European Conference on Software Maintenance and Reenginering, pp. 289–298 (2003)
36. Cook, T.D., Campbell, D.T.: Quasi Experimentation – Design and Analysis Issues for Field Settings. Houghton Mifflin Company (1979)
37. Baldassarre, M.T., Visaggio, G.: Description of a Paradigm for Empirical Investigations. Technical Report (2008), http://serlab.di.uniba.it/
38. Ardimento, P., Cimitile, M., Visaggio, G.: PROMETHEUS: From State of the Art to the State of Practices Technical Report (2008), http://serlab.di.uniba.it/
39. Wohlin, C., Runeson, P., Host, M., Ohlsson, M.C., Regnell, B., Wesslèn, A.: Experimentation in Software Engineering. Kluwer Academic Publishers, Dordrecht (2002)
40. Basili, V.R., Caldiera, G., Rombach, H.D.: Goal Question Metric Paradigm. In: Encyclopedia of Software Engineering, vol. 1, pp. 528–532. John Wiley & Sons, Chichester (1994)

An Ontological Investigation in the Field of Computer Programs

Pascal Lando, Anne Lapujade, Gilles Kassel, and Frédéric Fürst

MIS, Jules Verne University of Picardie, 33 rue Saint Leu, F-80039 Amiens, France
pascal.lando@u-picardie.fr, anne.lapujade@u-picardie.fr,
gilles.kassel@u-picardie.fr, frederic.furst@u-picardie.fr

Abstract. Over the past decade, ontology research has extended into the field of computer programs. The work has sought to define conceptual descriptions of programs in order to master the latter's design and use. Unfortunately, these efforts have only been partially successful. Here, we present the basis of a Core Ontology of Programs and Software (COPS) which integrates the field's main concepts. But, above all, we emphasize the method used to build the ontology. Indeed, COPS specializes the DOLCE foundational ontology ([10]) as well as core ontologies of domains (e.g. artefacts, documents) situated on a higher abstraction level. This approach enables us to take into account the "dual nature" of computer programs, which can be considered as both syntactic entities (well-formed expressions in a programming language) and artefacts whose function is to enable computers to process information.

Keywords: Knowledge engineering, ontological engineering, foundational ontologies, core ontologies, ontologies of computer programs.

1 Introduction

Over the last ten years or so, the field of computing programs has witnessed an increasing number of ontological investigations in several disciplines and with different objectives: in the philosophy of computer science, the goal is to attain better knowledge of the nature [3, 4] and semantics of computer programs [18], whereas in software engineering, formal descriptions seek to facilitate program maintenance [19, 13]. Such descriptions can also be used to orchestrate and automate the discovery of web services [14].

Our current work is in line with these efforts. In a first step it seeks to build a general or "core" ontology [6] of the domain of programs & software, and which will encompass the latter's main concepts and relations. This core ontology, (named COPS for "Core Ontology of Programs and Software"), will be used in a second step to conceptualize a sub-domain of computer programs, namely that of image processing tools. This step takes place within the project NeuroLOG (http:// neurolog. polytech. unice.fr) which aims at developing a distributed software platform to help members of the neuroimaging community to share images and image processing programs. This platform relies on an ontology integrating COPS as a component [17].

J. Filipe et al. (Eds.): ICSOFT/ENASE 2007, CCIS 22, pp. 371–383, 2008.

In this paper, we present not only our current content for COPS but more generally the methodological process that we used to build the ontology and the resulting structural features. The COPS ontology indeed specializes more abstract modules which strongly determine its structure, including the DOLCE foundational ontology [10] and the I&DA core ontology [5]. This type of design process is aimed at mastering two sorts of complexity (i) *conceptual* complexity, providing the ability to model complex objects (such as programs) at different abstraction levels and (ii) *modeling* complexity, providing the re-use of previously used & approved modules and also, the ability to design new modules by working in a distributed manner.

Two syntactic manifestations of COPS exist: one is coded in the semi-informal language from the OntoSpec method [17] and the other is coded in the formal web ontology language OWL. Within the NeuroLOG project, this latter expression is implemented into software under development that includes a semantic research tool called CORESE (designed as part of the ACACIA project at the INRIA institute, http://www.inria.fr/acacia/corese). However, due to space restrictions, we shall disregard these syntactic aspects in this paper and focus on the ontology's content.

The remainder of this article is structured in the following way. In section 2, we present the overall ontological framework of reference used (and the DOLCE and I&DA ontologies in particular). Section 3 details the COPS ontology's content and our design choices. In section 4, our work is compared with other efforts to design ontologies in the domain of computer programs.

2 Our Ontological Framework

As shown in Figure 1, the COPS ontology is integrated into a larger ontology composed of sub-ontologies which are situated at different levels of abstraction: a descending link between two sub-ontologies O_1 and O_2 means that the conceptual entities (concepts and relations) of O_2 are defined by specialization of the conceptual entities of O_1. The DOLCE foundational ontology and the different core ontologies make up the resource used by the OntoSpec methodology (http://www.laria.u-picardie.fr/IC/site/) to help structure application ontologies (which include all the concepts necessary for a particular application), such as that developed within the NeuroLOG project.

As a consequence of this overall structure, COPS's conceptualization depends on ontological commitments and modeling choices made upstream, i.e. in components situated at a higher level of abstraction. In this section, we present these main commitments and modeling choices by successively introducing the DOLCE ontology (2.1), the modeling of (participant) roles & artefacts (2.2) and the I&DA ontology (2.3).

2.1 DOLCE (Particulars)

DOLCE is a "foundational" ontology, which means that it comprises abstract concepts aimed at generalizing the set of concepts that we may encounter in the different domains of knowledge. In accordance with philosophically-grounded principles, DOLCE's domain – that of *Particulars* – is partitioned into four sub domains (*cf.* Fig. 2):

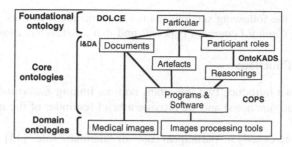

Fig. 1. Structure of the application ontology in the NeuroLOG project

- *Endurants* are entities "enduring in time" (e.g. the present article). Within *Endurants*, *Physical Objects* (e.g. a person, a pen or a sheet of paper) are distinguished from *Non-Physical Objects*, since only the former possess direct spatial *Qualities*. The domain of *Non-Physical Objects* covers social entities (e.g. the French community of researchers in knowledge engineering) and cognitive entities (e.g. your notion of knowledge engineering). To take plural entities into account (e.g. a family, a political group or the proceedings of a conference), the notion of *Collection* was recently introduced under *Non-Physical Objects* (Bottazzi *et al.*, 2006).
- *Perdurants* are entities "happening in time" (e.g. your reading of this article). *Perdurants* are generated by *Endurants* that *participateIn Perdurants* at *Time Intervals*. Among *Perdurants*, *Actions* "exemplify the intentionality of an agent": they correspond to *Accomplishments* that are controlled by an *Agent* (This latter notion is further defined in 2.2).
- *Endurants* and *Perdurants* have inherent properties (*Qualities*) that we perceive and/or measure (e.g. the weight of the paper copy of the article you may be holding or how long it takes you to read this article).
- These *Qualities* take a value (*Quale*) within regions of values which are *Abstracts* (e.g. 20 grams, 15 minutes).

These concepts are defined in DOLCE by means of rich axiomatization, which space restrictions prevent us from presenting. In particular, *Endurants* and *Perdurants* can be differentiated in terms of the dissimilar temporal behaviors of their parts. The interested reader is invited to refer to [10].

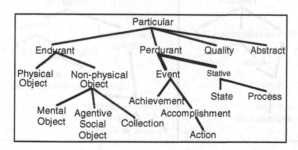

Fig. 2. An excerpt from DOLCE's hierarchy of concepts

Comment: in the following sections, names of conceptual entities will be noted in italics, with first *Capital Letters* for concepts and in a *javaLikeNotation* for relations.

2.2 Roles and Functions

In this section, we introduce two important notions linking *Endurants* to *Perdurants* (namely *roles* and *functions*) and we provide a brief reminder of the underlying modeling choices [2].

A *role* (or more exactly a "participant role" or "thematic role" [15] accounts for the the behaviour of *Endurants* when *participatingIn* (as defined by DOLCE) *Perdurants*. By way of an example, during the writing of an article, several entities participate in this *Action*: a person as an *Agent*, a pencil or a pen as an *Instrument* and the article itself as a *Result*. Here, the term "participant role" designates a category of concepts (e.g. *Agent, Instrument, Result*) which constitute a sub-ontology of *Endurants*, since the signature of the *participation* relation of DOLCE (notably the specification of its domain) constraints participants to be *Endurants* (cf. Fig. 3a). Formally, these concepts are defined by the introduction of primitive relations which specialize the relation *participatesIn* (e.g. *Agents control Actions, Patients areAffectedBy Actions*).

A *function* can be defined as the ability – assigned by agents to *Endurants* – to facilitate the performance of an *Action*, i.e. the ability of playing the role of *Instrument* in a *Perdurant*; in turn, this notion enables definition of the concept of an *Artefact* - an *Endurant* to which a function is assigned. According to the type of *Action* (the subontology of *Actions* on which COPS relies is discussed in 3.2), different types of *Artefacts* can be distinguished (cf. Fig. 3b): *Tools* are distinguished from *Cognitive Artefacts* according to whether the *Action* they can perform corresponds to modification of the physical world or the non-physical world. Of the latter, *Artefacts of Communication* enable communication of information to agents, whereas *Artefacts of Computation* allow computers to perform *Actions* as *Agents*.

In Figure 3, one can note that the concepts of *Author* and *Scientific Publication* encapsulate the type of entity and, respectively, the role and function assigned to the entity. This modeling choice, which is consistent with the most common paradigm for role modeling [16], leads to a tangled taxonomy.

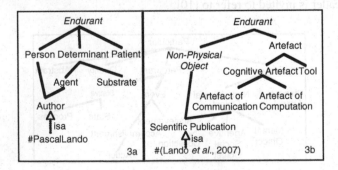

Fig. 3. Modeling of roles (a) and functions-artefacts (b)

2.3 I&DA (Inscriptions, Expressions and Conceptualizations)

I&DA is a core ontology in the domain of semiotics, and was initially built to classify documents according to their contents.

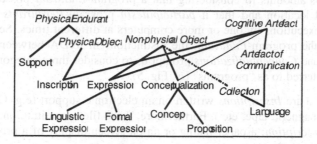

Fig. 4. The top level of I&DA's hierarchy of concepts

As shown in Figure 4, I&DA extends DOLCE by introducing three main concepts:

- *Inscriptions* are knowledge forms materialized by a substance and inscribed on a physical *Support* (e.g. a written text materialized by some ink on a sheet of paper). However, what counts more than the physicality of these *Physical Objects* is the fact that they represent another entity. Indeed, *Inscriptions* are intentional objects, which hold for other entities: *Inscriptions realize Expressions*.
- *Expressions* are non-physical knowledge forms *orderedBy* a *Language*. As *Non-physical Objects*, *Expressions* can be physically *realizedBy* many different *Inscriptions* (e.g. a text written on a sheet of paper can also be displayed on a computer screen). In turn, *Expressions* hold for other entities, namely contents that agents attribute to them: *Expressions express Conceptualizations*.
- Lastly, *Conceptualizations* are means by which agents can reason about (and refer to) a world. Within *Conceptualizations*, a functional distinction is made between *Propositions* (which are descriptions of situations) and *Concepts* (which serve to classify entities in a world).

The reader will note that, in order to account for documents, I&DA chooses to consider three distinct entities rather than three different views of the same entity. We shall see in the next section that this modeling choice has important repercussions on the structure of COPS.

3 COPS: A Core Ontology of Programs and Software

The COPS ontology indeed classifies a program as a document whose main characteristic is to allow a computer to perform information processing. In section 3.1, we first show how the ontological framework presented so far contributes to the definition of these particular documents. In sections 3.2 to 3.5, we then present several sub-ontologies dedicated to the different aspects of the notion of "program".

3.1 The Dual Nature of Programs

Firstly, the distinction made in DOLCE between *Endurants* and *Perdurants* prompts us to distinguish between the program (as an *Endurant*) and its executions (*Perdurants*). This amounts to considering that a program-*Endurant* possesses a spatial and temporal location and that it *participatesIn* program-*Perdurants* which correspond to its executions by one or more computers at different times. Secondly, when focusing on the program-*Endurant*, the distinctions in I&DA between *Inscriptions*, *Expressions* and *Conceptualizations* prompt us to consider three categories of entities commonly referred to as "programs" (*cf.* Fig. 5):

- *Files*, which are *Inscriptions* written on an electronic support (e.g. CDs, computer memory, magnetic tape, etc.). Furthermore, these files constitute only one type of program *Inscription*; a paper listing or an on-screen display of a program are also program *Inscriptions*.
- *Computer Language Expressions*, which are well-formed formulas (*isAWell-FormedFormulaOf* is a sub-relation of *isOrderedBy*) in a *Computer Language*. These expressions include *Programs*.
- *DataTypes* and *Algorithms*, which are *Conceptualizations* that represent the semantics of *Programs*. *DataTypes* are *Concepts* on which rely programming languages (e.g. variable, class, structure) and which are reflect to diversity of programming languages ([18]). *Algorithms* describe calculus steps in terms of these *DataTypes* (e.g. affecting a constant to the value of a variable, then adding another value, etc.).

This approach boils down to considering programs as *Expressions*, which is a consensual point of view in both computer science and philosophy. However, we consider that this purely syntactical description of a program is not enough to fully capture the nature of programs.

Indeed, programs have also a functional dimension, in that they allow computers to perform *Actions* (*Computations*). This functional dimension is present in expressions such as "sort program", "program for calculating the greatest common divisor of two numbers" or "image processing program". Programs are therefore also *Artefacts of Computation* (cf. Fig. 5). As commonly proposed in philosophy for the characterization of artefacts [9], we therefore end at a dual characterization of programs, considered to be both *Computer Language Expressions* and *Artefacts of Computation*.

In order to account for these dimensions of programs (and refine them), COPS proposes a sub-ontology of *Actions* (*cf.* 3.2) and a sub-ontology of *Languages* (*cf.* 3.3). We shall see in 3.4 that COPS's concept of *Program* integrates complementary constraints with regard to this first characterization.

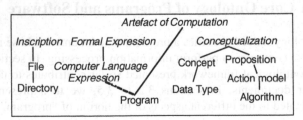

Fig. 5. The general structure of COPS

3.2 A Sub-ontology of Actions

In order to refine the functional dimension of *Programs* (and *Computer Language Expressions* in general) and specify what these latter allow a person or a computer to perform, COPS is endowed with an ontology of *Actions* (*cf.* Fig. 6).

Basically, we assimilate *Actions* to transformations taking place in a world and which are controlled (guided and monitored) by an *Agent*. According to this concept, *Actions* are distinguished according to (i) the world in which the transformation-*Action* occurs (physical (*Doing*) or non-physical (*Non-physical Action*)) and (ii) the *Agent* performing the *Action* (a human (*Human Action*) or a computer (*Computational Action*)).

The first semantic axis relies on a strong hypothesis dealing with the identity criteria of *Actions*, namely that *Actions* performed in separate worlds of entities are themselves distinct *Actions*. The worlds of entities considered in COPS converge with the common hierarchy of computer description levels, to which we add the "knowledge level" postulated by Newell (1982). This hypothesis prompts specialization of *Non-physical Actions* into *Symbolic Actions* (which, at the symbolic level, consist in transforming *Expressions* - e.g. *Executing a Program, Compiling a Program*) and *Conceptual Actions* (which, at the knowledge level, consist in transforming *Conceptualizations*). Of the latter, *Actions* involving *knowledge models* and which are taken into account by the CommonKADS methodology [2] (e.g. *Diagnosing a car's failure, Monitoring a patient*) are distinguished from *Actions* performed on *Data* and *Data Types* constituting the paradigms generated by the different programming languages (e.g. *Incrementing a Variable, Creating an Object*). The reader should note (cf. Fig. 6) that *Discourse Acts* are *Non-physical Actions*. The latter (considered as *Actions* which lead to a change in the state of knowledge of the addressee of the discourse) are used in COPS to account for *Actions* such as *Requests* (e.g. querying databases) or *Orders* for executing *Programs*.

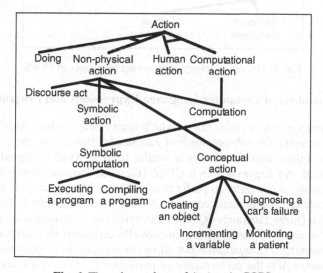

Fig. 6. The sub-ontology of *Actions* in COPS

3.3   A Sub-ontology of Computer Languages

In order to refine the syntactical dimension of *Expressions*, COPS includes an ontology of computer languages. Classically, one distinguishes between natural languages and formal languages. The formal languages of interest here are *Computer Languages, i.e.* those designed for interpretation by a computer (microprocessor) or a program.

Our conceptualization of *Computer Languages* (*cf.* Fig. 7) is based on the functions (the artefactual dimension) of the *Expressions* that they can *order*. The first category of computer languages is that of *General Purpose Computer Languages* (*GPCLs*), *i.e.* Turing-complete languages enabling the description of arbitrary *Computational Actions*. The second category is that of *Domain-Specific Computer Languages* (*DSCLs*), *i.e.* non-Turing-complete languages restricted to the writing of particular types of expressions (database queries, operating system commands, etc.). *Programming languages* are all *GPCLs* that are understandable by humans. *GPCLs* that are only understandable by computers or programs are *Low-level Computer Languages* (or low-level programming languages): *Machine Languages* (understandable by a processor) and *Byte-code Languages* (understandable by a virtual machine).

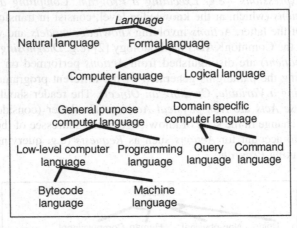

Fig. 7. The sub-ontology of computer languages in COPS

3.4   A Sub-ontology of Computer Language Expressions and Programs

The sub-ontology of *Expressions* (all of which are considered here to be well-formed *Expressions*) mirrors the sub-ontology of *Languages* that *order* these *Expressions*. The structure of these two taxonomies is similar (*cf.* Fig. 8) and is based on the functional dimension. An *Expression* in a GPCL (*General Purpose Computer Language Expression*) allows a computer to perform an arbitrary *Computation* (e.g. declaring a variable, calculating the greatest common divisor of two numbers). By contrast, an *Expression* in a *Query Language* (*Query Expression*) or an *Expression* in a *Command Language* (*Command Expression*) are functionally different: they do not allow computers to performing a *Computation* but allow (human) users to *Ordering* (which is a kind of *Discourse Act*) the performance of particular *Computations*, such as, for example, querying or modifying a set of data. These *Expressions* are therefore *Artefacts of Communication* and this functional distinction has repercussions for the definition of COPS's concept of *Program*.

On one side, we consider that a *Program* syntactically corresponds only to a particular type of *Expressions orderedBy* a *Programming Language*. Indeed, the peculiarity of *Expressions* qualified as *Programs* is that they can be either directly executed by a computer (after a compilation) or taken in charge by an interpreter. As an example, a program in the language C is composed of one or more functions, one of these functions being necessarily called "main". By contrast, *Expressions* such as a function or an instruction do not possess this entry point rendering the *Expression* executable or interpretable.

On another side, we consider that there exist executable or interpretable *Expressions* which are not *Programs*. Indeed, in the same lines as Eden and Turner [4], we only speak of *Programs* as *Expressions* being *orderedBy* Turing-complete languages (or *General Purpose Programming Languages*). Hence COPS does not consider a SQL query or a shell command, which are yet interpretable or executable, to be *Programs*. To sum up, we define a *Program* as an *Expression* in a Turing-complete language which can be interpreted or compiled and executed by an *Operating System* (*Executable Program*) or a *Virtual Machine* (*Byte-code Program*).

In addition, crosscutting relations link the different types of *Programs*: a *Source Code hasForExecutable* (or "can be compiled into several") *Executable programs*, and conversely an *Executable Program* or a *Byte-code Program hasForSourceCode* a *Source Code*.

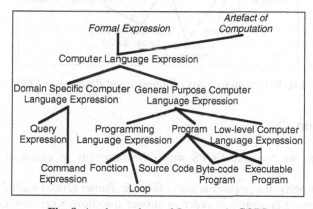

Fig. 8. A sub-ontology of *Programs* in COPS

Note that *Operating Systems* or *Virtual Machines* do not appear in COPS's sub-ontology of *Programs*. The reason is that they are not *Programs*, but they are rather made up of a set of *Programs*. They are therefore related to software, which are defined in COPS as collections of *Programs*, as presented in the next section.

3.5 A Sub-ontology of Software and Platforms

This sub-ontology of COPS models entities that are collections of *Programs* rather than single *Programs*. The concept *Library of Programs* (*cf.* Fig. 9) designates a *Collection* of *Programs* and, potentially, other documents (such as manuals).

By analogy with *Program*, *Software* is defined as both a *Library of Programs* and an *Artefact of Computation*. Since it must be executable, *Software* includes at least one *Executable* (or *Interpretable*) *Program*. *Software* includes *Compilers* (whose function is to allow a computer to translate a *Source Code* into an *Executable program*), *Interpreters* (whose function is to allow a computer to execute a *Source Code*) and *Operating Systems* (whose function is to allow a computer to execute *Executable Programs*). This function defines another class of *Artefacts* - the *Platforms*.

A *Platform* can be a purely material entity (*Hardware Platform*) or an entity that is partially made up of *Software* (*Software Platform*). *Software Platforms* include *Operating Systems* and *Computers* on which *Operating Systems* run.

Crosscutting relations link *Programs* and particular types of *Software*: a *Source Code isCompilableBy* particular *Compilers* and/or *isInterpretableBy* particular *Interpreters*; an *Executable Program runsOn* a particular *Operating System*; a *Byte-code Program runsOn* a particular *Virtual Machine*.

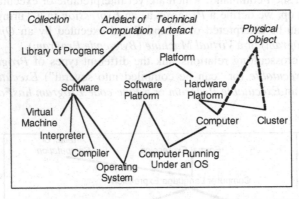

Fig. 9. The sub-ontology of *Software* and *Platforms* in COPS

4 Discussion

In this section, we compare COPS with other attempts to elaborate ontologies of programs.

In the philosophy of computer science, Eden & Turner (2007) recently began creating an ontology of programs in order to answer questions like: how can we display differences between hardware and software entities, and how can we distinguish a program from a program specification? Even though the ontological tools used in the present work are different from Eden & Turner's, it is interesting to compare the respective conceptualizations. For example, Eden & Turner (2007) define a program as a "well-formed expression in a Turing-complete programming language" - emphasizing the syntactical dimension of programs but setting aside their functional dimension (which exists in the COPS concept of *Program*). The fact that these two notions differ suggests that it would be useful to extend the COPS core ontology to other concepts.

The "web services" community has generated a variety of initiatives – METEOR-S, OWL-S, WSMO ([14]) – which seek to formally describe the discovery, evocation and orchestration (at different levels of automation) of such services. These efforts are currently far removed from COPS' aims because (i) the work emphasizes the operational nature of the descriptions and (ii) these descriptions concerning the function (the *Action* in COPS) realized by the service (e.g. booking a travel ticket) are situated on a meta level, which allows definition of the prerequisites for operation of the service (*e.g.* information about the travel has to be given) and the effects resulting from its execution (*e.g.* the ticket price is debited from a bank account). Within the framework of the NeuroLOG project, the functionalities targeted in terms of the evocation and orchestration of software tools are similar, which is why we plan to extend COPS to consider this level of description.

In the software engineering domain, Welty (2005) has suggested developing Comprehensive Software Information Systems (for software maintenance) by using an ontology which enables a detailed, conceptual description of software. This ontology could be considered as an extension of the COPS *Expression* sub-ontology, as it enables description at the code level and consideration of all the syntactical constructions available in programming languages. On the other hand, it supposes (strangely) that the entities playing *data* and *result* roles are real world entities (*e.g.* persons) and not conceptualizations modeling the real world. In COPS, we chose to follow [18] idea whereby program semantics are based on data and data types which model real world entities - for example (in the object paradigm), an *instance* which models an individual person or a *class* which models a set of persons.

Other work in the software engineering domain [13] led to publication of the Core Software Ontology (CSO) in order to better develop, administer and maintain complex software systems. The ontology-building approach is similar to ours, with re-use of the DOLCE high level ontology and core ontologies such as DnS [7]. COPS and CSO also share some modeling choices, such as the distinction between three entities (called *Inscriptions*, *Expressions* and *Conceptualizations* in COPS). However, we can note some different modeling choices. For example, in CSO (and assuming that every program can be a data item for another program), the *Data* concept subsumes the *Software* concept. In contrast, COPS assimilates the *Data* concept to a participant role (*cf.* 2.2) which can be played by arbitrary entities - *Programs*, for example. In fact, whereas CSO considers only one type of *Action* (namely "computational activities" whose participants are necessarily *Inscriptions* (in the sense of COPS) inscribed on some sort of hardware), COPS distinguishes several categories of *Actions* according to the nature of the participant entities (*cf.* 3.2). COPS' richer framework allows it to define a *Program Compilation* as an *Action* in which at least two *Programs* participate. Lastly, we can note that the functional dimension of programs is absent in CSO.

Those comparisons show that other core ontology proposals for the software domain do exist but that (i) the various efforts have not been coordinated and (ii) the existing ontologies display some important differences in terms of both range and structure.

5   Conclusions

In the present paper, we have presented the foundations of a core ontology of programs and software (COPS) derived by specializing the DOLCE foundational ontology and whose goal is to help structure more specific programming domains. In this connection, the next application of COPS within the NeuroLOG project, to help conceptualizing the domain of image processing tools, will provide an opportunity for evaluating the modeling choices made for the building of the ontology.

COPS' current conceptualization reveals a domain populated by entities having various nature. Indeed, there are temporal entities (program executions), physical entities (program inscriptions), plural entities (program collections), functional entities (program execution platforms) and, lastly, dual-nature (syntactic and functional) entities - the programs themselves. COPS' model-building feedback confirms the fact that ontological resource re-use (enabling modeling choices at several abstraction levels) is necessary for controlling the complexity of such domains.

In its current version, COPS only covers a part of this domain. Work in progress is extending the ontology in several directions. A first goal is to extend the programs semantics: links with processing (functions) only give an account of the "what", so it lacks the "how" - requiring us to take into account algorithms and data types which have only been positioned (cf. Fig. 5) and not precisely analyzed. A second goal is to enlarge COPS to program specifications: we plan to re-use the "problem resolution model" notion in OntoKADS [2] to extend COPS to the more general class of action models performed by computers using programs.

Acknowledgements. This work was funded in part by the NeuroLOG project (ANR-06-TLOG-024) under the French National Research Agency's Software Technologies program (http://neurolog.polytech.unice.fr).

References

1. Bottazzi, E., Catenacci, C., Gangemi, A., Lehmann, J.: From Collective Intentionality to Intentional Collectives: an Ontological Perspective. In: Cognitive Systems Research, Special Issue on Cognition, Joint Action and Collective Intentionality, vol. 7(2-3), pp. 192–208. Elsevier, Amsterdam (2006)
2. Bruaux, S., Kassel, G., Morel, G.: An ontological approach to the construction of problem-solving models. In: Clark, P., Schreiber, G. (eds.) 3rd International Conference on Knowledge Capture (K-CAP 2005), pp. 181–182. ACM, New York (2005) A longer version is published as LaRIA's Research Report 2005-03,
   http://hal.ccsd.cnrs.fr/ccsd-00005019
3. Colburn, T.R.: Philosophy and Computer Science. Explorations in Philosophy Series. M.E. Sharpe, New York (2000)
4. Eden, A.H., Turner, R.: Problems in the Ontology of Computer Programs. Applied Ontology 2(1), 13–36 (2007)
5. Fortier, J.-Y., Kassel, G.: Managing Knowledge at the Information Level: an Ontological Approach. In: Proceedings of the ECAI 2004 Workshop on Knowledge Management and Organizational Memories, Valencia, Spain, pp. 39–45 (2004)
6. Gangemi, A., Borgos, S. (eds.): Proceedings of the EKAW 2004 Workshop on Core Ontologies in Ontology Engineering, Northamptonshire (UK), vol. 118 (2004),
   http://ceur-ws.org
7. Gangemi, A., Mika, P.: Understanding the Semantic Web through Descriptions and Situations. In: Meersman, R., et al. (eds.) CoopIS 2003, DOA 2003, and ODBASE 2003. LNCS, vol. 2888, pp. 689–706. Springer, Heidelberg (2003)

8. Kassel, G.: Integration of the DOLCE top-level ontology into the OntoSpec methodology. LaRIA Research Report 2005-08 (2005), `http://hal.ccsd.cnrs.fr/ccsd-00012203`

9. Kroes, P., Meijers, A.: The Dual Nature of Thechnical Artifacts – presentation of a new research programme. Techné 6(2), 4–8 (2002)

10. Masolo, C., Borgo, S., Gangemi, A., Guarino, N., Oltramari, A., Schneider, L.: The WonderWeb Library of Foundational Ontologies and the DOLCE ontology. In: WonderWeb Deliverable D18, final report (vr. 1.0, 31-12-2003) (2003)

11. Newell, A.: The Knowledge Level. Artificial Intelligence 18, 87–127 (1982)

12. Niles, I., Pease, A.: Towards a standard upper ontology. In: Proceedings of the International Conference on Formal Ontology in Information Systems (FOIS 2001), pp. 2–9. ACM Press, New York (2001)

13. Oberle, D., Lamparter, S., Grimm, S., Vrandecic, D., Staab, S., Gangemi, A.: Towards Ontologies for Formalizing Modularization and Communication in Large Software Systems. Applied Ontology 1(2), 163–202 (2006)

14. Roman, D., Keller, U., Lausen, H., de Bruijn, J., Lara, R., Stollberg, M., Polleres, A., Feier, C., Bussler, C., Fensel, D.: Web Service Modeling Ontology. Applied Ontology 1, 77–106 (2005)

15. Sowa, J.F.: Knowledge Representation: Logical, Philosophical and Computational Foundations. Brooks/Cole (2000)

16. Steimann, F.: On the representation of roles in object-oriented and conceptual modelling. Data and Knowledge Engineering 35, 83–106 (2000)

17. Temal, L., Lando, P., Gibaud, B., Dojat, M., Kassel, G., Lapujade, A.: OntoNeuroBase: a multi-layered application ontology in neuroimaging. In: Proceedings of the 2nd Workshop: Formal Ontologies Meet Industry: FOMI 2006, Trento, Italy (2006)

18. Turner, R., Eden, A.H.: Towards a Programming Language Ontology. In: Dodig-Crnkovic, G., Stuart, S. (eds.) Computation, Information, Cognition – The Nexus and the Liminal, ch. 10, pp. 147–159. Cambridge Scholars Press, Cambridge (2007)

19. Welty, C.: An Integrated Representation for Software Development and Discovery. Ph.D. Thesis, RPI Computer Science Dept. (July 1995), `http://www.cs.vassar.edu/faculty/welty/papers/phd/`

Evaluation of Novel Approaches
to Software Engineering

Formal Problem Domain Modeling within MDA

Janis Osis, Erika Asnina, and Andrejs Grave

Faculty of Computer Science and Information Technology, Institute of Applied
Computer Systems, Riga Technical University, Latvia
{janis.osis,erika.asnina}@cs.rtu.lv, agrave@cs.rtu.lv

Abstract. The proposed approach called Topological Functioning Modeling for
Model Driven Architecture (TFM4MDA) uses formal mathematical foundations of Topological Functioning Model (TFM). It introduces the main feature
of MDA – Separation of Concerns by formal analysis of a business system,
enables mapping to functional requirements and verifying whether those requirements are in conformity with the TFM of the problem domain. By using a
goal-based method, a holistic behavior of the planned application can be decomposed in accordance with the goals. Graph transformation from the TFM to
a conceptual model (or a domain object model) enables establishing the definition of domain concepts and their relations. The paper also suggests a concept
of a tool for TFM4MDA, which is realized as an Eclipse plug-in.

1 Introduction

The main idea of the given work is to introduce more formalism into the problem
domain modeling within OMG Model Driven Architecture (MDA) [1] in the field of
object oriented software development. For that purpose, formalism of a Topological
Functioning Model (TFM) is used [2]. The TFM holistically represents complete
functionality of the system from computation-independent viewpoint. It considers
problem domain information separate from the application domain information captured in requirements and thus satisfies the main feature of MDA – Separation of
Concerns, therefore the TFM is an expressive and powerful instrument for a clear
presentation and formal analysis of system functioning and the environment the system works within.

This paper is organized as follows. Section 2 describes key principles and suggested solutions for a computation independent modeling as well as their weaknesses
in the object oriented analysis (OOA) within MDA. Section 3 discusses a developed
approach, i.e. *Topological Functioning Modeling for Model Driven Architecture
(TFM4MDA)*. TFM4MDA makes it possible to use a formal model (i.e. a TFM) as a
computation independent one without introducing complex mathematics. Besides
that, it allows verification of functional requirements at the beginning of analysis.
Section 4 describes the concept of a tool that partially automates the suggested approach. Conclusions establish further directions into the research of a *Computation
Independent Model (CIM)*.

J. Filipe et al. (Eds.): ICSOFT/ENASE 2007, CCIS 22, pp. 387–398, 2008.
© Springer-Verlag Berlin Heidelberg 2008

2 Constructing the CIM within MDA

MDA states that the CIM usually includes several distinct models that describe system requirements, business processes and objects; environment the system will work within, etc. OOA is a semiformal specification technique that contains use case modeling, class modeling, and dynamic modeling. Use cases are rather weak formalized approach that fragmentary describes the application domain. Their usage is not systematic in comparison with systematic approaches that enable identification of all system requirements. Fig. 1 shows several existing ways of creating behavior models (e.g. a use case model in case of OOA) and establishing of concepts and relations among them. One way is to apply assisting questions [3, 4], categories of concepts and concept relations [5] or goals [4, 6] in order to identify use cases and concepts from the description of the system (in a form of an informal description, expert interviewing, etc.). Another way is drafting a system requirements specification using some requirement gathering technique. Later these requirements are used for identification of use cases and creation of a conceptual model. The most complete way is to identify use cases and concepts, having knowledge of the problem domain as well as the system requirements specification [7]. All these ways use some mixture of information on both sides of the dashed line in Fig. 1.

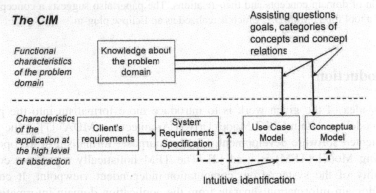

Fig. 1. Current ways of creating the CIM in OOA

Use case modeling starts with some initial estimate (a tentative idea) about where the system boundary lies. As an example we can mention the *Unified Process* [7], where use cases are driven by system requirements, the *B.O.O.M.* [8] that is IT project driven, and *Alistair Cockburn's approach* [6]. Use cases' fragmentary nature does not give any answer to questions about identifying all of system's use cases, conflicts among the use cases, gaps in the system's requirements, and how changes can affect behavior that other use cases describe [9].

First, a set of use cases as well as other means such as sets of UML activity diagrams, process diagrams, workflow diagrams and so on represents <u>decomposition</u> of the entire system's functionality. Second, we consider that modeling and understanding of the problem domain shall to be the primary stage in the software development, especially in case of embedded and complex business systems, whose failure can lead to huge losses. This means that decomposition of system's functionality must be applied as *a part of* a technique, whose first activity is a construction of a well-defined problem domain model. Such an approach is *TFM4MDA* which is discussed in this paper. This research can be considered as a step towards MDA completeness.

3 Topological Functioning Modeling for MDA

TFM4MDA is based on the formalism of the topological functioning model and uses some capabilities of universal category logic [2, 10, 11].

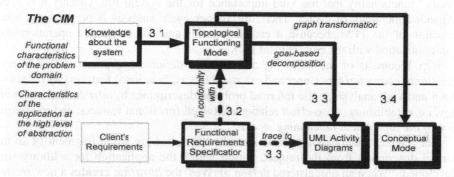

Fig. 2. Creating the CIM using TFM4MDA in OOA with <u>Separation of Concerns</u>

The main steps of TFM4MDA are illustrated by bold lines in Fig. 2. There are two <u>separate</u> branches at the beginning of the problem analysis: analysis on the (business or enterprise) system level, and analysis on the application system level. Having knowledge about a complex system that operates in the real world, a TFM of this system can be composed (Sect. 3.1). *The main idea is that the functionality determines the structure of the planned system.* This means that the TFM of the system controls and can be partially changed by functional requirements (Sect. 3.2). Then TFM functional features are associated to business goals of the system; this provides decomposition of system's functionality according to those goals in compliance to the problem domain realities. Moreover, after those activities functional requirements are not only in conformity with the business system functionality but also can be traceable to the system's activities, e.g. in UML Activity Diagrams (Sect. 3.3). Problem domain objects are identified from the TFM and described in UML Class Diagrams (Sect. 3.4).

3.1 Step 1: Construction of the Topological Functioning Model

The TFM has a solid mathematical base. It is represented in a form of a topological space (X, Θ), where X is a finite set of functional features of the system under consideration, and Θ is topology that satisfies axioms of topological structures and is represented in a form of a directed graph. The necessary condition for construction of a topological space is a meaningful and exhaustive verbal, graphical, or mathematical system description. The adequacy of a model describing the functioning of some concrete system can be achieved by analyzing mathematical properties of such abstract object [2].

TFM has topological (*connectedness, closure, neighborhood,* and *continuous mapping*) and functional (*cause-effect relations, cycle structure,* and *inputs* and *outputs*) characteristics. It is acknowledged that every business and technical system is a subsystem of the environment. Besides that the common thing for functioning of all systems (technical, business, or biological) should be *the main feedback*, whose visualization is an oriented cycle. Therefore, it is stated that at least one directed closed loop must be in every topological model of system functioning. It visualizes the "main" functionality that has vital importance for the system life. Usually it is even expanded hierarchy of cycles. Therefore, proper cycle analysis is necessary in construction of the TFM, because it enables careful analysis of system operation and communication with the environment [12].

Step 1 consists of the following activities: a) *definition of physical or business functional characteristics* (inner and external objects, functional features) by means of noun and verb analysis in the informal problem description; b) *introduction of topology*, i.e. establishing cause-effect relations between functional features; and c) *separation of the topological functioning model*.

For a better understanding let us consider the following small fragment of an informal description from the project, within which the application for a library was developed: "When an unregistered *person* **arrives,** the *librarian* **creates** a new *reader account* and a *reader card*. The *librarian* **gives out** the *card* to the *reader*. When the *reader* **completes** the *request for a book,* hi **gives** it to the *librarian*. The *librarian* **checks out** the requested *book* from a *book fund* to a *reader*, if the *book copy* is available in a *book fund*. When the *reader* **returns** the *book copy,* the *librarian* **takes** it **back** and **returns** the *book* to the *book fund*. He **imposes** the *fine* if the *term of the loan* is exceeded, the *book* is lost, or is damaged. When the *reader* **pays** the *fine,* the *librarian* **closes** the *fine*. If the *book copy* is hardly damaged, the *librarian* **completes** the *statement of utilization,* and **sends** the *book copy* to a *Disposer*."

Definition of physical or business functional characteristics consists of the following activities: 1) Definition of objects and their properties from the problem domain description, that is performed by noun analysis, i.e. by establishing as meaningful nouns and their direct objects as handling synonyms and homonyms; 2) Identification of external systems (objects that are not subordinated to the system rules) and partially-dependent systems (objects that are partially subordinated to the system rules, e.g. system workers' roles); and 3) Definition of functional features is preformed by verb analysis in the problem domain description, i.e. by founding meaningful verbs.

In the fragment, nouns are denoted by *italic*, verbs are denoted by **bold**, and action pre- (or post-) conditions are underlined. The identified objects (or concepts) are the following: a) inner objects are a librarian, a book copy (a synonym is a book), a reader, a reader card, a request for a book, a fine, a loan term, a statement of utilization, book fund, and b) external objects are a person, a reader, and a disposer.

Within TFM4MDA, each functional feature is a tuple $<A, R, O, PrCond, E>$, where A is an object's action, R is a result of this action, O is an object (objects) that receives the result or that is used in this action (for example, a role, a time period, a catalog, etc.), $PrCond$ is a set $PrCond = \{c_1, ..., c_i\}$, where c_i is a precondition or an atomic business rule (it is an optional parameter), and E is an entity responsible for action performing. Each precondition and atomic business rule must be either defined as a functional feature or assigned to the already defined functional feature. Two textual description forms are defined. The more detailed form is as follows:

```
<action>-ing the < result> [to, into, in, by, of, from]
a(n) <object>, [PrCond,] E
```

And the more abstract form is the following:

```
<action>-ing a(n) <object>, [PrCond,] E
```

The identified functional characteristics are described in the form <identifier: functional feature, [{precondition},] the responsible entity (where "Lib" denotes "librarian", "R" denotes "reader"), subordination ("in" is inner, "ex" is external)>.

The identified TFM functional features are as follows. **1:** Arriving a person, person, ex; **2:** Creating a reader account, {unregistered person}, Lib, in; **3:** Creating a reader card, Lib, in; **4:** Giving out the card to a reader, Lib, in; **5:** Getting the status of a reader, R, ex; **6:** Completing a request_for_book, R, in; **7:** Sending a request_for_book, R, in; **8:** Taking out the book copy from a book fund, Lib, in; **9:** Checking out a book copy, {completed request, book copy is available}, Lib, in; **10:** Giving out a book copy, Lib, in; **11:** Getting a book copy [by a registered reader], R, ex; **12:** Returning a book copy [by a registered person], R, ex; **13:** Taking back a book copy, Lib, in; **14:** Checking the term of loan of a book copy, Lib, in; **15:** Evaluating the condition of a book copy, Lib, in; **16:** Imposing a fine, {the loan term is exceeded, the lost book, or the damaged book}, Lib, in; **17:** Returning the book copy to a book fund, Lib, in; **18:** Paying a fine, {imposed fine}, R, in; **19:** Closing a fine, {paid fine}, Lib, in; **20:** Completing a statement_of_utilization, {hardly damaged book copy}, Lib, in; **21:** Sending the book copy to a Disposer, Lib, in; **22:** Utilizing a book copy, Disposer, ex.

Introduction of topology means establishing cause-and-effect relations between the identified functional features. Cause-and-effect relations are visualized as arcs of a digraph that are oriented from a cause vertex to an effect vertex.

Cause-effect relations form causal chains that sometimes are functioning cycles. All the *cycles* and *sub-cycles* should be carefully analyzed in order to completely identify existing functionality of the system. The *main cycle* (cycles) of system's functioning (i.e. functionality that is vitally necessary for system's life) must be found and analyzed before starting further analysis. In case of studying a complex system, a TFM can be separated into a series of subsystems according to identified cycles.

The result of the first two activities is represented in Fig.3a that illustrates a TFM composed for the mentioned fragment of library's functioning. Fig. 3a clearly shows that cause-and-effect relations forms functioning cycles, and illustrates the *main functional cycle* defined by experts, which includes the following ordered functional features *"17-8-9-10-11-5-12-13-14-15-17"* (bold arrows). These functional features describe book checking out and taking back. They are assumed to be main, because have a major impact on business system operating. A cycle that includes the functional features "5-6-7-17-8-9-10-11-5" illustrates an example of the first-order sub-cycle.

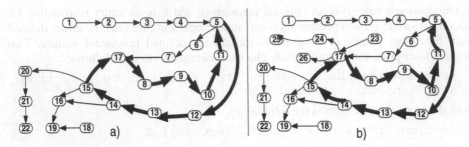

Fig. 3. Topological space of the library functioning (a), the modified topological space of the library functioning (b)

Separation of the topological functioning model is performed by applying the closure operation [2] over a set of system inner functional features.

A topological space is a system represented by Eq. (1). Where N is a set of system's inner functional features, and M is a set of functional features of other systems that interact with the system or of the system itself, which affect the external ones.

$$Z = N \cup M \qquad (1)$$

$$X = [N] = \bigcup_{\eta=1}^{n} X_\eta \qquad (2)$$

A TFM ($X \in \Theta$) is separated from the topological space of a problem domain by the closure operation over the set N as it is shown by Eq. (2). Where X_η is an adherence point of the set N and capacity of X is the number n of adherence points of N. An adherence point of the set N is a point, whose each neighborhood includes at least one point from the set N. The neighborhood of a vertex x in a digraph is the set of all vertices adjacent to x and the vertex x itself. It is assumed here that all vertices adjacent to x lie at the distance $d=1$ from x on ends of output arcs from x.

The example illustrates how the closure operation (2) is applied over the set N in order to get entire system functionality – the set X. The set of system's inner functional features $N=$ {2, 3, 6, 7, 8, 9, 10, 11, 12, 13, 14, 15, 16, 17, 19, 20}. The set of external functional features and system's functional features that affect the external environment $M=$ {1, 4, 5, 18, 21, 22}.

The neighborhood of each element of the set N is as follows: $X_2 = $ {2, 3}, $X_3=$ {3, 4}, $X_6=$ {6, 7}, $X_7=$ {7, 17}, $X_8=$ {8, 9}, $X_9 = $ {9, 10}, $X_{10} = $ {10,11}, $X_{11} = $ {11, 5}, $X_{12} = $ {12, 13}, $X_{13} = $ {13, 14}, $X_{14} = $ {14, 15, 16}, $X_{15} = $ {15,16, 17, 20}, $X_{16} = $ {16, 19}, $X_{17} = $ {17, 8}, $X_{19} = $ {19}, and $X_{20} = $ {20, 21}.

The union set is $X=$ {2, 3, 4, 5, 6, 7, 8, 9, 10, 11, 12, 13, 14, 15, 16, 17, 19, 20, 21}.

3.2 Step 2: Functional Requirement Conformity with the TFM

Step 2 is verification of whether functional requirements (hereafter *requirements*) are in conformity with the constructed TFM.

TFM functional features specify functionality that *exists in the problem domain*. In turn, requirements specify functionality that *must exist in the application*. Thus, requirements can be mapped onto TFM functional features. Mappings are described with arrow predicates — constructs borrowed from the universal categorical logic for computer science that is explored in details in [13].

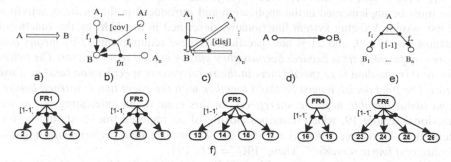

Fig. 4. Functional requirements mapping onto TFM functional features

Within TFM4MDA, five types of mappings and corresponding arrow predicates are defined:

- **One to One:** *An inclusion predicate* in Fig. 4a is used if the requirement A completely specifies what will be implemented in accordance with the feature B.
- **Many to One:** *A covering predicate* in Fig. 4b is used if requirements $(A_1, A_2,..., A_n)$ overlap the specification of what will be implemented in accordance with the feature B. In case of covering requirements, their specification should be refined. *A disjoint (component) predicate* in Fig. 4c is used if requirements $(A_1, A_2,..., A_n)$ together completely specify the feature B and do not overlap each other.
- **One to Many:** *Projection* in Fig. 4d is used if some part of the requirement A incompletely specifies some feature B_i. *Separating family of functions* in Fig. 4e is used if one requirement A completely specifies several features $(B_1, ..., B_n)$. It can be because: a) the requirement joins several requirements and can be split up, or b) features are more detailed than this requirement.
- **One to Zero:** One requirement specifies some new or undefined functionality. In this particular case it is necessary to define possible changes in the problem domain functioning.
- **Zero to One:** Specification does not contain any requirement corresponding to the defined feature (a missed requirement). Thus, it is mandatory to make a decision about implementation of the discovered functionality together with the client.

Results of this activity are both checked and conformed functional requirements and TFM, which describes needed system functionality and the environment it operates within.

For illustration, let us assume that the drafted requirements in our example are as follows. **FR1:** The system shall perform registration of a new reader; **FR2:** The system shall perform check out of a book copy; **FR3:** The system shall perform check in of a book copy; **FR4:** The system shall perform imposing of a fine to a reader; and **FR5:** The system shall perform handling of unsatisfied request (the description: the unsatisfied request should be added to the wait list; when a book copy is returned to the book fund, the system checks which requests can be satisfied and in success informs the readers by SMS).

The requirements maps onto the features as follows: "FR1"= {2, 3, 4}; "FR2"= {7, 8, 9}; "FR3"= {13, 14, 15, 17}; "FR4"= {16}; and "FR5" describes new functionality that must be implemented in the application and introduced in the business activities of the system. Existing system functionality described in the TFM by the functional features 18, 19, 20, and 21 is not specified by these requirements. *This means that more careful analysis is needed, because they can be missed requirements. The better way in this situation is to specify them in the requirements specification (and as a use case). The final decision must be taken together with the client that is warned beforehand about possible negative aftereffects.* In this context, the interesting one is the functional feature 19, which describes closing of an imposed fine. It should be implemented. Therefore, *"FR4" is modified as "The system shall perform imposing and closing of a fine to a reader".* Then, "FR4"= {16, 19}.

The new functionality introduced by "FR5" can be described by new identified objects (the system, a wait list and SMS), and additional functional features: **23:** Adding the request_for_book in a wait list, {unavailable book}, Lib, in; **24:** Checking the request_for_book in a wait list, {a book copy is returned to the book fund}, system, in; **25:** Informing the reader by SMS, {a request in the wait list can be satisfied}, system, in; **26:** Avoiding a request for a book, {book copy is not available}, system, in.

Introducing this functionality into the TFM, we must *recheck* all the previously identified cause-and-effect relations taking into account new functional features and possible changes in causes and effects. The set N= {2, 3, 6, 7, 8, 9, 10, 11, 12, 13, 14, 15, 16, 17, 19, 20, 23, 24}. The set M= {1, 4, 5, 18, 21, 22, 25, 26}. After the closuring, the set X= {2, 3, 4, 5, 6, 7, 8, 9, 10, 11, 12, 13, 14, 15, 16, 17, 19, 20 21, 23, 24, 25, 26}. A resulting model is represented in Fig. 3b.

The final correspondence between functional features and functional requirements is illustrated in Fig. 4f. All identified mappings of functional requirements onto the functional features have the type "one to many".

3.3   Step 3: Decomposition of the Holistic View on Functionality

Transition from an initial problem domain model to a CIM "output" model, i.e. a set of UML activity diagrams, goes as follows: 1) *Identification of system's users and their goals.* External entities establish business goals, but inner entities (humans, roles, etc.) either establish or implement goals. In the TFM, users are objects responsible for execution of functional features. Identification of *goals* is identification of the set of functional features necessary for the satisfaction of the goal; 2) *System's functionality decomposition in order to satisfy the identified goals* includes discovering functional features specified by requirements that are needed to achieve these goals. Identified parts of the TFM can be represented in an UML activity diagram. In that case, functional features must be mapped into activities, but cause-effect relations must be mapped into transition flows between these activities. By far, pre- and postconditions of transitions must also be presented and, if it is needed, refined. As previously mentioned, business rules, post and pre conditions were defined at the stage of definition of functional features. Therefore, they are known and can be described.

3.4   Step 4: Composing of a Conceptual Model

The last step is identification of a conceptual model (or a domain object model). After Step 3, the TFM shows functionality that persists together with logic to be implemented; hence, it includes all concepts that are necessary for proper functioning.

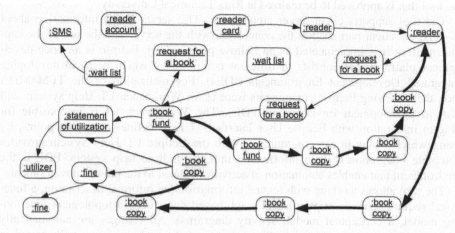

Fig. 4. The graph of domain objects

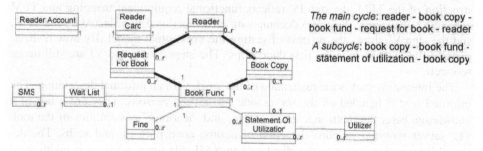

Fig. 5. The initial conceptual model

In order to obtain a conceptual model each TFM functional feature is detailed to the level where it uses only objects of one type. After that, this model must be transformed one-to-one into a graph of domain objects: Vertices with the same type of objects must be merged while keeping all relations to graph vertices with other types of objects. The result is a graph of domain objects with indirect associations. Concepts used in the main functional cycles are necessary in all cases.

In our example, the result of transformation from the TFM to the domain object graph is demonstrated in Fig. 4. In turn, the UML class diagram in Fig. 5 reflects the TFM after the gluing of all graph vertices with the same types of objects. This reflects the idea proposed in [2, 12] that the holistic domain representation by means of the TFM allows identifying of all necessary domain objects, and, even, allows to define their necessity for a successful implementation of the system.

4 Automation of TFM4MDA

TFM4MDA uses a complex graph-based constructs that require additional efforts. The main purpose of a tool for TFM4MDA is model management, i.e. model verification, traceability handling, step automation, etc. This section discusses the concept of the tool that is approved to be realized at Riga Technical University.

The tool supports client-server architecture. The server keeps information about models; the client part enables the connection with the server and the use of the kept information. It is implemented as an Eclipse plug-in [14]. Eclipse is an open development platform that consists of different components, which helps in developing Integrated Development Environments (IDEs). For realization of the TFM4MDA tool, the following Eclipse components were used: Workbench UI, Help system, and Plug-in Development Environment (PDE). The Workbench UI is responsible for plug-in integration with Eclipse User Interface (UI). It defines extension points, by using which a plug-in can communicate with the Eclipse UI. Help System provides complete integration of help information into the Eclipse help system. PDE is the environment that enables automation of activities related to the plug-in development.

The tool allows working with textual information (an informal description, a functional requirements description), and graph-based constructs (a topological functioning model, a conceptual model, activity diagrams). All changes are automatically propagated to the related models. The scheme of the tool activities is illustrated in Fig. 6. It describes the considered TFM4MDA steps. The first three steps reflect construction of the TFM, the step IV reflects functional requirement mapping and TFM enhancing, the step V illustrates decomposition of functionality reflected by the TFM, and the step VI shows the process of getting the conceptual model. By now realized parts of the tools include the first three steps. The steps IV, V and VI are still under research.

The interesting part is the realization of the work with an informal description. The informal text is handled on the server side for several reasons. They are a use of a knowledge base, the multi-user environment, and "learning" possibilities of the tool. The server program supports detection of nouns, noun phrases, and verbs. The detected information is sent to the client side in XML file form, where it is highlighted to the user in different ways (different colors, fonts, etc.). The tool provides convenient interface for handling this information and creation of TFM functional features. The topology introduced between TFM functional features is realized as a mix of their graphical and textual representation. The tool offers the user to join, split up and define cause-effect relations between these functional features using a tabular representation, but the result is also represented in the form of graphs.

The TFM4MDA tool provides a separate editor for each step. Each editor has relevant views that represent actual information. All automated steps that need user participation are realized as wizards that open corresponding editors. By now, there are three wizards constructed in the tool. The first wizard creates a system description file with the detected nouns, noun phrases and verbs by the Natural Language Processing Server (NLPS). The second one creates a topological space. And the third one creates the TFM.

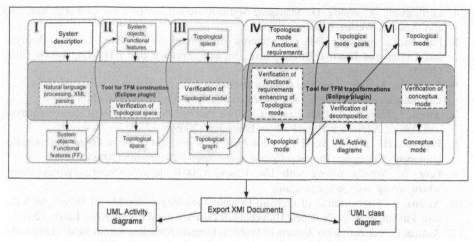

Fig. 6. The scheme of the tool supporting TFM4MDA

5 Conclusions

Application of TFM4MDA has the following advantages. First, careful cycle analysis can help to identify all (possible at that moment) functional and causal relations between objects in complex business systems. Implementation priorities of functional requirements can be ordered not only in accordance with the client's wishes, but in accordance with the functioning cycles. It makes it possible to take a decision about functional change acceptability before their realization in the application, and helps to check completeness of the functional requirements. Second, it does not limit the use of any requirement gathering techniques. Third, TFM4MDA provides more formal decomposition of functionality, and allows avoiding problems that arise due fragmentary nature of functionality representation such as incompleteness, conflicts among functioning parts and their affects on each other.

The tool partially automates the steps of TFM4MDA described above. But this approach still requires human participation. Therefore, the further research is related to enhancing TFM4MDA with capabilities of natural language handling in order to make it possible to automate more steps of this approach and to decrease human participation in decision making.

References

1. Miller, J., Mukerji, J. (eds.): OMG: MDA Guide Version 1.0.1 (2003), http://www.omg.org/docs/omg/03-06-01.pdf
2. Osis, J.: Formal computation independent model within the MDA life cycle. In: International Transactions on Systems Science and Applications, vol. 1(2), pp. 159–166. Xiaglow Institute Ltd., Glasgow (2006)
3. Jacobson, I., Christerson, M., Jonsson, P.: Object-Oriented Software Engineering: A Use Case Driven Approach. Addison-Wesley, Reading (1992)

4. Leffingwell, D., Widrig, D.: Managing Software Requirements: a use case approach, 2nd edn. Addison-Wesley, Reading (2003)
5. Larman, C.: Applying UML and Patterns: An Introduction to Object-Oriented Analysis and Design and Iterative Development, 3rd edn. Prentice Hall PTR, Englewood Cliffs (2005)
6. Cockburn, A.: Structuring Use Cases with Goals, http://alistair.cockburn.us/crystal/articles/sucwg/structuringucwithgoals.htm
7. Arlow, J., Neustadt, I.: UML2 and the Unified Process: Practical Object-Oriented Analysis and Design. Addison-Wesley, Pearson Education (2005)
8. Podeswa, H.: UML for the IT Business Analyst: A practical Guide to Object-Oriented Requirements Gathering. Thomson Course Technology PTR, Boston (2005)
9. Ferg, S.: What's Wrong with Use Cases? (2003), http://www.ferg.org/papers/ferg-whats_wrong_with_use_cases.html
10. Asnina, E.: Formalization of Problem Domain Modeling within Model Driven Architecture. PhD thesis, Riga Technical University, RTU Publishing House, Riga, Latvia (2006)
11. Asnina, E.: Formalization Aspects of Problem Domain Modeling within Model Driven Architecture. In: Databases and Information Systems, 7th International Baltic Conference on Databases and Information Systems, Communications, Materials of Doctoral Consortium, pp. 93–104. Technika, Vilnius (2006)
12. Osis, J.: Software Development with Topological Model in the Framework of MDA. In: Proceedings of the 9th CaiSE/IFIP8.1/EUNO International Workshop on Evaluation of Modeling Methods in Systems Analysis and Design (EMMSAD 2004) in connection with the CaiSE 2004, vol. 1, pp. 211–220. RTU, Riga (2004)
13. Diskin, Z., Kadish, B., Piessens, F., Johnson, M.: Universal Arrow Foundations for Visual Modeling. In: Anderson, M., Cheng, P., Haarslev, V. (eds.) Diagrams 2000. LNCS (LNAI), vol. 1889, pp. 345–360. Springer, Heidelberg (2000)
14. Eclipse – an open development platform, http://www.eclipse.org

Model Based Testing for Agent Systems

Zhiyong Zhang, John Thangarajah, and Lin Padgham

School of Computer Science, RMIT, Melbourne, Australia
{zhzhang,johthan,linpa}@cs.rmit.edu.au

Abstract. Although agent technology is gaining world wide popularity, a hindrance to its uptake is the lack of proper testing mechanisms for agent based systems. While many traditional software testing methods can be generalized to agent systems, there are many aspects that are different and which require an understanding of the underlying agent paradigm. In this paper we present certain aspects of a testing framework that we have developed for agent based systems. The testing framework is a model based approach using the design models of the Prometheus agent development methodology. In this paper we focus on model based unit testing and identify the appropriate units, present mechanisms for generating suitable test cases and for determining the order in which the units are to be tested, present a brief overview of the unit testing process and an example. Although we use the design artefacts from Prometheus the approach is suitable for any plan and event based agent system.

1 Introduction

Agent systems are increasingly popular for building complex applications that operate in dynamic domains, often distributed over multiple sites. While the dream of theory based verification is appealing, the reality is that these systems are reliant on traditional software testing to ensure that they function as intended. While many principles can be generalised from testing of object oriented systems [1], there are also aspects which are clearly different and that require knowledge of the underlying agent paradigm.

For example in many agent systems paradigms (including BDI - Belief, Desire, Intention [2]) there is a concept of an *event* which triggers selection of one of some number of identified plans, depending on the situation. If one of these plans is actually never used, then this is likely to indicate an error. The concepts of event and plan, and the relationships between them are part of typical agent designs, and can thus be used for model based testing of agent systems. Effective testing of an agent system needs to take account of these kinds of relationships.

In this paper, we describe some of the aspects of a framework we have developed to automatically generate unit test cases for an agent system, based on the design models. The testing framework includes components that generate the order in which the units are to be tested, generate inputs for creating test cases, automate the test case execution, augment the system code to enable the testing to be performed, and a test agent that activates the testing process, gathers the results and generates a report that is easily understood.

J. Filipe et al. (Eds.): ICSOFT/ENASE 2007, CCIS 22, pp. 399–413, 2008.

We base our approach on the notion of model based testing ([3,4]) which proposes that testing be in some way based on design models of the system. There are a number of agent system development methodologies, such as Tropos [5], Prometheus [6], MASE [7] and others, which have well developed structured models that are potentially suitable as a basis for model based testing. In our work we use the Prometheus models. The models that are developed during design provide information against which the implemented system can be tested, and also provide an indication of the kind of faults that one might discover as part of a testing process.

There has been some work by others on testing agent systems in recent years. However, they have either focused on testing for the properties of abstract BDI-agents [8], or performed *black box* testing of the system [9].

In this paper, we focus on *unit testing* the components of a single agent. Unlike more traditional software systems, such as those based on Object-Oriented principles, where the base units are classes that are called via method invocation, the units in agent systems are more complex in the way they are called and are executed. For instance, plans are triggered by events, an event may be handled by more than one plan, plans may generate events that trigger other plans either in sequence or in parallel and so on. A testing framework for agent based systems must take these details into consideration in identifying the appropriate units and developing appropriate test cases.

In the sections ahead, we first identify what the natural units for an agent based system are, and how we use the model to determine the various test cases and their expected outcomes. We then provide an overview of the testing process and provide details on the reasoning that is done regarding dependencies between units, the necessary ordering of test cases, and the way in which inputs are generated for the various test cases. We provide a brief example from the evaluation with a case study, and then conclude with a discussion that identifies related and future work.

2   Test Units

The type of testing that we perform is *fault-directed* testing, where we intend to reveal faults in the implementation through failures [1, p.65]. This is in contrast to *conformance-based* testing, which tests whether the system meets the business requirements[1].

In order to perform fault-directed testing we require knowledge about the failures that can occur within the design paradigm (often called the fault model). In this section, we identify the units to be tested and identify possible points of failure for each unit that are independent of the implementation. We begin by examining the Prometheus design artefacts to identify suitable units for testing. Figure 1 outlines the components of an agent within the Prometheus methodology[2]. An agent may consist of plans, events and belief-sets, some of which may be encapsulated into capabilities. Percepts and incoming messages are inputs to the agent, while actions and outgoing messages are outputs from the agent.

---

[1] We expect to eventually also add conformance based testing, using use cases and other artefacts from the models developed at the requirements analysis stage, rather than the detailed design models being used here.

[2] Other agent oriented methodologies use similar constructs.

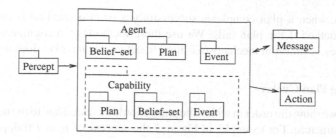

Fig. 1. Agent Component Hierarchy in Prometheus

Beliefsets are essentially the agent's knowledge about the environment and therefore constitute the situations in which testing must be done. The basic units of testing then are the plans and the events. Percepts and messages are also treated as events in agent development tools like JACK [10] and similar systems, and we also use this same generalisation.

We now discuss informally appropriate fault models for testing *plans* and *events*.

2.1 Testing Plans

A plan in its simplest form consists of a *triggering event*, a *context condition*, which determines the applicability of the plan with respect to the agent's beliefs about the current state of the world, and a *plan body* which outlines a sequence of steps. These steps may be subtasks, activated by posting events that are handled by the agent itself or external message events, which will be handled by another agent.

When we consider a plan as a single unit we test for the following aspects:

- Does the plan get triggered by the event that it is supposed to handle?
  If it does not, then there could be two possible reasons. The first is that some other plan always handles it, and the other is that there could be an inconsistency between the design and code and no plan actually handles that particular event[3].
- Is the context condition valid?
  The context condition for a plan is optional. The absence of a context condition denotes that the plan is always applicable. However, if the designer includes a context condition, then it should evaluate to true in at least one situation and not in all.
- Does the plan post the events that it should?
  Events are posted from a plan to initiate sub-tasks or send messages. If some expected events are never posted, we need to identify them as this may be an error.
- Does the plan complete?
  While it is difficult to determine whether a plan completes successfully or not, we can at least determine whether the plan executed to completion. If the plan does not complete then there is an error[4]. In implementation systems like JACK[10], for

---

[3] Here we can only check if the design matches the code, and can not check, for example, if the plan handling a particular event is correct or sensible.

[4] When deployed, a plan may well fail due to some change in the environment after the time it was selected. However, in the controlled testing situation where there are no external changes, then a plan that does not complete properly (due to failure at some step) should not have been selected.

example, when a plan completes successfully a success method is invoked, or a failure method if the plan fails. We use these methods to recognize when a plan completes. A time-out mechanism is used to detect when a plan does not complete.

2.2 Testing Plan Cycles

In Section 4 we show the order in which the plans should be tested due to the dependencies in the plan structure. For example, in Figure 2 the success of plan $P0$ depends on the success of either plan $P2$ or plan $P3$. These dependencies may on some occasions be cyclic. For example, there is a cyclic dependency between plans $P0$, $P2$ and $P1$. In this special case we cannot test each plan individually as they are dependent on each other. Hence, such plans are considered as a single unit which we shall term *cyclic plans*.

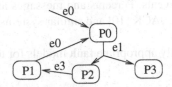

Fig. 2. Plan Dependencies

Each plan in the cycle is tested for the aspects discussed above, and in addition the following aspects are tested with respect to the cycle that they form:

- Does the cycle exist at run-time?
  If the cycle never occurs at run-time then the developer of the system should be notified, as the cycle may have been a deliberate design decision[5].
- Does the cycle terminate?
  Using a pre-defined maximum limit for the number of iterations in the cycle, we can determine if the cycle exceeds that limit and warn the user if it does.

2.3 Testing Events

An event as we generalized previously is either a percept, a message, or an event within the agent. The purpose of the event is to trigger the activation of a plan. Each event unit is tested for the following:

- Is the event handled by some plan?
  If the event does not trigger a plan, it could be due to two reasons. The first is if there is no plan that handles that particular event (which is easily checked by the compiler). The second is if the context conditions of all the plans that handle the event are false. This is a test for coverage.
- Is there more than one plan applicable for the event?
  If at design time the developer has indicated that only one plan is expected to be applicable, then the existence of multiple applicable plans for a given situation (referred to as overlap) is an error.

---

[5] Alternatively the cycle can be detected at design time and the developer asked whether it is expected to occur at runtime. This information can then be used in testing.

Mistakes in specification of context conditions in plans, leading to unexpected lack of coverage, or unexpected overlap, are common causes of error in agent programming. Consequently it is a good idea to warn the user if this occurs (though they can also specify that it is expected in which case no warning need be generated).

3 Testing Process: Overview

The unit testing process consists of the following steps:

- Determination of the order in which the units are to be tested.
- Development of test cases with suitable input value combinations.
- Augmentation of the code of the system under test with special testing code to facilitate the testing process.
- Testing, gathering of results, analysis and generation of an appropriate test report.

All of the above steps are automated and can be performed on a partial implementation of the system if needed. This supports the *test as you go* approach of unit testing.

In this section we briefly discuss the process of testing each type of unit. In the sections to follow we present the method for determining the order in which the units are to be tested and a mechanism for generating practically feasible input combinations for the test cases for each unit. Due to space limitation we do not discuss the implementation of augmenting the code of the system under test or the process of the report generation.

3.1 The Testing Framework

Figure 3 shows an abstract view of the testing framework for a plan unit. It has two distinct components, the *test-driver* and the *subsystem under test*. The test-driver component contains the test-agent, testing specific message-events that are sent to and from the test-agent, and a plan (test-driver plan) that initiates the testing process. This plan is embedded into the subsystem under test as part of the code augmenting process. The *subsystem under test* is the portion of the system that is needed for testing of the relevant unit and includes the necessary data and beliefsets, the *supporting hierarchy* of the *key plans* and the *key units*. The supporting hierarchy of a key plan is the events and plans on which it is dependent for full execution. For testing a plan, the key units are the plan itself, and its triggering event. For testing an event the key units are the event and all plans for which that event is a trigger. For testing a plan cycle the key units are all plans in the cycle and their triggering events.

Figure 3 illustrates the steps in the testing process for a plan: the *test-agent* generates the test cases, and runs each test case by sending an activation message to the *test-driver plan*; the *test-driver plan* sets up the input and activates the subsystem under test that executes and sends information (via specially inserted code) back to the *test-agent*; when testing is complete the *test-agent* generates a report which addresses the questions we have discussed related to each unit in section 2.

Figure 4 shows a similar process for an event, testing for coverage and overlap.

Plans that form a cyclic dependency (a cyclic plan set) need to be tested together. In addition to testing each plan in the set, in the same way as a single plan, the specific

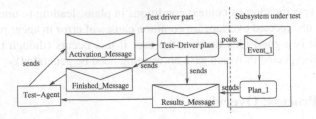

Fig. 3. Abstract Testing Framework for a Plan

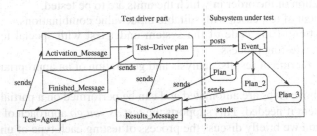

Fig. 4. Abstract Testing Framework for an Event

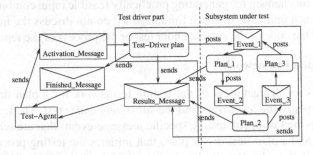

Fig. 5. Abstract Testing Framework for Cyclic Plans

questions about the cyclic dependency need to be assessed: does the cycle occur at run-time? does the cycle terminate? Figure 5 shows the test-driver activating just one plan of the cycle (*Plan_1*), this however must be done in turn for each plan of the cycle.

3.2 Automated Code Augmentation

In order to execute the various test cases generated, the code of the system under test is augmented to include special code for testing purposes. This augmentation is automated. The first step in the process is to copy the original source code into testing directories. There is a separate directory for each test case as the code modifications depends on the test case. Into each test system the *test driver* component (see figure 3) is added and testing specific code is inserted into the code of the *key units* (discussed above). These augmented test systems are then compiled so that the test cases maybe executed.

```
package conference_travel;
/** BEGIN - import statements for testing
import au.edu.rmit.cs.prometheus.test.*;
import tm_agent.*;
/** END - import statements for testing
import conference.bookings.*;

public agent Book_Conference_Agent extends Agent {
    /**BEGIN Test Code -  add the capability for testing*/
    #has capability cap_AUT_Testing  cap_aut_test;
    /**END Test Code -  add the capability for testing*/
    #has capability Transport cap;
    #posts event Lookfor_Transport m_ev_trans;

    public Book_Conference_Agent(String name) { super(name); }
    public void lookforTransport(Long nBudget) {.....}
    ......
    /*************** BEGIN Test Code - Recording methods ******************/
    /* BIT code: this method informs the TM_Agent the CC value of the Plan
     * pName. It is invoked in two places: 1.the beginning of "body()" method
     * 2. BIT_Start_Plan */
    public boolean BIT_record_CCResult(PlanTestParams p,String pName,
                                        boolean ccVal){...}
    /* BIT code: this method informs the TM_Agent about the message received
     * by the PUT. It is invoked in the "context()" method of PUT */
    public boolean BIT_record_PlanReceivedMsg(PlanTestParams params,
                                        String strPlanName){...}
    /* BIT code: this method informs the TM_Agent when messages are sent out
     * by the PUT. It is invoked before the "@post/@send/@subtask" calls in
     * the body of the PUT */
    public boolean BIT_record_MsgOut(PlanTestParams params, Event outEvent,
                                     String strPlanName, int sendType) {...}
    ....
    /*************** END Test Code - Recording methods ******************/
```

Fig. 6. Example of an Augmented code for an Agent

```
....

#reasoning method body()
{
    /** BEGIN Test Code - report the true CC value to the TM_Agent*/
    m_in_agency.BIT_getPlanTestParams().setActuallyTriggeredPlanName(
                                        BIT_getCurrPlanName());
    bit_self.BIT_record_CCResult(m_in_agency.BIT_getPlanTestParams(),
                                 BIT_getCurrPlanName(), true);
    /** END Test Code - report the true CC value to the TM_Agent*/
    ....
```

Fig. 7. Example of an Augmented code for a Plan body

For example, Figure 6 shows an example of the code of an agent from a conference travel booking system implemented in JACK [10], where the agent under test is the *Book_Conference_Agent*. The code of that agent is modified to include: the relevant packages for testing (via import statements), the testing capability, and reasoning methods required to record various aspects of testing such as the messages received and sent out by its plans when they are tested. Figure 7 shows an example of a plan under test that is modified to include, among others, code at the start of the plan body to record the success of the context condition of that plan and that is has begun execution.

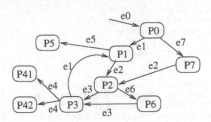

Fig. 8. Testing order

```
PROCEDURE getOrder(PlanNode N)
    IF tested(N) THEN
        terminate the procedure
    stack.push(N) // Store the current path explored
    FOR EACH child Nᵢ of N
        IF Nᵢ is the ancestor of any Plan in the stack
            THEN testqueue.add(CT(Nᵢ, . . . , N))
        ELSE getOrder(Nᵢ)
    FOR EACH child-set N(Nᵢ, Nⱼ, . . .) that share
    the same trigger e
        testqueue.add(ET(e))
    testqueue.add(PT(N))
    stack.pop(N)
END PROCEDURE
```

Fig. 9. Testing order: Algorithm

4   The Order of Testing

Recall that an event may be handled by one or more plans and each plan may post sub-tasks. The success of the top level plans is partly dependent on the success of the plans triggered by the sub-tasks (if any). The order of testing is, therefore, bottom-up where we test a unit before we test any other unit that depends on it. For example, from Figure 8 we test plan $P7$ before we test plan $P0$. The complicating factor is the presence of cyclic dependencies. Plans that form cyclic dependencies are to be tested together as a single unit as previously described.

In order to determine the order of testing we apply the following steps. We use Figure 8 as an example design and abbreviate the following: Plan Test - PT; Event Test - ET; Cyclic Plans Test - CT.

1. We perform a modified depth-first search outlined in Figure 9, which performs a typical depth-first search but also identifies cyclic dependencies as well as plans that share the same trigger event (for testing coverage and overlap). The order returned by this algorithm for our example is: PT(P5), CT(P3, P1, P2), PT(P41), PT(P42), ET(e4), PT(P3), CT(P6, P3, P1, P2), PT(P6), PT(P2), PT(P1), PT(P7), PT(P0).
2. From the above order, we can eliminate all unit test of plans that are part of any cyclic dependency as they will be tested when the cyclic plans are tested. The resulting ordered set is: PT(P5), CT(P3, P1, P2), PT(P41), PT(P42), ET(e4), CT(P6, P3, P1, P2), PT(P7), PT(P0).

3. In the order above the cyclic plans are not in the correct order as they must be tested only when all of its plans' children have been tested. For instance P41 is a child of P3. This re-ordering is a trivial operation and when complete reveals: PT(P5), PT(P41), PT(P42), ET(e4), CT(P3, P1, P2), CT(P6, P3, P1, P2), PT(P7), PT(P0).
4. The final step is to combine cyclic dependencies that overlap. By overlap we mean cycles that have at least one plan in common. In our example one cycle is a subset of the other hence when merged the resulting final order of testing is: PT(P5), PT(P41), PT(P42), ET(e4), CT(P6, P3, P1, P2), PT(P7), PT(P0).

5   Test Case Input Generation

The variables that we consider as test parameters are those within the context conditions or body[6] of the plans to be tested and the variables of the *entry-event*. The entry-event is the initial trigger event when testing a plan, or is the event itself when testing an event. The variables within the event may be used within the plan body. We need to generate practical combinations of these variables to adequately test the plans and events.

There are 3 steps in generating value combinations that form different test cases:

1. Variable extraction from the design documents.
2. Generation of Equivalence Classes, which is a heuristic for reducing the size of the input range. While the concept of Equivalence Classes is not novel, we have adopted our own techniques in applying the heuristic.
3. Generating the input combinations from the equivalence classes using a heuristic to reduce the number of different combinations.

5.1   Extraction of Variables

The detailed description of plans and events in our design documents contains a list of variables and their types. For variables in context conditions we also have a list of conditions that must be satisfied for the context condition to return True. We call values that satisfy these conditions **valid** variables for the context condition. Following are some examples of such variables and their associated conditions.

stock, int, $\geq 0; \leq 200;$
price, float, $>0.0;$
bookName, string, $!=null;$
supplier, SupplierType, $==$ *"Amazon",* $==$ *"Powells"*

In our testing framework we define four basic variable types: *integer, float, string* and *enumerated*. Other types are considered as special cases of these four basic ones. For example, *boolean* is considered as a special case of *enumerated*, and *double* is a special case of *float*. The definition of the enumerated types must be contained within the design. For example, the enumerated type *SupplierType* may be defined as:

[EnumType, SupplierType, { *"Amazon",*
"Angus&Robertson", "Powells", "Dymocks" }*].*

[6] We have not yet implemented the use of variables in the body other than those in the context condition and the event. However the information is available from the design and follows the same principles.

5.2 The Generation of ECs

It is not possible to create test cases for every valid value of a variable since some domains are infinite, such as $(0.0, +\infty)$ Additionally we wish to test with some invalid values. Even for non-infinite domains the number of test values may be extremely large. To address this issue we use the approach of *equivalence partitioning* [11, p.67] to obtain a set of representative values. An *Equivalence Class* (EC) [1, p.401] is a set of input values such that if any value is processed correctly (or incorrectly), then it is assumed that all other values will be processed correctly (or incorrectly). We consider the open intervals and the boundary values of the variable domains to generate ECs, as the former gives equivalent valid values and the latter are *edge* values that should be checked carefully during testing. We also consider some invalid values.

An EC that we define has five fields:

1. *var-name*: The name of the variable.
2. *Index*: A unique identifier.
3. *domain*: An open interval or a concrete value.
4. *validity*: Whether the domain is valid or invalid.
5. *sample*: A sample value from the domain: if the domain is an open interval (e.g. $(0.0, +\infty)$), it is a random value of this interval (e.g 778); if the domain is a concrete value (x=3), it is this value.

Table 1 gives the equivalence classes for the example variables above.

When generating the ECs for a particular variable we use the following rules (we refer to Table 1):

- One EC is generated for each boundary value of the variable. The *sample* value of that EC is the boundary value. E.g., for variable 'stock', EC-2 and EC-4 are created using the boundary values.
- For an *integer* or *float* variable, one EC is generated for each open interval between two neighbouring boundary values. The *sample* value is a random value in this interval. E.g., for variable 'stock', EC-1 EC-3, and EC-5 are generated using boundary value intervals.
- For a *string* variable, one EC is generated to represent the domain of valid values. The *sample* value is a random string that is not a valid value. E.g., for variable 'bookName' EC-2 is such an EC.
- For an *string* variable, one EC is generated to accommodate the NULL value.
- For an *enumerated* variable, one EC is generated for each value of the enumerated type.

The generated ECs for the sample variables given above are displayed in Table 1.

5.3 Reducing the Size of the Test Set

It is straightforward to generate the set of all possible combinations of variable ECs, which could then be used for the value combinations for test cases. However the number of the combinations may still be quite large. In our example, there are 120 combinations of all the variables. This number can be reduced further by using the approach of combinatorial design [12]. This approach generates a new set of value combinations that

Table 1. ECs of all variables

variable	index	domain	valid	sample
stock	EC-1	(-∞, 0)	no	-823
	EC-2	0	yes	0
	EC-3	(0.0, 200)	yes	139
	EC-4	200	yes	200
	EC-5	(200, +∞)	no	778
price	EC-1	(-∞, 0.0)	no	-341.0
	EC-2	0.0	yes	0.0
	EC-3	(0.0, +∞)	yes	205.0
book Name	EC-1	NULL	no	NULL
	EC-2	not NULL	yes	"random"
supplier	EC-1	"Amazon"	yes	"Amazon"
	EC-2	"Angus& Robertson"	yes	"Angus& Robertson"
	EC-3	"Powells"	yes	"Powells"
	EC-4	"Dymocks"	yes	"Dymocks"

Table 2. List of EC value combinations

index	stock	price	bookName	supplier
1	139	205.0	"random"	"Amazon"
2	200	205.0	"random"	"Amazon"
...	...	...	...	...
23	139	205.0	NULL	"Powells"
24	200	205.0	"random"	"Powells"

cover all n-wise (n≥2) interactions among the test parameters and their values in order to reduce the size of the input data set. Hartman and Raskin have developed a software library called CTS (Combinational Testing Service)[7] which implements this approach. We do not expand on these techniques as we do not modify them by any means. Using this software we are able to reduce the set of 120 combinations to a smaller set of 24 combinations. We then use the sample value from each EC to obtain the concrete test data for each test case that will be run. Table 2 shows some sample value combinations from the reduced list, where each combination represents the input to a unique test case. Whether or not this method is used can be determined depending on the number of total combinations. It is also possible to differentiate between valid and invalid data, reducing the number of combinations for invalid data, while using all possibilities for valid data to ensure that all options through the code are exercised.

6   Case Study

As a first step in evaluating our testing framework we took a sample agent system, systematically introduced all types of faults discussed in section 2 into the system and used it as input to the testing framework. The testing framework successfully uncovered each of these faults in the automated testing process.

---

[7] http://www.alphaworks.ibm.com/tech/cts

Table 3. ECs of all variables

variable	index	domain	valid	sample
BookID	EC-1	$(0, +\infty)$	yes	11
	EC-2	$(-\infty, 0)$	no	-2
	EC-3	0	yes	0
Number	EC-1	$(0, +\infty)$	yes	8
Ordered	EC-2	$(-\infty, 0)$	no	-9
	EC-3	0	no	0
Urgent	EC-1	yes	yes	yes
	EC-2	no	yes	no

The sample system that we used, was the *Electronic Bookstore* system as described in [6]. This is an agent-based system dealing with online book trading, containing agents such as *Customer Relations, Delivery Manager, Sales Assistant* and *Stock Manager* agents. We used the *Stock Manager* agent as the agent under test (AUT), and specifically edited the code to introduce all identified types of faults. The testing framework generator automatically generated the testing framework for the testable units of the *Stock Manager* agent, and then executed the testing process for each unit following the sequence determined by the testing-order algorithm. For each unit, the testing framework ran one test suite, which was composed of a set of test cases, with each case having as input one of the value combinations determined.

For example, as discussed earlier, one kind of fault that can occur is that a particular subtask is never posted from a plan, despite the fact that the design indicates it should be. In the *Stock Manager* the plan *Out of stock response* had code that, when the book is needed urgently and the number of ordered books is less than 100, checks if the default supplier currently has stock and if not posts the subtask *Decide supplier*. We modified the body of the code for *Out of stock response* so that a condition check would always be false, thus leading to the situation that the *Decide supplier* subtask event would in fact never be posted.

The plan *Out of stock response* had as its trigger event *No stock* which included the boolean variable *Urgent*. The context condition of this plan was:

(BookID $\geq$ 0 AND NumberOrdered > 0).

Within the body of the plan we had the code :

```
IF       Urgent=YES AND NumberOrdered < 100
THEN   postEvent(Decide supplier)
ENDIF
```

To introduce the fault into the system we modified the IF condition above to be :

IF Urgent=YES AND NumberOrdered < 0

which will of course result in *Decide supplier* never being posted. The input arguments for the test are then *BookID, NumberOrdered* and *Urgent*, with the following specifications:

BookID, int, $\geq$ 0
NumberOrdered, int, > 0
Urgent, boolean

Table 4. List of Equivalence Class combinations

index	BookID	Number Ordered	Urgent	Validity
1	EC-1 (11)	EC-1 (8)	EC-1 (yes)	valid
2	EC-3 (0)	EC-1 (8)	EC-1 (yes)	
3	EC-1 (11)	EC-1 (8)	EC-2 (no)	
4	EC-3 (0)	EC-1 (8)	EC-2 (no)	
5	EC-1 (11)	EC-2 (-9)	EC-1 (yes)	invalid
6	EC-1 (11)	EC-2 (-9)	EC-2 (no)	
7	EC-1 (11)	EC-3 (0)	EC-1 (yes)	
8	EC-1 (11)	EC-3 (0)	EC-2 (no)	
9	EC-2 (-2)	EC-1 (8)	EC-1 (yes)	
10	EC-2 (-2)	EC-1 (8)	EC-2 (no)	
11	EC-2 (-2)	EC-2 (-9)	EC-1 (yes)	
12	EC-2 (-2)	EC-2 (-9)	EC-2 (no)	
13	EC-2 (-2)	EC-3 (0)	EC-1 (yes)	
14	EC-2 (-2)	EC-3 (0)	EC-2 (no)	
15	EC-3 (0)	EC-2 (-9)	EC-1 (yes)	
16	EC-3 (0)	EC-2 (-9)	EC-2 (no)	
17	EC-3 (0)	EC-3 (0)	EC-1 (yes)	
18	EC-3 (0)	EC-3 (0)	EC-2 (no)	

This gives the equivalence classes as shown in table 3, giving 18 possible combinations of values shown in table 4, which is reduced to 9 if the combinatorial testing reduction is used.

This error was discovered by the testing system by analysing the results of the test suite and observing that *Decide supplier* was never posted. The following is an example of the warning message that is supplied to the user:

Type of Fault: *Subtask never posted*
WARNING: The event *Decide supplier* is never posted in any test case. Value combinations used in test suite were:[8]

BookID=11 NumberOrdered=8 Urgent=yes
BookID=0 NumberOrdered=8 Urgent=yes
BookID=11 NumberOrdered=8 Urgent=no
BookID=0 NumberOrdered=8 Urgent=no

If some other value combination would result in posting of event *Decide supplier*, please provide these values to the testing system.

7 Discussion

The need for software testing is well known and accepted. While there are many software testing frameworks for traditional systems like Object-Oriented software systems, there is little work on testing Agent-Oriented systems. In particular to the best of our knowledge there is no testing framework that is integrated into the development methodology.

---

[8] Only valid values are provided as invalid values would not cause the plan to run and are hence irrelevant for this error.

In this paper we present part of a framework for testing agent systems that we have developed, which performs model based unit testing. We have identified as units, plans, events that are handled by multiple plans, and plans that form cyclic dependencies. We have presented an overview of the testing process and mechanisms for identifying the order in which the units are to be tested and for generating the input that forms test cases.

There has been some work on testing agent based systems in recent years (e.g [8,9]). The former provides an approach to compare the properties of the agent and the observable behaviours with the specification of the system, by building a behavioral model for the system using extended state machines. The latter studied how to build a state-based model for an agent-based system using extended Statecharts, and then proposed an approach to generate test sequences. Both of the above work is based on *conformance testing*, which tests if the system meets the business requirements and are restricted to *black-box* testing. In contrast to these approaches, our work looks at *fault directed testing* which tests the internal processes of the system and not the business requirements. Our approach is also integrated with the design methodology and supports testing at early stages of development.

There are other work on multi-agent testing that defines agents as test units (e.g [13,14]). We however, explore the internals of an agent and choose plans and events as test units.

While we obtain and use more structural information than standard black box testing, we are limited in the information we use as we obtain this information from the design. Hence, implementation specific structure is not considered. The testing framework is also reliant on the implementation following the design specification.

Although we have completed the implementation of the testing framework using JACK Intelligent Systems [10], and done some preliminary evaluation as discussed in the previous section, further evaluation is required. For this purpose we intend to use programs developed by post-graduate students as part of an agent programming course.

In this work we have only addressed unit testing, in future work we will extend this work to include *integration* testing. To this end, we expect to build on existing work (e.g. [15,16]). The former described a debugger which, similar to this work, used design artefacts of the Prometheus methodology to provide debugging information at run-time. Their approach of converting protocol specifications to petri-net representations is of particular relevance to our future work on integration testing. The latter presented a unit testing approach for multi-agent systems based on the use of *Mock-Agents*, where each Mock-Agent tests a single role of an agent under various scenarios.

As future work we also look to embed the testing functionality into the Prometheus Design Tool (PDT) [17]. PDT is a tool for developing agent systems following the Prometheus methodology, and includes automated code generation which we hope to extend to generate testing specific code.

Acknowledgements. We would like to acknowledge the support of the Australian Research Council and Agent Oriented Software, under grant LP0453486.

References

1. Binder, R.V.: Testing Object-Oriented Systems: Models, Patterns, and Tools. Addison-Wesley Longman Publishing Co., Inc., Boston (1999)
2. Rao, A.S., Georgeff, M.P.: BDI Agents: From Theory to Practice. In: Lesser, V. (ed.) The First International Conference on Multi-Agent Systems, San Francisco, pp. 312–319 (1995)
3. Apfelbaum, L., Doyle, J.: Model Based Testing. In: The 10th International Software Quality Week Conference, CA, USA (1997)
4. El-Far, I.K., Whittaker, J.A.: Model-Based Software Testing. In: Encyclopedia of Software Engineering, pp. 825–837. Wiley, Chichester (2001)
5. Bresciani, P., Perini, A., Giorgini, P., Giunchiglia, F., Mylopoulos, J.: Tropos: An Agent-Oriented Software Development Methodology. Autonomous Agents and Multi-Agent Systems 8, 203–236 (2004)
6. Padgham, L., Winikoff, M.: Developing Intelligent Agent Systems: A practical guide. Wiley Series in Agent Technology. John Wiley and Sons, Chichester (2004)
7. DeLoach, S.A.: Analysis and design using MaSE and agentTool. In: Proceedings of the 12th Midwest Artificial Intelligence and Cognitive Science Conference (MAICS 2001) (2001)
8. Zheng, M., Alagar, V.S.: Conformance Testing of BDI Properties in Agent-based Software Systems. In: APSEC 2005: Proceedings of the 12th Asia-Pacific Software Engineering Conference (APSEC 2005), Washington, pp. 457–464. IEEE Computer Society Press, Los Alamitos (2005)
9. Seo, H.S., Araragi, T., Kwon, Y.R.: Modeling and Testing Agent Systems Based on Statecharts. In: Núñez, M., Maamar, Z., Pelayo, F.L., Pousttchi, K., Rubio, F. (eds.) FORTE 2004. LNCS, vol. 3236, pp. 308–321. Springer, Heidelberg (2004)
10. Busetta, P., Rönnquist, R., Hodgson, A., Lucas, A.: JACK Intelligent Agents - Components for Intelligent Agents in Java. Technical report, Agent Oriented Software Pty. Ltd., Melbourne, Australia (1999)
11. Patton, R.: Software Testing, 2nd edn. Sams, Indianapolis (2005)
12. Cohen, D.M., Dalal, S.R., Fredman, M.L., Patton, G.C.: The AETG system: An Approach to Testing Based on Combinatorial Design. Software Engineering 23, 437–444 (1997)
13. Caire, G., Cossentino, M., Negri, A., Poggi, A., Turci, P.: Multi-Agent Systems Implementation and Testing. In: The Fourth International Symposium: From Agent Theory to Agent Implementation, Vienna (2004)
14. Rouff, C.: A Test Agent for Testing Agents and their Communities. In: Proceedings on Aerospace Conference, vol. 5, p. 2638. IEEE, Los Alamitos (2002)
15. Padgham, L., Winikoff, M., Poutakidis, D.: Adding Debugging Support to the Prometheus Methodology. Engineering Applications of Artificial Intelligence, special issue on Agent-Oriented Software Development 18, 173–190 (2005)
16. Coelho, R., Kulesza, U., von Staa, A., Lucena, C.: Unit Testing in Multi-Agent Systems using Mock Agents and Aspects. In: Proceedings of the 2006 International Workshop on Software Engineering for Large-Scale Multi-Agent Systems, pp. 83–90 (2006)
17. Thangarajah, J., Padgham, L., Winikoff, M.: Prometheus design tool. In: The 4th International Joint Conference on Autonomous Agents and Multi-Agent Systems, Utrecht, The Netherlands, pp. 127–128 (2005)

A Metamodel for Defining Development Methodologies

Manuel Bollain and Juan Garbajosa

System and Software Technology Group, Technical University of Madrid (UPM)
E.U. Informatica, Km. 7 Carretera de Valencia, E-28031 Madrid, Spain
mbollain@eui.upm.es, jgs@eui.upm.es

Abstract. The concept of software product is often associated to software code; process documents are, therefore, considered as by-products. It is also often the case that customers demand first and foremost "results" leaving documentation in second place. Development efforts are then focused on code production at the expense of document quality and corresponding verification activities. As discussed within this paper, one of the root problems for this is that documentation in the context of methodologies is often described with insufficient level of detail. This paper presents a metamodel to address this problem. It is an extension of ISO/IEC 24744, the metamodel for methodologies development. Under this extension, documents can become the drivers of methodology activities. Documents will be the artifact which method engineers should focus on for methodology development, defining their structure and constraints. Developers will put their effort into filling sections of the documents as the way to progress in process execution; in turn, process execution will be guided by those documents defined by the method engineers. This can form the basis for a new approach to a Document-Centric Software Engineering Environment.

1 Introduction

In software production, work products are both programs and documents. Methodologies describe the activities and tasks for producing such documents and programs but in many cases, organizations have problems following the defined process (if any). Time schedule delays force developers to skip over document production, and related activities, trying to release an "operational" product in time. This is a common situation that pushes companies to adhere to process maturity models, and attain a maturity level. Maturity models may guide efforts to cope with the documentation issue, do not provide a complete solution. This is true even in the case of defining a specific maturity model for documentation process as in [1] and even more so in scenarios with tight time constraints or high levels of pressure. Technicians like to solve technical problems, and associated documentation is often considered as the "boring" part of the work. For all these reasons, documentation quality is often neglected. We can also consider applying agile methods and therefore reduce the effort in documentation production, while maintaining some control over documentation quality and completeness. However, the ongoing debate still remains the same: "Working software over comprehensive documentation" [2].

One of the root problems is that the existing predefined processes and methodologies consider documents as one of their *natural outputs*. Very often they simply provide

J. Filipe et al. (Eds.): ICSOFT/ENASE 2007, CCIS 22, pp. 414–425, 2008.

guidelines on how to perform the process considering proper documents as resulting from performing good process practices. The problem with this assumption is that, being true, the granularity level is too coarse. It is necessary to establish, as formally as possible, the existing relationship between documents and other software process elements at a level of detail sufficient to consider the production of documents not simply as a result of activities but as a part of the process itself.

Within this paper, the concepts of process and methodology as described in [3] will be followed, and applied in the context of ISO 24744:2007 emerging standard [4]. Following this standard *A document is a durable depiction of a fragment of reality.*However it is possible to be more specific and consider that a document will be structured into sections and subsections. It is possible to describe models in terms of a document; UML class model is an example when described using OCL. It is just necessary to describe the model in a textual fashion. In our approach, all activities must correspond to a product, that is, a document; we adopt a strategy in which all process outputs can be represented as documents - this is a document-centric perspective of the software process as in [5] and [6]. In this way all the activities will produce a "touchable" result in form of a document; this includes source code. The approach to this research work is based on specifying the relationship between documents and the rest of the elements of the software process.

It is common that standards define activities as lists of tasks in which the verb *to document* participates at the beginning or the end of the list. For example, in ISO 12207 [7], we can go to section 5.3.4.1 that reads *The developer shall establish and **document** software requirements, including the quality characteristics specifications, described below [...].*

There are several standards and guidelines collections for the software development activities with detailed tables of contents, such as the International Organization for Standardization, IEEE Standards Collection - Software Engineering or the European Cooperation on Space Standardization (ECSS). These documents can be easily adopted in a document-centric methodology. Thus, the software engineers effort is less when defining a methodology that must include a previously normalized documentation.

As discussed above, in many cases, the description of an activity includes the act of documenting its execution results. This takes us to a situation in which, in practice, the tools used for assisting in an activity have some sort of utility intended for report or document generation. This report or document will contain the results of the activity, usually understood as the software product. This makes activity execution and documentation two different issues. If we accept that all software products will be documents, corresponding to previously defined templates, and that activities will be defined in accordance with those templates, the act of document creation will mean performing the corresponding activities or tasks. With this approach we found the following advantages:

– All the project information is within documents, following the template established by the method engineer and there is no need for further work to document the performed activities. This will require defining modelling formalisms in terms of documents as outlined in [6]. Stakeholders models adopted in software process models such as [8] and [9] can also be partially supported with this approach. Defining

roles for different participants or stakeholders could be achieved using the different points of view of each stakeholder for different documents or document sections.
- Documentation preparation is timely. This is ISO/IEC 12207, section 6.4.2.7 compliant. As progress in the development process is through documents fulfillment, it is not necessary to verify if documentation preparation is timely.

This paper is structured as follows: this section, 1, is an introduction. Related work is analysed in section 2. The necessary standard extensions for achieving a document-centric methodology definition, the outcome classes and types and the new relationships will be explained in section 3. How to make use of this extension for defining a document-centric methodology will be developed in section 4, and finally, some conclusions, advantages of the approach, pitfalls and future work will be presented in section 5.

2 Related Work

The Document-centric approach has been addressed in [5] and more recently in [10]. Reference [5] introduces a document-centric approach and concludes that a document driven approach is feasible and useful. However, the approach presented in [5] differs from this paper in the sense that it is based on how documents can be translated from a user-oriented form into machine equivalent forms; that is, it is more oriented to document processing and how to take advantage of this. Instead, the approach presented within this paper is based on a study of process metamodelling. Reference [10] introduces some ideas in line with this work but they are not developed in detail. Reference [11] presents an environment in which documents play an essential role, however, documents are simply an element necessary to support web applications. In [12], XML documents are simply project information containers, that is, documents and data are the same. Our approach makes it possible to support traceability and configuration management on the document itself. In all the cases, the source code is still independent of the project documents.

The use of documents to drive the software process is presented in [6] and [13]; these can be considered precedents for this research, although the theoretical background is not provided.

There are several techniques for defining and supporting process. Process definition usually lies on Process Modelling Languages (PML) based on some linguistic paradigms such as [14] Petri nets or logical languages, several of them discussed in [15]. Others are component-based software process support [16], role-based approaches [17], coordination rules [18]. Some interesting reflections can also be found in [19]. Even process support could be based on tools integration mechanisms, as in [20].

In [3], several approaches to process metamodelling are discussed: a well accepted modelling language such as UML [21] deal with modelling issues but neglect process, while widespread methodological frameworks such OPEN [22] or Extreme Programming [23] emphasize the process side and are less detailed when it comes to work product. At the start of 2007, ISO/IEC 24744:2007 [4] Metamodel for development methodologies was published; it offers a balance between process and product in the methodology definition issue. This standard has been the basis for defining our metamodel in which process and product will be closely related.

Other related works include document extraction from central repositories like in [24], [25] and [26]. Document content is based on project information, but just as a result of querying the project repository. Also Hypertext linking management in HTML documents has been studied in [27] and [28], where part of the project data, such as traceability information, is in the document, but the document info is still the result of querying a repository.

3   A Metamodel for Supporting Documents as *Full* Products

ISO/IEC 24744 standard [4], has a three domain context: the metamodel domain, the methodology domain and the endeavour domain, as shown in figure 1. According to ISO/IEC 24744 an endeavour is an information-based domain development effort aimed at the delivery of some product or service through the application of a methodology.

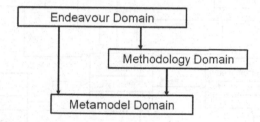

Fig. 1. Three domains layers conforming to ISO/IEC 24744:2007

Metamodels are useful for specifying the concepts, rules and relationships used to define methodologies. A methodology is the specification of the process to follow together with the work products to be used and generated, plus the consideration of the people and tools involved, during an Information-based domain development effort. A methodology specifies the process to be executed, usually as a set of related activities, tasks and/or techniques, together with the work products to be manipulated (created, used or changed) at each moment and by whom, possibly including models, documents and other inputs and outputs. In turn, specifying the models that must be dealt with implies defining the basic building blocks that should be used to construct these models.

Most metamodelling approaches define a metamodel as a model of a modelling language, process or methodology that developers may employ. Following this conventional approach, classes in the metamodel are used by the method engineer to create instances (i.e. objects) in the methodology domain and thus generate a methodology. However, these objects in the methodology domain are often used as classes by developers to create elements in the endeavour domain during methodology enactment. This apparent contradiction is solved by conceiving a metamodel as a model of both the methodology and the endeavour domains. Thus, we find a dual-layer modelling, one layer in the methodology domain and other in the endeavour domain. Modelling a methodology element in both layers should be performed by means of *Clabjets*- a dual entity that is both a class (for the endeavour layer) and an object (for the methodology layer) at the same time ISO/IEC 24744 is an open and flexible standard and has been designed to be extended. Therefore the approach to define in detail the existing relation between documents, tasks and other process elements was to extend ISO/IEC 24744 to provide the method engineer with a new tool. ISO/IEC 24744 classes are *immutable*. New attributes and relationships are only allowed to be introduced in new extended classes.

ISO/IEC 24744 contains a relationship between producer, task and document, as shown in figure 2, but this is not enough for defining a direct relationship between them. ISO/IEC 24744 classes will be extended in order to obtain a direct relationships between producer (role) and product (document). The same type of relationship between document, task, technique and tool is required. The following paragraphs present how classes have been extended.

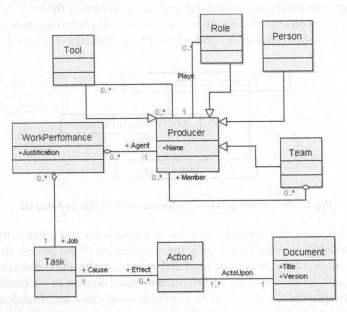

Fig. 2. Relationships between producers, tasks and documents in the standard

Extension of Role Class. This is a subclass of Producer. In ISO/IEC 24744 it is possible to define the roles to be played by persons, teams or even tools. There is an indirect relationship between producers and products through Workperformance, Task and Action classes. It is possible to specify which producers have been involved in creating different products, but is not possible to state explicitly which participant is involved in the creation of a particular product. In our approach, a direct relationship for setting this participation is needed. At the same time, it is necessary to set the producers role type in relation to the product. To achieve this, a new attribute called Type is created with two possible values: *producer* or *consumer*. The new extended class is called Figure and it is a subclass of the ISO/IEC 24744 class Role. Both extended classes in the methodology domain and in the endeavour domain are shown in figure 3.

Extension of Document Class. As presented in figure 4, Document class and its parent class, WorkProduct are the main classes in our approach. It is important to highlight

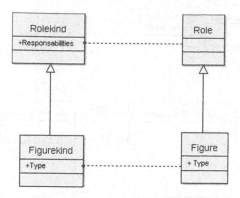

Fig. 3. Role Class Extension

that a Document could have a parent document. This hierarchy allows a *subdocuments* schema that could be used to support document sections and subsections. This provides the metamodel with a mechanism for defining the methodology documentation at any level of detail, depending on the required document granularity. The Document class extension is necessary for setting the following new relationships:

- Relationship with producers, that is, with Figure class, as previously discussed.
- Relationship with tasks, techniques and tools needed for document or subdocument (section) development. In ISO/IEC 24744 there is a direct relationship between tasks and products (documents) through the Action class, but it is not possible to state the same type of relationship with the applicable techniques. Although this direct relationship is not necessary in the methodology domain, we consider it essential in the endeavour domain for our approach. The same case occurs with the *tools* relationship: there is no direct relationship in ISO/IEC 24744, but we consider it necessary for achieving a well defined document-centric methodology .
- Relationship between documents for setting up constraints among them. In the methodology domain, the different types of documents and the constraints among them should be defined. A document could have constraints that involve other documents that are subdocuments of the first one.

Extension of Constraint Class. The Standard Constraint class is a condition that holds or must hold at a certain point in time. Constraints are often used to declaratively characterize the entry and exit conditions of actions. This class applies only in the methodology domain, making no sense in the endeavour level. Initially, it is related to the ActionKind class, setting conditions form the execution of its related actions. In our approach, the constraints will be applied to documents: the document production will depend on conditions applied to other documents (or subdocuments). From this, it is necessary to extend of the Constraint class and to set a new relationship.

Extension of TaskTechniqueMapping Class. the TaskTechniqueMapping ISO/IEC 24744 class establishes the relationship between tasks and the applicable techniques in both the methodology and the endeavour domains. However, our aim, as discussed

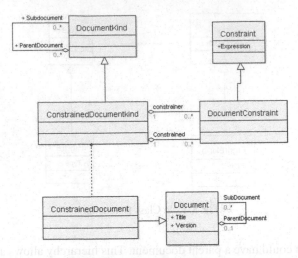

Fig. 4. Document Class Extension

above is to provide the method engineer with the possibility of assigning tasks to documents and corresponding techniques with a direct relationship. At the same time, we consider useful the possibility of assigning which tools should be involved in certain tasks execution, following certain techniques, related to these documents. For this, we need to extend the TaskTechniqueMapping ISO/IEC 24744 class in the new Document-TaskTechniqueToolMapping class. This makes it possible directly to assign the document to the corresponding tasks for completing it, the applicable techniques in each case, and the tools to be used. Figure 5 shows the classes involved in this mapping. In our approach, this mapping is necessary in both methodology and endeavour domains in order to provide support not only for the methodology definition, but also for the Document-Centric Software Engineer Environments.

Extension of Tool Class. ISO/IEC 24744 tool class is a subclass of the Producer class and it is related to other producers that are assisted by it. As Producer subclass, it has the same relationship with products as described for Role class. As previously discussed, in our approach, tools are related in a direct way with tasks and techniques required for developing a document. The relationship between tools and assisted producers is established, in an indirect way, through the document in which they take part. The relationship between Tool class and DocumentTaskTechniqueToolMapping could be achieved by the extension of the Tool class into a new class called AssignedTool.

The metamodel extension, in the methodology domain, is shown in figure 6. The ConstrainedDocumentkind is the core of our proposal: Figures are related to documents and will follow the tasks, use the techniques and tools attached to the documents under the corresponding documents constraints.

The metamodel extension in the endeavour domain is shown in figure 7. It is very similar to the corresponding methodology domain point of view. The main difference lies in the lack of constraints in this domain.

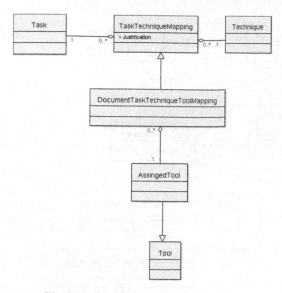

Fig. 5. Task,Technique and tool mapping

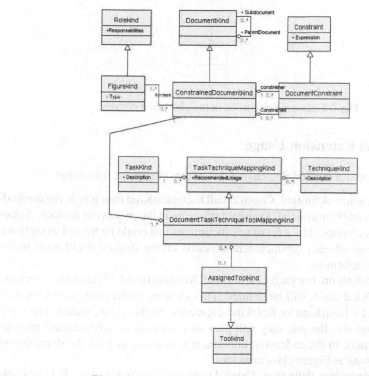

Fig. 6. Metamodel extension in the methodology domain

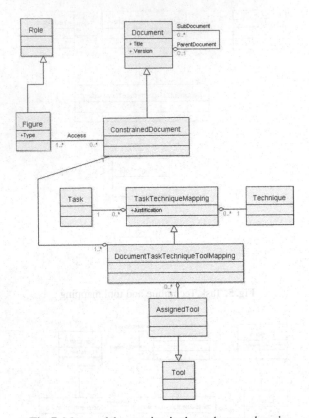

Fig. 7. Metamodel extension in the endeavour domain

4   Metamodel Extension Usage

The steps for defining a methodology using this extension are the following:

- Document structure definition. ConstrainedDocumentKind structure in the method-ology domain and ConstrainedDocument structure in the endeavour domain. A doc-ument kind is composed by a set of subdocuments that could be treated as sections. The detail level of each subdocument depends on the desired detail level in the methodology definition.
- FigureKind definition. For each ConstrainedDocumentKind, FigureKind, an exten-sion of RoleKind class, will be defined. This element could grant access on doc-ument kinds to TeamKind or Rolekind elements. At this point, stakeholders will be defined and also the role they will play as document or subdocument produc-ers or consumers. In the endeavour domain, it is possible as well, the definition of concrete persons as Figure class instances.
- Document constraints definition. Using DocumentConstraint class, it is possible the documents (or subdocuments) to define constraints such as creation conditions or documents precedence. For example, a Unit Test Report document could be pro-duced only if a Unit Test Plan document is approved and the first draft of Source

Code document is baselined. This constraint could only be defined in the methodology domain. Though working at document level can be considered a coarse granularity, it is necessary to recall that this kind of constraint can equally be defined at any document substructure level.

- Document tasks, techniques and tools assignment. Any document development will involve tasks, techniques and tools that will be carried out by the figures with producer grants on this document. For example, a Unit Test Report document can be assigned to a producer Figure called *Unit Test Team*. The *Run Unit Test* task, using the *Black Box* technique, with the *Test Case Generator* tool support are also assigned to this document.

Therefore, tasks precedence, responsibilities distribution, applicable techniques and tool definition are determined by complete document definition, and consequently, documents are the drivers of methodology definition and process execution.

5   Conclusions and Future Work

This paper has presented an approach that is characterised by two issues: firstly, documents become not by-products, but the key product developers produced; secondly, it enables the possibility to define methodologies that use constraints on documents templates to drive software processes. For this, the standard ISO/IEC 24744:2007 Software Engineering - Metamodel for Development Methodologies has been extended in order to provide a modelling baseline to support the approach.

While the approach looks promising on the one hand, one of the identified limitations is related to the need for detailed document templates. A balance between the approach and the required information for a cost-effective project is required. Another issue is that software/system models should be specified in the form of documents. For instance, in the case UML, OCL obviously will facilitate this task.

As future work, it is planned to extend this metamodel in order to support configuration management. This will imply that for any type of document it will be necessary to identify at which level configuration control will be performed. For instance, for a software requirement document the configuration item could be each requirement. This will be a methodological decision that should be taken in each case by the method engineer.

Automation may be a key issue to support the approach described within this paper. Therefore it is planned to set up the basis to define a software engineering environment architecture in which documents are the central issue and support all the concepts introduced within the paper. Documents will be used as the integration mechanism for tools and services using XML schemes to define the document templates. These XML schemes will be the result of applying some mapping rules to the proposed metamodel. A Software Engineering Environment will manage the XML schemes and documents and will trigger the execution of the defined tools for performing the tasks assigned to different documents or subdocuments. A prototype was already developed few years ago as presented in [13].

Acknowledgements. Authors are indebted to Cesar Gonzalez for the support and advice provided while working with ISO/IEC 24744. The work reported herein has been partially supported by Spanish "Ministerio de Educacion, Ciencia y Cultura" within the AGMOD TIC2003-08503 and the OVAL/PM TIC2006-14840, the project VUL-CANO FIT-340503-2006-3 from "Ministerio de Industria, Turismo y Comercio". The DOBERTSEE project, European Space Agency ESA/ESTEC Contract No. 15133/01/ NL/ND, has provided part of the necessary background to produce this work.

References

1. Visconti, M., Cook, C.R.: Software system documentation process maturity model. In: ACM Conference on Computer Science, pp. 352–357 (1993)
2. Boehm, B.W.: Get ready for agile methods, with care. IEEE Computer 35, 64–69 (2002)
3. Gonzalez-Perez, C., Henderson-Sellers, B.: Templates and resources in software development methodologies. Journal of Object Technology 4(4) (2005)
4. ISO: ISO/IEC 24744:2007 Software Engineering – Metamodel for Development Methodologies. International Organization for Standarization (2007)
5. Luqi Zhang, L., Berzins, V., Qiao, Y.: Documentation driven development for complex real-time systems. IEEE Transactions on Software Engineering, 936–952 (2004)
6. Bollain, M., Alarcon, P.P., Garbajosa, J., Amador, J.: A low-cost document-centric software/system engineering environment. In: Proceedings of the 16th International Conference Software & Systems Engineering and their Applications, Paris (2003)
7. ISO: ISO/IEC 12207:1995 Information technology - Software life cycle processes (1995)
8. Sharp, H., Finkelstein, A., Galal, G.: Stakeholder identification in the requirements engineering process. In: DEXA Workshop, pp. 387–391 (1999)
9. Robinson, W.N., Volkov, V.: A meta-model for restructuring stakeholder requirements. In: ICSE, pp. 140–149 (1997)
10. Rausch, A., Bartelt, C., Ternité, T., Kuhrmann, M.: The V-Modell XT Applied – Model-Driven and Document-Centric Development. In: 3rd World Congress for Software Quality, Online Supplement. Number 3-9809145-3-4, International Software Quality Institute GmbH, vol. III, pp. 131–138 (2005), http://www.isqi.org/isqi/deu/conf/wcsq/3/proc.php
11. Nguyen, T.N., Munson, E.V.: The software concordance: a new software document management environment. In: Proceedings of the 21st annual international conference on Documentation, pp. 198–205. ACM Press, New York (2003)
12. Nguyen, T.N., Munson, E.V., Boyland, J.T.: Configuration management in a hypermedia-based software development environment. In: Proceedings of the fourteenth ACM conference on Hypertext and hypermedia, pp. 194–195. ACM Press, New York (2003)
13. Alarcón, P.P., Garbajosa, J., Crespo, A., Magro, B.: Automated integrated support for requirements-area and validation processes related to system development. In: IEEE INDIN. IEEE Computer Society Press, Los Alamitos (2004)
14. Cass, A.G., Lerner, B.S., Stanley, M., Sutton, J., McCall, E.K., Wise, A., Osterweil, L.J.: Little-jil/juliette: a process definition language and interpreter. In: ICSE 2000: Proceedings of the 22nd international conference on Software engineering, pp. 754–757. ACM Press, New York (2000)
15. Fuggetta, A.: Software process: a roadmap. In: Proceedings of the conference on The future of Software engineering, pp. 25–34. ACM Press, New York (2000)
16. Gary, K., LindqKuist, T., Koehnemann, H., Derniame, J.: Component-based software process support. In: 13th IEEE International Conference on Automated Software Engineering (ASE 1998), pp. 196–199 (1998)
17. Kopka, C., Wellen, U.: Role-based views to approach suitable software process models for the development of multimedia systems. In: ISMSE, pp. 140–147 (2002)

18. Ciancarini, P.: Modeling the software process using coordination rules. In: WETICE, pp. 46–53 (1995)
19. Barnes, A., Gray, J.: Cots, workflow, and software process management: An exploration of software engineering tool development. In: Proceedings of the 2000 Australian Software Engineering Conference, p. 221. IEEE Computer Society Press, Los Alamitos (2000)
20. Pohl, K., Weidenhaupt, K., Haumer, P., Jarke, M., Klamma, R.: PRIME - toward process-integrated modeling environments. ACM Trans. Softw. Eng. Methodol. 8, 343–410 (1999)
21. OMG: OMG Unified Modelling Language Specification. Version 1.4. Object Management Group (2001)
22. Atkinson, C., Kune, T.: Meta-level independent modelling. In: 14th European Conference on Object-Oriented Programming International workshop on model engineering (2000)
23. Beck, K.: Extreme Programming Explained: Embrace Change. Addison-Wesley Professional, Reading (1999)
24. Gray, J., Scott, L., Liu, A., Harvey, J.: The first international symposium on constructing software engineering tools (coset 1999). In: Proceedings of the 21st international conference on Software engineering, pp. 707–708. IEEE Computer Society Press, Los Alamitos (1999)
25. Henrich, A.: Document retrieval facilities for repository-based system development environments. In: Proceedings of the 19th annual international ACM SIGIR conference on Research and development in information retrieval, pp. 101–109. ACM Press, New York (1996)
26. Singh, H., Han, J.: Increasing concurrency in object-oriented databases for software engineering environments. In: DASFAA, pp. 175–184 (1997)
27. Devanbu, P., Chen, Y.F., Gansner, E., Muller, H., Martin, J.: Chime: customizable hyperlink insertion and maintenance engine for software engineering environments. In: Proceedings of the 21st international conference on Software engineering, pp. 473–482. IEEE Computer Society Press, Los Alamitos (1999)
28. Oinas-Kukkonen, H.: Flexible case and hypertext. ACM Comput. Surv. 31, 7 (1999)

Author Index

Communications
in Computer and Information Science

For information about Vols. 1–2

please contact your bookseller or Springer

Printed in the United States
By Bookmasters